T0262202

Encyclopedia of Salmonella: Diverse Topics

Volume III

Encyclopedia of Salmonella: Diverse Topics
Volume III

Edited by **Alan Klein**

New York

Published by Callisto Reference,
106 Park Avenue, Suite 200,
New York, NY 10016, USA
www.callistoreference.com

Encyclopedia of Salmonella: Diverse Topics
Volume III
Edited by Alan Klein

© 2015 Callisto Reference

International Standard Book Number: 978-1-63239-292-3 (Hardback)

This book contains information obtained from authentic and highly regarded sources. Copyright for all individual chapters remain with the respective authors as indicated. A wide variety of references are listed. Permission and sources are indicated; for detailed attributions, please refer to the permissions page. Reasonable efforts have been made to publish reliable data and information, but the authors, editors and publisher cannot assume any responsibility for the validity of all materials or the consequences of their use.

The publisher's policy is to use permanent paper from mills that operate a sustainable forestry policy. Furthermore, the publisher ensures that the text paper and cover boards used have met acceptable environmental accreditation standards.

Trademark Notice: Registered trademark of products or corporate names are used only for explanation and identification without intent to infringe.

Printed in the United States of America.

Contents

Preface

This book presents new developments and researches in the field of salmonella from across the globe. Salmonella comprises of two species (enterica and bongori) and is said to be an intensely variegated genus which infects a variety of hosts. This group is made up of 2579 serovars which fascinates researchers and draws their attention towards its different characteristics. Salmonella causes problems of zoonoses and also leads to food borne illness. Diseases caused by salmonella are becoming a matter of concern for developed and developing countries because of their impact on economy and other important sectors. Antimicrobial resistance in salmonella makes it difficult to reveal different mechanisms involved and this problem seems to increase further. This book focuses on environmental interactions of salmonella, problem of antimicrobial resistance and genetic aspects of salmonella. Internationally acclaimed researchers and practitioners in this field have made their valuable contributions to this book.

The information shared in this book is based on empirical researches made by veterans in this field of study. The elaborative information provided in this book will help the readers further their scope of knowledge leading to advancements in this field.

Finally, I would like to thank my fellow researchers who gave constructive feedback and my family members who supported me at every step of my research.

Editor

Part 1

Environmental Interactions

Invasion and Survival of Salmonella in the Environment: The Role of Biofilms

Cynthia L. Sheffield and Tawni L. Crippen
United States Department of Agriculture, Agriculture Research Service
Southern Plains Agricultural Research Center
USA

1. Introduction

Bacteria compose the majority of living biomass on Earth and play a vital role in the recycling of elements critical to sustaining life. We are discovering that they often exist as interlinked, multispecies colonies termed biofilms. They are all around us, on us, and in us. In fact, over 99% of microorganisms on Earth live as biofilms. They play a critical role in the ecology of the earth and the sustainability of life. For many years, studies of bacterial physiology focused primarily on the planktonic state neglecting the bacteria within the biofilm. The biofilm state is now recognized as the predominant form in which bacteria endure the stresses of the environment (An and Parsek, 2007; Hall-Stoodley et al., 2004; Hoffman et al., 2005; Karatan and Watnick, 2009; Stoodley et al., 2001)

Bacterial biofilms have long been recognized as participants in tooth decay, slippery rock surfaces, and contaminated water. Now these colonies are being investigated as perpetrators of persistent low-level food contamination which threaten animal and human health. Bacteria existing as biofilms are capable of surviving for extended periods in various environments, such as water, animal manure, and a range of agricultural soil types. For example, human pathogens can attach to and colonize the surfaces of plants and form biofilms on plant tissues (Annous et al., 2006). These biofilms are problematic because they are extremely hearty and difficult to remove by simple washing techniques. Causing, foodborne illnesses associated with human consumption of contaminated fresh fruits and vegetables (Fett and Cooke, 2003; Sivapalasingam et al., 2004). Living in biofims is advantageous for bacteria as it increases survival chances when confronted with unpredictable environmental stresses such as: temperature changes, desiccation, ultraviolet rays, etc.

In recent years, bacterial biofilms have been increasingly linked to food safety issues worldwide. The culprits of three recent foodborne illness outbreaks in cantaloupe melons, apples, and leafy greens have been identified as pathogenic bacteria existing in biofilms (Annous et al., 2009). They have also been implicated as the cause of many chronic infections in humans and are frequently associated with implanted devices, such as catheters, prosthetics, and contact lenses (Prouty et al., 2002). There is increasing interest in biofilms found on mucosal surfaces, such as the colon, particularly with respect to their role in disease processes (Macfarlane and Macfarlane, 2006).

There are numerous definitions of biofilms but all share the common threads of a concept involving an assemblage of microorganisms in which some of the bacteria adhere to the surface and exude an extracellular polymeric substance (EPS) that forms a matrix for further cellular attachment. The matrix is comprised of proteins, polysaccharides, extra-cellular DNA, and the various organisms involved. Biofilms can range from simple single species monolayer matrices, to complex multi-organism communities and sometimes even involve higher level organisms such as nematodes and larvae (Cloete et al., 2009).

Initially, the term "biofilm" was used informally among scientists for many years. It first appeared in a scientific journal in 1977 (Montana State University, http://www. biofilm.montana.edu/node/2930). Early researchers examining the phenomenon of microbes attaching to surfaces include Windogradsky, Cholodny, and Conn in the 1930's (Lappin-Scott, 1999). An important observation made by these scientists, was that bacteria which grew attached to a surface (in this case glass slides immersed in soil slurry) were phenotypically different from those cultured from the water phase of soil slurry (Lappin-Scott, 1999). Henrici studying freshwater bacteria observed that "for the most part water bacteria are not free floating but grow attached to the surfaces" (Lappin-Scott, 1999). These early researchers described how bacteria that were attached to surfaces exhibited diverse populations and developed into "microbial films". ZoBell's research from the early 1930's, focused on the role of bacteria in biofouling (the unwanted accumulation of microorganisms on surfaces) (Lappin-Scott, 1999). In fact ZoBell & Allen (1935), report the first apparatus specifically designed to examine bacterial attachment to surfaces. It was a carrier that held 16 glass slides and was designed to be lowered into the ocean where marine microbes could attach to the glass. Using this apparatus, ZoBell & Allen found a greater diversity of bacteria in the biofilm "lawn" on the slide than that which could be cultured from the sea water.

The bacteria found in biofilms are phenotypically distinct from their planktonic form. These changes include alterations in the regulation of large suites of genes (Hall-Stoodley et al., 2004; Karatan and Watnick, 2009). The transformation from planktonic existence to biofilm formation is a complex process, often triggered by various alterations in the surrounding environment. Bacteria in biofilms exhibited: protein profiles that more closely resemble those of exponentially growing planktonic cells (Mikkelsen et al., 2007); significant differences in the genes that are expressed (Teplitski et al., 2006; Trevors, 2011); and significant differences in the degree of resistance to antibiotics and disinfectants (Brooun et al., 2000; Ryu and Beuchat, 2005).

Bacteria living within biofilms can exhibit 1000 times more resistance to antimicrobials than their planktonic peers. The close proximity of fellow bacteria within this community allows for the increased incidence of gene transfer; resulting in increased genetic diversity, including augmented antimicrobial resistance. Biofilms impart increased levels of protection against environmental stresses, such as depleted nutrient, moisture and oxygen levels; inhospitable surrounding pH and salinity; excessive shear forces and UV exposure, and even metal toxicity. Additionally, life in a biofilm protects against attacks by a host immune system's protective proteins and signaling molecules, phagocytes, antibiotics and disinfectants (Jefferson, 2004; Mara and Horan, 2002).

Even after more than 80 years of research, there are still many unanswered question about the formation, function, maturation and eventual death of biofilms. Biofilms are typically attached and sessile. However, they have become ubiquitous in the environment because, portions can detach and relocate to other hospitable surroundings. There is widespread scientific interest in investigating the molecular mechanisms underlying life in these intriguing bacterial communities that are able to inhabit such diverse environments.

2. Biofilm development

Despite the years of research into the mechanism of bacterial attachment, there remain many basic facets of the process that are still a mystery. The nuances of the attachment are difficult to elucidate. What is known is that the multifaceted process involves a complete alteration in life style of the bacteria involved. A generalized model for bacterial transformation from a planktonic to biofilm existence can be made (Lemon et al., 2008). This model contains five major phases: attachment; formation; micro-colony development; maturation; and finally detachment/dispersal of the biofilm. Each phase can be described by key features and triggers unique to that phase of development and will be discussed in the remainder of this section.

Cell attachment occurs in five stages. The first stage is a reversible stage where cells lightly attach to the surfaces. It is followed by a second, more permanent stage, where the cells affix themselves securely by forming an adhesive exopolymeric compound. Then in stage three, the biofilm begins to expand by the recruitment of cells into micro-colonies. In stage four, the mature biofilm is characterized by the development of a three-dimensional structure containing cells packed in clusters with channels forming to aid in the movement of nutrients and molecules to cells beneath the colony surface. In the fifth and final stage, the cells detach which facilitates dispersal and the initiation of new similar biofilms at more favorable locations. It is important to note that cell division is uncommon in mature biofilms, and energy is used predominantly to produce exopolysaccharides (Watnick and Kolter, 2000).

Bacteria within biofilms exhibit a range of phenotypes; some of these do not exist in the planktonic phase. These phenotypes include: freely suspended naked cells (resuming their planktonic state); cells reversibly attached to a surface; cells irreversibly attached to a surface and not encapsulated by EPS; embedded attached cells surrounded by EPS matrix or deeply embedded attached cells within a the three dimensional microbial stack; embedded cells sloughed into suspension; and planktonic daughter cells (Parry, 2004).

Quorum sensing allows bacterial cells to communicate resulting in a cohesiveness of function that benefits an entire population and allow the community to operate as a living system (Smith and Chapman, 2010). The channels between cell clusters deliver water and nutrients to each cell and facilitate waste removal. These structures combined with strong adhesive properties and sophisticated cell-cell communication make biofilms highly resistant to conventional cleansing agents such as biocides and disinfectants. Not surprisingly, once biofilms form, they are difficult to eliminate.

2.1 Attachment

Surface attachment offers distinct advantages for bacteria which depend on the diffusion of nutrients and wastes for their well-being. Most natural aqueous environments contain only dilute substances which can be used for metabolism and growth. On the other hand, natural surfaces tend to collect and concentrate nutrients by charge-charge or hydrophobic interactions; which provide bacteria exposure to more concentrated foodstuffs. Biofilms are initiated when individual motile bacteria localize onto a surface and begin major physiological alterations. This initial attachment is reversible but encourages aggregation and attachment of more planktonic bacteria and other organisms. During this phase the attraction is mediated by weak forces, such as van der Waals, acid-base and simple electrostatics processes.

2.2 Formation

Permanent formation and expansion of the biofilm occurs when the initial transient attachment is reinforced by the production of cell surface adhesive compounds, pilli and fimbriae (Kaplan, 2010). The complex transition from transient to permanent attachment is associated with the formation of a monolayer via the up-regulation of genes responsible for the production of an extracellular matrix composed of exopolysaccharides, and extracellular DNA. Bacterial motility is lost by removal of cell flagellum by protease and replacement with a holdfast protrusion composed of oligomers of N-acetylglucosamine. The holdfast is composed of a strong adhesive polysaccharide that ensures a tight bond to the surface (Karatan and Watnick, 2009). In some strains of bacteria cell wall bound surface proteins called biofilm-associated protein (BAP) begin to be expressed and promote cell to cell interactions and the development of the extracellular matrix (Lasa and Penades, 2006).

Further development of the biofilm is promoted by the production of molecules which cause potassium leakage and trigger the activation of a membrane kinase (Lopez et al., 2009). In addition, the transcription of flagellar genes is repressed when the monolayer stage is achieved. Transcription of a large number of methyl-accepting chemotaxis genes are activated in the monolayer stage. Studies suggest that chemotaxis proteins influence monolayer formation. One possibility is that flagellar rotation pausing, which plays a role in the response to chemoattractants, also enhances the transition to permanent attachment (Karatan and Watnick, 2009). Some of the different components involved in the formation of the matrix include pilli and extracellular DNA (Banas and Vickerman, 2003; Kachlany et al., 2001; Petersen et al., 2005).

2.3 Micro-colony development

Now that the bacteria are sessile and biofilm formation is initiated, the bacteria actively multiply and communicate via quorum sensing signals. Once the quorum sensing threshold is achieved, exopolysaccharide production begins and micro-colonies develop through a variety of mechanisms. *Pseudomonas aeruginosa* use flagella and pili-mediated twitching motility to redistribute across the surface. *Escherichia coli* utilize fimbriae, flagella and pili for the same purpose. Others spread and generate micro-colonies through cell division, where the daughter cells spread outward and upward (Cloete et al., 2009).

2.4 Maturation

Maturation results in the formation of pillars and masses of tightly packed cells intermixed with fluid filled channels allowing for the exchange of nutrients, oxygen, and waste products between the biofilm and the surrounding liquid (Cloete et al., 2009). EPS is a key component of the biofilm matrix and may be composed of a number of sugar monomers such as glucose, galactose, mannose and xylose and some non-carbohydrate substitutes (such as acetate, pyruvate, succinate, and phosphate). Most EPS molecules are neutral or polyanionic in nature, which aids in immune evasion and tolerance toward antibacterial agents. Enzymatic alteration of EPS is thought to significantly change its physicochemical properties and consequently the entire structure. Some examples of polymeric biofilm matrix constituents include the glucan polysaccharides produced by *Streptococcus mutans* (Banas and Vickerman, 2003), proteinaceous fimbriae produced by *Aggregatibacter actinomycetemcomitans* and *Porphyromonas gingivalis* (Kachlany et al., 2001; Lamont et al., 2002); extracellular, double-stranded DNA in biofilms produced by *A. actinomycetemcomitans, S. mutans,* and *Streptococcus intermedius* (Inoue et al., 2003; Petersen et al., 2004; Petersen et al., 2005) and a wide variety of proteins, glycoproteins, glycolipids, and enzymes.

Mature biofilms are intricate structures containing sectors with distinctive microenvironments that differ in cell densities, oxygen and nutrient levels, and pH ranges. As a result, the metabolic and reproductive functionality of the bacteria located in these distinct sectors are quite divergent (Kaplan, 2010). Metabolically dormant cells located in the interior of the colony are often more resistant to the actions of antimicrobial agents that target actively growing cells near the exterior (Fux et al., 2005).

2.5 Detachment and dispersal

The fifth and final phase of a biofilm lifecycle is detachment and dispersal. Growth and detachment are interdependent. Under robust conditions, the detachment rate has been shown to increase with increasing growth rates (Gjaltema et al., 1997). This phase leads to colonization of new areas offering fresh resources, which is critical for long-term survival. This phase is also important in the dissemination of infection and therefore, disease transmission in clinical and public health settings. As with all aspects of the biofilm lifecycle, the processes surrounding detachment and dispersal are very intricate; involving a wide variety of environmental and physiological triggers and signal transduction pathways (Karatan and Watnick, 2009). Individual bacteria employ somewhat different methods of dispersal, which can be divided into three discrete stages: (a) detachment of cells from the colony; (b) relocation of cells to an alternative site; and (c) re-attachment of the cells to a new substrate site (Kaplan, 2010). Rochex et al. (2009) found that one dominant species often comprises most of the weakly cohesive, thick top layer of the biofilm; while a more diverse population comprises the strongly cohesive, thin basal layer. These findings suggest that determining species diversity may be an important parameter in understanding detachment and dispersal.

2.5.1 Key factors for detachment

Both biochemical and physical factors participate in the major processes facilitating biofilm detachment; those being erosion; sloughing; abrasion, grazing, and human intervention Numerous biochemical factors involved in detachment are: the production of EPS-degrading enzymes; lytic bacteriophage activation; expression of phosphodiesterases; and quorum-sensing signaling. Physical detachment factors are : microbiologically generated gas bubbles; the presence of cross-linking cations; nutrient limitations; metabolite accumulations; changes in osmolarity; high cell density growth; and fluidic shear factors (Thormann et al., 2006).

2.5.2 Erosion, sloughing and abrasion

Erosion and sloughing are two mechanisms of spontaneous biofilm cellular detachment. The distinction between erosion and sloughing has a considerable effect on bacterial species competition within biofilms and thus morphology (Telgmann et al., 2004). Erosion is the continual detachment of single cells or small fragments from the biofilm at low levels over the course of formation. Researchers have noted that the rate of erosion from the biofilm increases with increased matrix thickness and fluidic shear forces at the cell-liquid interface. An increase in the flow velocity causes the hydrodynamic boundary layer to decrease, resulting in amplified turbulence at the biofilm surface. Sloughing is the swift, massive loss of large chunks of biofilm greater than or equal to the overall thickness. Sloughing is a more random occurrence than erosion and is thought to result from nutrient or oxygen depletion within the structure and is more commonly observed in thicker systems (Donlan, 2002).

Erosion and sloughing occur when local shear forces overwhelm the cohesiveness of the biofilm. Overall cohesiveness is strongly influenced by the composition and the structure of the polymeric matrix, which is dependent on the formation history, the environmental growth conditions and the developmental stage of the biofilm. The resulting strength of biofilm attachment is contingent on cell density, composition of extracellular polymeric substances, and levels of specific compounds, such as the calcium. Fast growing organisms with high initial cell growth rates favor the development of protrusions and the formation of a heterogeneous biofilm structure. Shear forces more easily erode these protrusions (Telgmann et al., 2004).

Abrasion is the loss of biofilm due to collision of particles from the surrounding fluid with the exposed surface. Biofilms in fluidized beds, filters, and particle-laden environments such as surface waters are often subjected to abrasion (Donlan, 2002). Work by Rochex et al. (2009) demonstrated that abrasion characteristics, such as particle collision frequency and pressure strongly affect biofilm detachment rates. Experiments by Gjaltema et al. (1997) have shown that inter-particle collisions cause an on-going abrasion of the biofilm

2.5.3 Grazing and human intervention

A key mortality factor in the control of bacteria within biofilms is grazing. Grazing is the uptake and killing of bacteria by phagocytic protozoa and metazoa in close association with biofilms. These biofilm-associated protozoa exhibit three modes of predation: 1) planktonic, but swimming close to the biofilm surface; 2) surface attachment on biofilm, but feeding on suspended prey; and 3) feeding directly on biofilm as prey. Protozoans benefit from this association as demonstrated by their increased numbers and taxa diversity when associated with a biofilm community compared to the surrounding plankton environment (Boenigk and Arndt, 2000).

Protozoa exhibit a sizeable diversification of morphologies consequent to developing a variety of means to capture and engulf their bacterial prey. However, they are commonly grouped into flagellates, ciliates, and amoebae. All three free-living groups efficiently graze on bacteria exposed on the biofilm surface. Flagellates and ciliates contain feeding types primarily focused on suspended bacteria with only a few that preferably feed on surface-bound bacterial prey (Parry, 2004). For instance, the flagellate *Rhynchomonas nasuta* feed on attached *Pseudomonas* spp. at rates between 13 and 120 bacteria per flagellate per hour (Boenigk and Arndt, 2000). Sibille et al. (1998) found that a mixed population of flagellates could consume on average 12 suspended bacteria per flagellate per hour. The ciliate *Euplotes* spp. grazes on adherent *Vibrio natriegens* and *Pseudomonas fluorescens* at rates of 120 and 882 bacteria per ciliate per hour, respectively (Lawrence and Snyder, 1998), while Ayo et al. (2001) found that in general ciliates showed a grazing rate of ≤20 free swimming bacteria per ciliate per hour.

Amoebae protozoans feed almost exclusively on surface-bound bacteria (Parry, 2004). Amoebae species such as *Hartmanella cantabrigiensis*, *Platyamoeba placida*, *Saccamoeba limax*, *Vahlkampfia avara* eat attached *Escherichia coli* at rates of 15 to 440 bacteria per amoeba per hour (Heaton et al., 2001).

Many predators are selective and remove only a subset of the microbial community thus altering the biofilm community structure (Parry, 2004). Morphological differences in biofilm structure correlate with predation. Without the pressure of predation a flat, compact structure results. Conversely in the presence of predators, an open and heterogeneous structure results.

Metazoa (rotatoria, nematoda, and oligochaeta) are the main group of higher level predators responsible for grazing. Their grazing on biofilms initially decreases microbial biomass, and unless grazing pressure is severe, the secondary microbial community that develops will have increased rates of metabolic activity and growth. Total microbial biomass will be greater and the turnover rates of both the substrates and microorganisms will increase. The diversity of the community structure will decrease as the biofilm community shifts towards faster growing organisms.

Bacterial predators, such as *Bdellovibrio bacteriovorus; Micavibrio* spp.; and *Hyphomicrobium* spp. also play a vital role in the life and death of biofilms. *Bdellovibrio bacteriovorus* is a gram-negative, aerobic bacterium that preys upon a wide variety of other gram negative bacteria, including *E. coli*; which, in simple biofilms, can devastate a community altogether (Dashiff et al., 2010). Additionally, *Micavibrio* spp. is also a gram-negative, aerobic bacterium that also preys on bacteria and biofilm structures. Unlike *Bdellovibrio* which penetrate their prey, *Micavibrio* attach to the outside surface and eventually lyse their host bacteria. *Bdellovibrio* and *Micavibrio* spp. have been shown to be extremely host specific; for example, *Micavibrio aeruginosavorus* strain ARL-13 preys only on *Pseudomonas aeruginosa*. In static and flow cell experiments, *M. aeruginosavorus* not only modified *P. aeruginosa* biofilm structure, but also decreased bacterial viability. The alterations were likely caused by increased cell-cell interactions brought about by the presence of the predator (Donlan, 2002).

Human intervention involves both mechanical action and the use of disinfectants. Any type of brush or scouring pad provides the agitation required to disrupt the biofilm structure. Once the community has been physically disrupted the addition of a surfactant and disinfectant is required to complete the destruction process. In the case of contact lens , Wu et al. (2011) found that *Staphylococcus aureus* or *Pseudomonas aeruginosa* biofilms required rubbing and rinsing with multipurpose disinfecting solutions followed by tissue-wiping and air-drying to remove them from the surface. *Listeria monocytogenes*, an important foodborne pathogen, has the ability to form persistent biofilm matrices in food processing environments. Soni & Nannapaneni, (2010) determined that a cocktail of different bacteriophages may be essential for their removal. Lequette et al. (2010) found that solubilization of polysaccharidases and proteases in a buffer containing surfactants, along with dispersing and chelating agents, enhanced their efficiency of removing biofilms by targeting several components of EPS of *Bacillus* spp. and *Pseudomonas* spp..

Biofilms have been extensively studied in the dental industry. Periodontitis is a chronic bacterial infectious disease whose hallmark is the presence of a bacterial biofilm at the gum line. The condition necessitates thorough removal of the biofilm for therapy. However, debridement using hand instruments or oscillating scalers is both technically demanding and time consuming, and may lead to severe root damage over time (Petersilka, 2011). Air-polishing with glycine powder proved to be an easy, safe and effective means of biofilm removal from teeth (Petersilka, 2011).

3. Quorum sensing

For many years, bacteria were believed to exist as individual cells that existed to find nutrients and multiply. The discovery of intercellular communication among bacteria led to the realization that bacteria are capable of coordinated activity that was once thought to be restricted to higher organisms (reviewed in (Waters and Bassler, 2005). The ability to behave collectively has obvious advantages, for example, the ability to migrate to a more suitable environment or better nutrient supply and to adopt new modes of growth, such as biofilm formation, which may afford protection from harmful environments. This intercellular communication is called quorum sensing. The mechanism used for quorum sensing is the process of recognition of and response to small molecules, called autoinducers, secreted by the bacteria themselves. The process of biofilm creation in a variety of bacteria has been shown to specifically involve quorum sensing. These autoinducers are used by bacteria to regulate their behavior according to population density. The phenomenon relies on the principle that when a single bacterium releases autoinducers into the environment, the concentration is too low to be detected. However, when sufficient bacteria are present, autoinducer concentrations reach a threshold level that allows the bacteria to sense a critical mass and respond by the activation or repression of target genes (de Kievit and Iglewski, 2000). Quorum sensing manifests itself as a synchronization of individual behavior into cooperative group activity, often resulting in a change of phenotype within a population once bacterial densities have reached a threshold level. The specific threshold level can be different for each population. Examples of density-dependent changes include the turning on of bioluminescence within *Vibrio fischeri*, conjugal transfer in *Agrobacterium tumefaciens*, swarming in *Serratia liquefacians*, production of virulence factors in *Burkholderia cepacia* and *Pseudomonas aeruginosa*, and biofilm formation in numerous species including *Pseudomonas aeruginosa*, *Pantoea stewartii* and *Vibrio cholera* (Bottomley et al., 2007; Davies et al., 1997; Nadell et al., 2008; Ward et al., 2004).

4. Biofilm and virulence

Many bacterial pathogens including *Listeria monocytogenes*, *Salmonella* spp., *Shigella* spp., *Staphylococcus aureus*, *Escherichia coli*, and *Enterobacter* spp., utilize a biofilm strategy to survive inhospitable conditions and to cause disease. Tamayo et al. (2010) found that pathogenic *Vibrio cholera* in both dispersed and intact biofilms vastly out-competed planktonic populations. Huang et al. (2008) found that that *Streptococcus mutans* utilizes the general secretory pathway to secrete virulence factor proteins and the level of SecA, the key factor in the general secretory pathway, was influenced significantly by biofilm formation. PrfA is the critical virulence transcription factor that regulates the switch from extracellular, flagellum-propelled bacterium to intracellular pathogen in *L. monocytogenes*. Lemon et al. (2010) reported the first evidence that PrfA has a significant positive impact on extracellular biofilm development. Mutants lacking *prfA* were defective in surface-adhered biofilm indicating that PrfA positively regulates biofilm establishment and has a role in modulating the life-style of *L. monocytogenes*. This could provide selective pressure to maintain this critical virulence regulator when *L. monocytogenes* is outside host cells. The human-enteropathogenic species *Yersinia enterocolitica* and *Y. pseudotuberculosis* and the highly virulent plague bacillus *Y. pestis*, represent ideal species to study how bacteria adapt from different environments and evolve to be highly virulent. The work of Hinchliffe et al. (2008) found that several alleged virulence determinants of the *Yersinia* species, regulated by a phosphorelay, also regulated proteins involved in biofilm formation, motility, mammalian cell adhesion and stress survival. *Escherichia coli* are one of the first colonizers of the gastrointestinal tract of newborns and a normal component of the gastrointestinal flora of almost every human being. Found in concentrations up to 10^8 cells ml^{-1} it is a major source for the spread of potentially pathogenic *E. coli* to susceptible sites via the fecal route. Adherence and invasion of intestinal epithelial cells mediated by type 1 fimbriae is a feature of *E. coli* strains isolated from lesions of Crohn's disease.

Salmonella enterica serovar Enteritidis has emerged as one of the most important foodborne pathogens for humans. It is often associated with consumption of contaminated produce, poultry meat and eggs. The spiA gene within S. enterica serovar Typhimurium encodes an outer-membrane component of the SPI-2 type III secretion system that is essential for virulence in host cells. Dong et al. (2011) found that that the spiA gene is also critical to biofilm formation. Biofilm cells, from *Listeria monocytogenes* and *S. enterica* serovar Typhimurium , which survived disinfection, seem to develop a stress response and become more virulent, which may compromise food safety and increase public health risk (Rodrigues et al., 2011a). Legendre et al. (2011) showed that adhered *S. enterica* serovar Enteritidis bacteria were more resistant to antibacterial agents than their planktonic counterparts. Xu et al. (2010) found that the enterotoxin production and invasion ability of biofilm *S. enterica* serovar Typhimurium cells is enhanced under acidic stress conditions. Further, cells of *S. enterica* serovar Typhimurium, collected from a biofilm, showed increased adhesive ability within the spleens of mice. The invasion of *S. enterica* serovar Typhimurium into the intestinal epithelial cells is the crucial step in pathogenesis. Wilson et al. (2007) reported that *S. enterica* serovar Typhimurium samples grown during the weightlessness of space flight exhibited enhanced virulence in a mouse infection model, along with extracellular matrix accumulation consistent with a biofilm.

5. Biofilm development in Salmonella

Scientific understanding of the formation process of biofilms by Salmonella is insufficient and replete with opportunities for further exploration. While some generalities can be made, each species has its own idiosyncrasies relating to the influence of local environmental conditions, gene expression and protein production and secretion. Some of these differences will be discussed in the following section.

In recent years, outbreaks Salmonellosis have often been traced back to contaminated plant sources (CDC, 2011). Lately it has been determined that contamination of plants with Salmonella is not superficial, but due to specific attachment of the bacteria to plant tissues by surface molecules (Barak et al., 2005; 2002). Salmonella uses extracellular matrix components, such as thin aggregative fimbriae and polymers (cellulose and O-antigen capsules) to colonize the plants, forming a biofilm, which is ultimately consumed by and causes illness in humans (Barak et al., 2007). The determination that Salmonella specifically attaches with biofilm formation, challenges the public concept that cleaning vegetables by simply rinsing with water is adequate for bacterial removal. These surface molecules appear to aid this pathogen in the utilization of plants as vectors for spreading and increase the risk of contamination of fresh produce.

Iturriaga, et al. (2007) found that during growth of tomatoes in greenhouses or during postharvest handling, higher humidity promotes biofilm development on the surface of the fruit. These biofilms provide a protective environment for pathogens and reduce the effectiveness of sanitizers and other inhibitory agents used to clean the fruit prior to consumption. S. enterica serovar Montevideo was shown to grow on tomato surfaces under a wide range of temperature and relative humidity combinations even when external nutrients were scarce. These findings reinforce the importance of maintaining fruits and vegetables under proper storage conditions to reduce the incidence of Salmonella biofilm development.

Fifteen S. enterica serotypes, Anatum; Baidon; Caracase; Cubana; Give; I 13,23,d-; Isangi; Montevideo; Muenchen; Newport; Onderstepoort; Senftenberg; Teko; Wandsbek and Weltevideo, found to form biofilms, were identified from various foods, spices and water sample (Xia et al., 2009). Pulse Field Gel Electrophoresis showed that eight out of the 15 serotypes had patterns indistinguishable from patterns of strains from human clinical samples or foods (US PulseNet National database); indicating that the isolates could potentially infect humans and cause salmonellosis.

Patel & Sharma, (2010) investigated the ability of five S. enterica serovars to attach to and colonize intact and cut lettuce (Iceberg, Romaine) and cabbage surfaces. They found that biofilm formation was significantly affected by the serovars used. Generally, S. enterica serovars Tennessee and Thompson showed significantly more biofilm formation than serovars Braenderup, Negev, and Newport; and were thus classified as strong biofilm producers according to the criteria suggested by Stepanovic et al. (2004). The criteria states, that strong biofilm producer had four times the optical density (OD) cutoff, which is three standard deviations above the mean OD of the negative control. Understanding the attachment mechanisms of Salmonella to vegetables may be useful in developing new intervention strategies to prevent contamination.

Kim and Wei, (2009) demonstrated that the knockout of the yjcC gene, encoding putative diguanylate cyclase/phosphodiesterase, in S. enterica serovar Typhimurium DT104 enhanced biofilm formation by the mutant in meat and poultry broths and on contact surfaces. This work also showed that biofilm formation by S. enterica serovar Typhimurium DT104 could be affected by the type of food products, since the yjcC mutant produced greater biofilms in meat and poultry broths than in vegetable broths. Therefore, the prevention of bacterial biofilm formation on food contact surfaces is critical for controlling cross-contamination of S. enterica serovar Typhimurium DT104 in food processing.

Bhowmick, et al. (2011)found the existence of an alternative biofilm regulatory pathway in *S. enterica* serovar Weltevreden from seafood isolates. This is the most prevalent serovar associated with seafood. While human illness caused by this serovar is rare in Europe and United States, it has been reported in Asia. In *S. enterica* serovar Typhimurium the gcpA gene plays a critical role in biofilm formation under low nutrient conditions. In *S. enterica* serovar Weltevreden deletion of the gcpA gene resulted in its inability to produce cellulose and failure to produce biofilm on polystyrene substrate. This indicated that in the case of *S. enterica* serovar Weltevreden, gcpA is critical for activating cellulose synthesis and biofilm formation. The characterization of genes involved in biofilm formation will help in defining critical control points within the process that may be manipulated to control for or possible eliminate the development of biofilms in certain environments.

Small RNAs (sRNA) are non-coding RNA molecules, 50-250 nucleotides in length, produced by bacteria. Kint et al. (2010) showed that biofilm formation is influenced by the sRNA molecule in various *S. enterica* serovar Typhimurium mutants. The sRNA was encoded in the same region as the quorum sensing synthase luxS. Quorum sensing represents a coordinated gene expression response in bacteria, stimulated by local population density. Autoinducer-2 (AI-2) is considered a universal signaling molecule in quorum sensing that is widespread in bacteria, and the LuxS enzyme is required for AI-2 synthesis. Quorum sensing plays an important role in biofilm formation and survival (see section 3). MicA is a family of small RNA molecules highly conserved in several Enterobacteriaceae. These sRNA's are reported to be a regulatory mechanism necessary for biofilm formation in many bacterial species and whose balanced expression level is essential for mature Salmonella biofilm formation.

The *ydcI* gene is differentially regulated in response to conditions of low fluid shear force that increase bacterial virulence and alter other phenotypes in *S. enterica* serovar Typhimurium (Jennings et al., 2011). They found that the *S. enterica* serovar Typhimurium strain in which *ydcI* expression is induced; invaded cells at a level 2.8 times higher than that of the wild type strain. Further, induction of *ydcI* resulted in the formation of a biofilm in stationary cultures, indicating that the *ydcI* gene encodes a conserved DNA binding protein involved with aspects of prokaryotic biology related to stress related biofilm production and possibly virulence. Further, these studies indicate that the *S. enterica* serovar Typhimurium *ydcI* gene is conserved across genera and has auto-regulated expression. When induced, it alters the interactions of *S. enterica* serovar Typhimurium host cells and expedites biofilm formation.

Human-to-human transmission of *S. enterica* serovar Typhimurium makes this a pathogen of global concern. Random transposon mutants of this serovar were screened for impaired adherence and biofilm formation on cholesterol-coated surfaces; 49 mutants with this phenotype were found (Crawford et al., 2010). It was determined that genes involved in flagellum biosynthesis and structure primarily mediated the attachment to cholesterol. In addition, the presence of the flagellar filament enhanced binding and biofilm formation in the presence of bile. This improved understanding of the early events during biofilm development, specifically how Salmonella bind to cholesterol, provides potential therapeutic targets for alleviating asymptomatic gallbladder carriage of *S. enterica* serovar Typhimurium.

6. Biofilm and Salmonella survival

Salmonella enterica serovar Enteritidis is a significant biofilm-forming pathogen. The survival of Salmonella on equipment and instruments in the food industry might be one of the most important contributing factors to food contamination and the subsequent foodborne infection. Further, the biofilm formation ability of foodborne pathogens has attracted much attention in the medical field and food industry due to its potential risks, including transfer of antimicrobial resistance and virulence factors (Xu et al., 2010).

Hasegawa et al. (2011) found that the ability of Salmonella strains to survive in the presence of acetic acid and rice vinegar paralleled their ability to form biofilms. Thus, Salmonella with a high biofilm-formation capability might be more difficult to kill in a food production setting. Salmonella cells embedded in these matrices show reduced susceptibility to trisodium phosphate, desiccation, and chlorination. Further, the connection between biofilm-forming ability and risk of foodborne outbreaks has been suggested in Salmonella. The work of Vestby et al. (2009) showed a correlation between persistence and biofilm establishment of Salmonella thus this may be an important factor for its longevity in the factory environment. These Salmonella strains appear to be a greater risk to human health via food contamination by surviving for longer periods (Iibuchi et al., 2010).

Mangalappalli-Illathu et al. (2008) found significant differences in the pattern and degree of resistance between planktonic and biofilm *S. enterica* serovar Enteritidis cells to benzalkonium chloride (BC). They established that the biofilm phenotype resulted in an early, more efficient adaptive response, and produced a higher proportion of adapted individuals than the planktonic phenotype. Once adapted, these cells were better able to survive BC than the planktonic cells. It is worth mentioning that disrupted BC adapted biofilm cells seem to have a better likelihood to attach, multiply, and form biofilms in BC-containing environments if the concentration is sublethal. The presence of these BC adapted *S. enterica* serovar Enteritidis biofilm cells presents a potential problem in environments such as health care facilities, the food industry, and households.

The presence of *S. enterica* serovars in animal feed ingredients is a well-known problem, resulting in contamination that vectors Salmonella infections in livestock farms. Dual-species biofilms favored Salmonella growth compared to Salmonella in mono-species biofilms, where biomass increased 2.8-fold and 3.2-fold in the presence of Staphylococcus and Pseudomonas, respectively (Habimana et al., 2010). Thus contamination with Salmonella in the presence of other bacteria will only exacerbate the problem of dissemination of Salmonella.

Fresh fruits and vegetables have been increasingly associated with outbreaks of foodborne illness. Salmonella contamination was higher on members of the Brassicaceae family (radish, turnip, and broccoli) than on lettuce, tomatoes, and carrots when sown and grown in contaminated soil. Vegetables that had soft rot exhibited twice the Salmonella contamination as did healthy produce. This could be stress related or possibly because the vegetables are already immunocompromised (Barak and Liang, 2008). Biofilm formation on plant tissue enabled foodborne pathogens to survive in the harsh phyllosphere and decreased the efficacy of commonly used sanitizers (Critzer and Doyle, 2010). Lapidot and Yaron, (2009) demonstrated that *S. enterica* serovar Typhimurium could be transferred from irrigation water to the edible parts of parsley plants. This work also revealed that *S. enterica* serovar

Typhimurium formed aggregates at a depth of 8 to 32 µm beneath the leaf surface. Penetration was most likely achieved through the roots or the phyllosphere. They further determined that, curli and cellulose, both components involved in the formation of biofilms, play a major role in the transfer or survival of S. enterica serovar Typhimurium in the plant. Incidences of salmonellosis caused by eating fresh produce continue to increase. This appears to be the result of S. enterica serovar Typhimurium attaching to and colonizing plants, rather than incidental contamination. S. enterica serovar Typhimurium that preferentially colonize roots use a hydrolase for swarming or biofilm production on plants; this multicellular behavior of S. enterica serovar Typhimurium has emerged as central to plant colonization (Barak et al., 2009).

A series of studies from our lab provided a molecular-based characterization of both the biofilm and planktonic populations from continuous-flow culture community. These studies examined the ability of S. enterica serovar Typhimurium to colonize a defined microfloral community established to model chicken ceca at day-of-hatch, 7 and 14 days old. The bacterial communities were allowed to equilibrate biofilm and planktonic populations for 3 weeks prior to introduction of S. enterica serovar Typhimurium. The one common factor relating to successful invasion of the community was the presence of S. enterica serovar Typhimurium within the biofilm. If the introduced S. enterica serovar Typhimurium could invade and sequester within the biofilm, then colonization appeared long-term. However if it only invaded the planktonic portion, then it was unable to gain a foothold and did not persist within the community (Crippen et al., 2008; Sheffield et al., 2009a, b).

7. Salmonella biofilms in the environment

Salmonella causes an estimated 93.8 million human infections and 155,000 deaths annually worldwide (Majowicz et al., 2010). The U.S. Centers for Disease Control and Prevention (CDC) have estimated that over 1.4 million cases of infection and 600 deaths related to salmonellosis may occur every year, accounting for about 31% of all food-related deaths in the USA (Wang et al., 2010). Poultry, poultry products, red meat, pork, wild game, and vegetables are all possible vehicles of transmission to humans.

7.1 Poultry

There are many avenues for Salmonella persistence in large scale poultry houses; one is to develop biofilms. Poultry feed has been demonstrated to be a leading source of Salmonella introduction into a poultry production facility (Park et al., 2011). Further, containers used in transporting live poultry between production and processing units have also been incriminated as primary sources of contamination for processed poultry products (Ramesh et al., 2002).

7.2 Non-poultry food animals

In developed countries, the production of food animals (i.e. cattle and hogs) is often limited to highly concentrated rearing facilities, also known as concentrated animal feeding operation (CAFO). This provides a conduit for the spread of Salmonella serovars to a large number of individuals within the herd, as demonstrated by the S. enteritidis pandemic in the 1990s, which affected both developed and developing countries (Hendriksen et al., 2011).

Additionally, where wild game is still a key food source, the incidence of Salmonella in feces is upwards of 22% of the wild boar and 48% of the wild rabbit populations in some areas (Vieira-Pinto et al., 2011). This demonstrates the potential for the exchange of bacterial pathogens between wild and domestic animals, which is cause of concern for the welfare of both the wild and the domestic populations, as well as for the humans in contact with them.

Antimicrobial resistance gene-bearing organisms that move from nutritionally rich to more dilute environments, such as when inadvertently washed from CAFO's into the surrounding watershed, survive longer in biofilms (Engemann et al., 2008). Additionally, antimicrobial resistance genes readily transfer into biofilms, which can then be transferred into the surrounding environment, in particular aquatic systems. These organisms are then accessible to wild fauna also utilizing the environment. Many studies have been performed investigating wild animals acting as reservoirs of disease for domestic animals. However, the influence of domestic animals serving as a reservoir of diseases transferable to wildlife is rarely considered. Domestic stock, particularly ungulates, have introduced many diseases into wildlife populations, sometimes with catastrophic results for that population and wildlife conservation on the whole (Mathews, 2010).

7.3 Processed foods

Food processing or handling equipment may provide a niche in which pathogenic bacteria such as *S. enterica* can grow rapidly into highly hydrated biofilms resulting in cross-contamination from food processing surfaces to food products. This cross-contamination can potentially lead to foodborne illnesses. Such cross-contamination of food products has been observed from the use of inadequately cleaned/sanitized processing equipment. Some examples include pumps, containers, or tanks first used for handling raw food materials and subsequently used for processed food products without first undergoing proper sanitation procedures (Jun et al., 2010). Predictably, the food industry has increased interest in chemical, physical, and biological interventions that mitigate food-borne pathogens on these products (Ha and Ha, 2011). *Salmonella* spp. is one of the most commonly isolated pathogens associated with fresh produce (Wong et al., 2011). Penteado & Leitao, (2004) demonstrated that low acid fruits are good substrates for the survival and growth of *S. enterica* serovar Enteritidis, a known biofilm forming pathogen.

8. Salmonella biofilm control measures

Salmonella enterica is a major cause of bacterial food-borne diseases worldwide, and serovars, such as Typhimurium, can cause a localized self-limiting gastroenteritis in humans. In immunocompromised people, Salmonella infections are often fatal if they are not treated promptly with antibiotics (Janssens et al., 2008). While Salmonella infections are most commonly treated using fluoroquinolones (e.g., ciprofloxacin) and extended spectrum cephalosporins (e.g., cefotaxime), there are disturbing reports regarding the development of resistance against these antimicrobials. Further, Salmonella is able to form biofilms on a variety of biotic and abiotic surfaces, where they are a double threat in that they allow the Salmonella to survive and spread in the environment outside the host (Janssens et al., 2008). The Salmonella found in these biofilms show an even higher tolerance to antibiotics than most Salmonella and according to the National Institutes of Health; approximately 80% of persistent bacterial infections in the United States are caused by biofilms (NIH, 1997).

Therefore, the need for alternative strategies to combat the spread of bacterial biofilm related infections is emerging (Janssens et al., 2008).

8.1 Chemical control

Salmonella in biofilms is less susceptible to disinfectants than planktonic Salmonella (Wong et al., 2010); therefore the eradication of biofilm sequestered pathogens is more challenging. S. enterica can itself form biofilms that are relatively resistant to chemical sanitizing treatments. The use of glutaraldehyde, formaldehyde, and peroxygen at a concentration of 1.0% in field conditions is insufficient to eradicate Salmonella biofilms (Marin et al., 2009). However, Rodrigues et al. (2011a) and Wong et al. (2011) showed sodium hypochlorite to be one of the most effective disinfectants against biofilms; with the ability to eradicate biofilms at concentrations as low as 3.125 mg per ml. Rodrigues et al. (2011a) also found that bacterial cells from biofilms, which survived disinfection, appeared to develop a stress response and/or become more virulent. The main finding of this work is the worrying fact that, even at concentrations that lead to significant reduction in biofilm biomass, disinfectants may actually enhance virulence within the surviving cells. Adding to this is the fact that the biofilm forms of Salmonella have significantly increased antibiotic resistance properties compared to their planktonic forms (Papavasileiou et al., 2010). These studies confirm that the biofilm form of Salmonella is not only more difficult to remove during sanitation procedures, but has an increased potential to compromise food safety and potentiate public health risk.

In further work, Rodrigues et al. (2011b) examined the adhesion, formation and viability within biofilms of S. enterica serovar Enteritidis on regular (granite, marble, stainless steel) and triclosan-impregnated kitchen bench stones (Silestones). Triclosan is a polychlorophenoxy phenol compound with broad spectrum antimicrobial activity that works by targeting lipid biosynthesis and inhibiting cell growth. Salmonella cells adhered equally well (4 to 5 log CFU per cm^2) to all surfaces, with the exception of silestone, which exhibited a potential for bacteriostatic activity. Less S. enterica serovar Enteritidis biofilms formed on impregnated silestones and cell viability was one to two logs lower than on other materials (Rodrigues et al., 2011b).

Hasegawa et al. (2011) observed a positive relationship between acid tolerance and biofilm-formation capability in Salmonella by examining the ability of strains to survive and form biofilms in the presence of acetic acid and rice vinegar. It has been suggested that a positive relationship exists between biofilm formation and increased risk of foodborne outbreaks. Therefore, when developing strategies for the prevention of Salmonella contamination of foods it is important to consider the biofilm-formation capability of each particular strain (Hasegawa et al., 2011).

Rosenberg et al. (2008) demonstrated that biofilm formation can be prevented through controlled release of nature-derived antimicrobials, such as salicylate-based poly (anhydride esters). The inhibition of the biofilm appeared to be caused by the irreversible interaction of salicylic acid molecules with the cells. The inhibition was not caused by interference with attachment but rather, via another mechanism essential for biofilm development that remains to be elucidated.

Another promising area of biofilm control is the use of essential oils from a variety of plants. The efficacy of essential oils from the leaves of *Myrcia ovata* Cambess for antimicrobial activity and prevention of the formation of microbial biofilms by *Enterococcus faecalis* was examined (Candido et al., 2010). The essential oil from this plant is commonly used in Brazil for the treatment of gastric illnesses. This oil showed antimicrobial activity against *E. faecalis*, *E. coli*, *P. aeruginosa*, *S. choleraesuis*, *Staphylococcus aureus*, *Streptococcus pneumoniae* and *Candida parapsilosis*. Further, at a concentration as low as 0.5 % it appreciably reduced the formation of biofilm by *E. faecalis* (Candido et al., 2010).

8.2 Predation

Protozoa are important participants within microbial food webs; however protozoan feeding preferences and their effects with respect to bacterial biofilms are not very clear. Work by Chabaud et al. (2006) demonstrated that protozoan grazing had a substantial effect on the removal of pathogenic coliforms in septic effluent and in the presence of a biofilm. Coliform survival was 10 times lower in a septic effluent with protozoa than without them. Further, removal of the bacteria within the biofilm was 60% higher in the presence of protozoa.

A landmark study examined the predatory range of *Myxococcus virescens* and *Myxococcus fulvus*, on a variety of human pathogens, including *Staphylococcus aureus*, *Mycobacterium phlei*, *Shigella dysenteriae*, *Vibrio cholerae*, *Proteus* X, and several Salmonella isolates (Mathew and Dudani, 1955). With the exception of M. phlei, all of the examined pathogenic species were completely or partially lysed, indicating that deciphering the predatory mechanism utilized by *Myxobacteria* species is of practical importance to improve our understanding of how to treat bacterial infectious diseases.

In 1983 Lambina and colleagues (Lambina et al., 1983) isolated a new species (*Micavibrio* spp.) of exoparasitic bacteria with an obligatory parasitic life cycle. They are gram negative, small curved rod shaped (0.5 x 1.5 mm), bacteria with a single polar flagellum. A titer as low as 10 plaque forming units per well of *M. aeruginosavorus* was sufficient to produce a 78% reduction in a *P. aeruginosa* biofilm after 30 min exposure in a static assay (Kadouri et al., 2007).

Dopheide et al. (2011) examined the grazing interactions of two ciliates, the free-swimming filter feeder *Tetrahymena* spp. and the surface-associated predator *Chilodonella* spp., on biofilm-forming bacteria. They found that both ciliates readily consumed cells from both *Pseudomonas costantinii* and *Serratia plymuthica* biofilms. They also found that both ciliates used chemical cues to locate biofilms. Further, using confocal microscopy they discovered that *Tetrahymena* spp. had a major impact on biofilm morphology, forming holes and channels throughout *S. plymuthica* biofilms and reducing *P. costantinii* biofilms to isolated, grazing-resistant microcolonies. Grazing by *Chilodonella* spp. resulted in the development of less-defined trails through *S. plymuthica* biofilms and caused *P. costantinii* biofilms to become homogeneous scatterings of cells (Dopheide et al., 2011).

Bdellovibrio spp. are small, predatory bacteria that invade and devour other gram-negative bacteria. Under dilute nutrient conditions, bdellovibrio prevented the formation of simple bacterial biofilms and destroyed established biofilms (Nunez et al., 2005). During the active prey-seeking period of its life cycle, it moved through water or soil searching for prey. Once it encountered a prey cell, bdellovibrio attached to the prey bacterium's surface, broke the outer membrane, and killed the prey cell by halting its respiration and growth. During the growth period, this predator utilized the prey's macromolecules for fuel and the carcass provided a protected, nutrient-rich habitat for development. Once the prey resource was exhausted, bdellovibrio divided into multiple progeny that lyse the remains of the prey and swim away to pursue new prey. Depending on the prey and the environmental conditions, its life cycle takes roughly 3–4 h (Berleman and Kirby, 2009; Nunez et al., 2005). While many predatory bacteria have been identified, most have been studied only superficially. Predation behavior has evolved a number of times. Examples of predatory bacteria are found in diverse genera, within the *Proteobacteria, Chloroflexi*, and *Cytophagaceae* (Berleman and Kirby, 2009). Dashiff et al. (2010) has demonstrated that predatory bacteria, *Bdellovibrio bacteriovorus* and *Micavibrio aeruginosavorus*, are able to attack bacteria from a variety of genus, including *Acinetobacter, Aeromonas, Bordetella, Burkholderia, Citrobacter, Enterobacter, Escherichia, Klebsiella, Listonella, Morganella, Proteus, Pseudomonas, Salmonella, Serratia, Shigella, Vibrio* and *Yersinia*. Further, predation occurred on single and multispecies planktonic cultures, as well as on monolayer and multilayer biofilms. Finally, *Bdellovibrio bacteriovorus* and *Micavibrio aeruginosavorus* have the ability to reduce many of the multidrug-resistant pathogens associated with human infection (Dashiff et al., 2010).

8.3 Radiation

Niemira & Solomon, (2005) found that while the radiation sensitivity of Salmonella is isolate specific, the biofilm associated cells of *S. enterica* serovar Stanley were significantly more sensitive to ionizing radiation than the respective planktonic cells. The dose of radiation value required to reduce the population of *E. coli* O157:H7 by 90% (D10) was highly dependent on the isolate. One isolate exhibited significantly ($P < 0.05$) higher D10 values for planktonic cells than those observed for biofilm cells indicating a significantly increased sensitivity to irradiation for cells in the biofilm habitat. However, for another isolate of *E. coli* O157:H7 exhibited exactly the opposite results. It appears that culture maturity had a more significant influence on the irradiation efficacy of planktonic cells than on biofilm-associated cells of *E. coli* O157:H7 (Niemira, 2007).

9. Future outlook

Current research investigating Salmonella biofilms covers efforts to fully understand the multifaceted process of biofilm development and the intricate relationships between biofilms and virulence, and to develop more effective and environmentally friendly control methods. In the following section we will discuss some of the most recent work reported in these areas.

Shah et al. (2011) have found an association between the pathogenicity of *S. enterica* serovar Enteritidis strains and the differential production of type III secretion system proteins during the production of biofims. In addition several factors including motility, fimbriae, biofilm production, and the presence of large molecular mass plasmids can augment pathogenicity. Such research will provide more insights into molecular basis of *S.* Enteritidis virulence and thus delineate a new direction for the reduction of virulence in *S.* Enteritidis. Based on recent finding, solid murine tumors might represent a unique model to study biofilm formation *in vivo*. Crull et al. (2011) found that systemic administration of *S. enterica* serovar Typhimurium to tumor bearing mice resulted in preferential colonization of the tumors by Salmonella and retardation of tumor growth. Ultrastructural analysis of these tumors did not detect the Salmonella intracellularly, but revealed that the bacteria had formed biofilms. This model could provide the means for further clarification of the biofilm development process. Research by Sha et al. (2011) utilized the high resolution tool, Rep-PCR, to differentiate closely related microbial strains among Salmonella. This methodology could provide more discriminatory information essential to pin pointing bacterial sources, which is critical to maintaining food safety and public health in the future.

Perez-Conesa et al. (2011) tested eugenol and carvacrol delivered within surfactant micelles at concentrations of 0.9 and 0.7%, respectively. Eugenol is a component of essential oils primarily from clove, nutmeg, cinnamon, and bay leaf; and carvacrol is a predominant phenol found in wild oregano oil. These oils decreased viable counts of 48 hr biofilms of pure *E. coli* O157:H7 or *L. monocytogenes* on stainless steel surfaces by 3.5 to 4.8 logs of CFU per cm2, respectively, within 20 minutes of exposure. Thus, micelle-encapsulated eugenol and carvacrol appear to be good vehicles to deliver hydrophobic antimicrobials through the exopolymeric structure to cells embedded within biofilms. Potentially, these oils could be used in combination with other treatments to diminish biofilm formation on food and food contact surfaces.

The pathogenicity of several significant human pathogens has been linked to the activity of AI-2 quorum sensing signaling, which is also involved with the development of biofilms (Roy et al., 2011). The ubiquitous nature of AI-2 makes it an excellent target as a potential antimicrobial therapy against a broad spectrum of pathogens. Additionally, as AI-2 is not essential for cell growth or survival, interference with its synthesis and processing will probably not stimulate development of resistance. However, as with any single piece of the biofilm pathogenicity puzzle, it is unlikely that quorum sensing quenching drugs will be the "magic bullet" for the treatment of bacterial infections. Therefore, according to Roy et al. (2011) a mixed therapy of quorum sensing quenchers and traditional antibiotics appears to be a promising approach for the future. Finally, it is important that our understanding of signaling molecules be increased, thereby allowing the identification of potential new antimicrobial therapies.

Many questions remain to be answered on the path to understanding the complicated processes involved in the development and expansion of biofilms in human, animal and environmental settings. What specific factors, both biotic and abiotic, govern the initiation and continuation of the biofilm process? What impact does quorum sensing have on the initiation and differential development of the unique biofilm characteristics? What influences the ability of Salmonella to form biofilms and the development of virulence and antibiotic resistance? The final question is how to use this knowledge to manage the environment, and components involved in the biofilm development process to reduce their negative impact on human and animal health.

10. References

An, D., and Parsek, M. R. (2007). The promise and peril of transcriptional profiling in biofilm communities. *Current Opinions in Microbiolology* 10, 292-296.
Annous, B. A., Fratamico, P. M., and Smith, J. L. (2009). Scientific status summary. *Journal of Food Science* 74, R24-37.
Annous, B. A., Solomon, E. B., and Niemira, B. A. (2006). Biofilms on fresh produce and difficulties in decontamination. *Food Quality Magazine* April/May 2006.

Ayo, B., Santamaria, E., Latatu, A., Artolozaga, I., Azua, I., and Iriberri, J. (2001). Grazing rates of diverse morphotypes of bacterivorous ciliates feeding on four allochthonous bacteria. *Letters in Applied Microbiology* 33, 455-60.

Banas, J. A., and Vickerman, M. M. (2003). Glucan-binding proteins of the oral streptococci. *Critical Reviews in Oral Biology and Medicine* 14, 89-99.

Barak, J. D., Gorski, L., Liang, A. S., and Narm, K. E. (2009). Previously uncharacterized *Salmonella enterica* genes required for swarming play a role in seedling colonization. *Microbiology* 155, 3701-3709.

Barak, J. D., Groski, L., Naraghi-Arani, P., and Charkowski, A. O. (2005). *Salmonella enterica* virulence genes are required for bacterial attachment to plant tissue. *Applied and Environmental Microbiology* 71, 5685-5691.

Barak, J. D., Jahn, C. E., Gibson, D. L., and Charkowski, A. O. (2007). The role of cellulose and O-antigen capsule in the colonization of plants by *Salmonella enterica Molecular Plant-MIcrobe Interactions* 20, 1083-1091.

Barak, J. D., and Liang, A. S. (2008). Role of soil, crop debris, and a plant pathogen in *Salmonella enterica* contamination of tomato plants. *PLoS One* 3, e1657.

Barak, J. D., Whitehand, L. C., and Charkowski, A. O. (2002). Differences in the attachment of *Samonella enterica* serovars and *Escherichia coli* O157:H7 to alfalfa sprouts. *Applied and Environmental Microbiology* 68, 4578-4763.

Berleman, J. E., and Kirby, J. R. (2009). Deciphering the hunting strategy of a bacterial wolfpack. *FEMS Microbiology Reviews* 33, 942-957.

Bhowmick, P. P., Devegowda, D., Ruwandeepika, H. A., Fuchs, T. M., Srikumar, S., Karunasagar, I., and Karunasagar, I. (2011). gcpA (*stm1987*) is critical for cellulose production and biofilm formation on polystyrene surface by *Salmonella enterica* serovar Weltevreden in both high and low nutrient medium. *Microb Pathog* 50, 114-122.

Boenigk, J., and Arndt, H. (2000). Comparative studies on the feeding behavior of two heterotrophic nanoflagellates: the filterfeeding choanoflagellate *Monosiga ovata* and the raptorial-feeding kinetoplastid *Rhynchomonas nasuta*. *Aquatic MIcrobial Ecology* 22, 243-249.

Bottomley, M. J., Muragila, E., Bazzo, R., and Carfi, A. (2007). Molecular insights into quorum sensing in the human pathogen *Pseudomonas aeruginosa* from the structure of the virulence regulator LasR bound to its autoinducer. *The Journal of Biogogical Chemistry* 282, 13592-13600.

Brooun, A., Lui, S., and Lewis, K. (2000). A dose-response study of antibioticr esistance in *Pseudomonas aeruginosa* biofilms. *Antimicrobioal Agents and Chemotherapy* 44, 640-646.

Candido, C. S., Portella, C. S. A., Laranjeira, B. J., da Silva, S. S., Arriaga, A. M. C., Santiago, G. M. P., Gomes, G. A., Almeida, P. C., and Carvalho, C. B. M. (2010). Effects of Myrica Ovata Cambess essential oil on planktonic growth of gastrointestinal microorganism and biofilm formation of Enterococcus faecalis. *Brazilian Journal of Microbiology* 41, 621-627.

CDC (August, 18, 2011) Centers for Disease Control and Prevention. Salmonella Outbreaks In:*Salmonella Homepage*, September 1, 2011, Available from: <http://www.cdc.gov/salmonella/outbreaks.html>.

Chabaud, S., Andres, Y., Lakel, A., and Le Cloirec, P. (2006). Bacteria removal in septic effluent: influence of biofilm and protozoa. *Water Research* 40, 3109-3014.

Cloete, T. E., Thantsha, M. S., Maluleke, M. R., and Kirkpatrick, R. (2009). The antimicrobial mechanism of electrochemically activated water against Pseudomonas aeruginosa and Escherichia coli as determined by SDS-PAGE analysis. *Jounal of Applied Microbiology* 107, 379-384.

Crawford, R. W., Reeve, K. E., and Gunn, J. S. (2010). Flagellated but not hyperfimbriated *Salmonella enterica* serovar Typhimurium attaches to and forms biofilms on cholesterol-coated surfaces. *J Bacteriol* 192, 2981-90.

Crippen, T. L., Sheffield, C. L., Andrews, K., Dowd, S. E., Bongaerts, R. J., and Nisbet, D. J. (2008). Planktonic and biofilm community characterization and Salmonella resistance of 14-day-old chicken cecal microflora-derived continuous-flow cultures. *Journal of Food Protection* 71, 1981-1987.

Critzer, F. J., and Doyle, M. P. (2010). Microbial ecology of foodborne pathogens associated with produce. *Current Opinions in Biotechnology* 21, 125-130.

Crull, K., Rohde, M., Westphal, K., Loessner, H., Wolf, K., Felipe-López, A., Hensel, M., and Weiss, S. (2011). Biofilm formation by *Salmonella enterica* serovar Typhimurium colonizing solid tumours. *Cellular Microbiology* 13, 1223–1233.

Dashiff, A., Junka, R. A., Libera, M., and Kadouri, D. E. (2010). Predation of microbial ecology of foodborne pathogens associated with produce *Micavibrio aeruginosavorus* and *Bdellovibrio bacteriovorus*. *Journal of Applied Microbiology* 110, 431-444.

Davies, D. G., Parsek, M. R., Pearson, J. P., Iglewski, B. H., Costerton, J. W., and Greenberg, E. P. (1997). The involvement of cell-to-cell signals in the development of a bacterial biofilm. *Science* 280, 295-298.

de Kievit, T. R., and Iglewski, B. H. (2000). Bacterial quorum sensing in pathogenic relationships. *Infection and Immunity* 68, 4839-4849.

Dong, H., Peng, D., Jiao, X., Zhang, X., Geng, S., and Liu, X. (2011). Roles of the Spia gene from Salmonella enteritidis in biofilm formation and virulence. *Microbiology* 157, 1798-1805.

Donlan, R. M. (2002). Biofilms: Microbial life on surfaces. *Emerging and Infectious Diseases* 8, 881-890.

Dopheide, A., Lear, G., Stott, R., and Lewis, G. (2011). Preferential feeding by the ciliates Chilodonella and Tetrahymena spp. and effects of these protozoa on bacterial biofilm structure and composition. *Applied and Environmental Microbiology* 77, 4564-4572.

Engemann, C. A., Keen, P. L., Knapp, C. W., Hall, K. J., and Graham, D. W. (2008). Fate of tetracycline resistance genes in aquatic systems: migration for water column to peripheral biofilms. *Environmental Science and Technology* 42, 5131-5136.

Fett, W. F., and Cooke, P. H. (2003). Reduction of *Escherichia coli* O157:H7 and *Salmonella* on laboratory-inoculated alfalfa seed with commercial citrus-related products. *Journal of Food Protection* 66, 1158-1165.

Fux, C. A., Costerton, J. W., Stewart, P. S., and Stoodley, P. (2005). Survival strategies of infectious biofilms. *Trends in Microbiology* 13, 34-40.

Gjaltema, A., Vinke, J. L., van Loosdrecht, M. C. M., and Heijen, J. J. (1997). Biofilm abrasion by particle collisions in airlift reactors. *Water Science and Technology* 36, 2221-2228.

Ha, J. H., and Ha, S. D. (2011). Synergistic effects of sodium hypochlorite and ultraviolet radiation in reducing the levels of selected foodborne pathogenic bacteria. *Foodborne Pathogens and Disease* 8, 587-591.

Habimana, O., Moretro, T., Langsrud, S., Vestby, L. K., Nesse, L. L., and Heir, E. (2010). Micro ecosystems from feed industry surfaces: a survival and biofilm study of Salmonella versus host resident flora strains. *BMC Veterinary Research* 6, 48.

Hall-Stoodley, L., Costerton, J. W., and Stoodley, P. (2004). Bacterial biofilms: from the natural environment to infectious diseases. *Nature Reviews Microbiology* 2, 95-108.

Hasegawa, A., Hara-Kudo, Y., and Kumagai, S. (2011). Survival of *Salmonella* strains differing in their biofilm-formation capability upon exposure to hydrochloric and acetic acid and to high salt. *Journal of Veterinary Medical Science.*

Heaton, K., Drinkall, J., MInett, A., Hunt, A., and Parry, J. D. (2001). Amoeboid grazing on surface associted prey. *In* "Biofilm Community Interactions: Chance or Necessity?" (P. Gilbert, D. G. Allison, M. Brading, J. Verran and J. Walker, eds.), pp. 293-301. Bioline Press, Cardiff.

Hendriksen, R. S., Vieira, A. R., Karlsmose, S., Lo Fo Wong, D. M., Jensen, A. B., Wegener, H. C., and Aarestrup, F. M. (2011). Global monitoring of Salmonella serovar distribution from the world health organization global foodborne infections network country data bank: results of quality assured laboratories from 2001 to 2007. *Foodborne Pathogens and Disease* 8, 887-900.

Hinchliffe, S. J., Howard, S. L., Huang, Y. H., Clarke, D. J., and Wren, B. W. (2008). The importance of the Rcs phosphorelay in the survival and pathogenesis of the enteropathogenic Yersiniae. *Microbiology* 154, 1117-1131.

Hoffman, L. R., D'Argenio, D. A., MacCoss, M. J., Zhang, Z., Jones, R. A., and Miller, S. I. (2005). Aminoglycoside antibiotics induce bacterial biofilm formation. *Nature* 436, 1171-1175.

Iibuchi, R., Hara-Kudo, Y., Hasegawa, A., and Kumagai, S. (2010). Survival of Salmonella on a polypropylene surface under dry conditions in relation to biofilm-formation capability. *Journal of Food Protection* 73, 1506-1510.

Inoue, T., Shingaki, R., Sogawa, N., Sogawa, C. A., Asaumi, J., Kokeguchi, S., and Fukui, K. (2003). Biofilm formation by a fimbriae-deficient mutant of Actinobacillus actinomycetemcomitans. *Microbiology and Immunology* 47, 877-881.

Iturriaga, M. H., Tamplin, M. L., and Escartin, E. F. (2007). Colonization of tomatoes by Salmonella montevideo is affected by relative humidity and storage temperature. *Journal of Food Protection* 70, 30-34.

Janssens, J. C., Steenackers, H., Robijns, S., Gellens, E., Levin, J., Zhao, H., Hermans, K., De Coster, D., Verhoeven, T. L., Marchal, K., Vanderleyden, J., De Vos, D. E., and De Keersmaecker, S. C. (2008). Brominated furanones inhibit biofilm formation by *Salmonella enterica* serovar Typhimurium. *Applied and Environmental Microbiology* 74, 6639-6648.

Jefferson, K. K. (2004). What drives bacteria to produce a biofilm? *FEMS Microbiology Letters* 236, 163-173.

Jennings, M. E., Quick, L. N., Soni, A., Davis, R. R., Crosby, K., Ott, C. M., Nickerson, C. A., and Wilson, J. W. (2011). Characterization of the Salmonella enterica serovar Typhimurium ydcI gene, which encodes a conserved DNA binding protein required for full acid stress resistance. *JOURNAL OF BACTERIOLOGY* 193, 2208-2217.

Jun, W., Kim, M. S., Cho, B.-K., Millner, P. D., Chao, K., and Chan, D. E. (2010). Microbial biofilm detection on food contact surfaces by macro-scale fluorescence imaging. *Journal of Food Engineering* 99, 314-322.

Kachlany, S. C., Planet, P. J., DeSalle, R., Fine, D. H., and Figurski, D. H. (2001). Genes for tight adherence of Actinobacillus actinomycetemcomitans: From plaque to plague to pond scum. *Trends in Microbiology* 9, 429-437.

Kadouri, D., Venzon, N. C., and O'Toole, G. A. (2007). Vulnerability of pathogenic biofilms to *Micavibrio aeruginosavorus*. *Applied and Environmental Microbiology* 73, 605-614.

Kaplan, J. B. (2010). Biofilm dispersal: Mechanisms, cinical implications, and potential therapeutic uses. *Journal of Dental Research* 89, 205-218.

Karatan, E., and Watnick, P. (2009). Signals, regulatory networks, and materials that build and break bacterial biofilms. *Microbiology and Molecular Biology Reviews* 73, 310-347.

Kim, S. H., and Wei, C. I. (2009). Molecular characterization of biofilm formation and attachment of Salmonella enterica serovar typhimurium DT104 on food contact surfaces. *Journal of Food Protection* 72, 1841-1847.

Kint, G., De Coster, D., Marchal, K., Vanderleyden, J., and De Keersmaecker, S. C. J. (2010). The small regulatory RNA molecule MicA is involved in *Salmonella enterica* serovar Typhimurium biofilm formation. *BMC Microbiology* 10, 276-283.

Lambina, V. A., Afinogenova, A. V., Romay-Penobad, Z., Konovalova, S. M., and Andreev, L. V. (1983). New species of exoparasitic bacteria of the genus *Micavibrio* infecting Gram-positive bacteria. *Mikrobiologiya* 52, 777-780.

Lamont, R. J., El-Sabaeny, A., Park, Y., Cook, G. S., Costerton, J. W., and Demuth, D. R. (2002). Role of the *Streptococcus gordonii* Sspb protein in the development of Porphyromonas gingivalis biofilms on Streptococcal substrates. *Microbiology* 148, 1627-1636.

Lapidot, A., and Yaron, S. (2009). Transfer of *Salmonella enterica* serovar Typhimurium from contaminated Irrigation water to parsley is dependent on curli and cellulose, the biofilm matrix components. *Journla of Food Protection* 72, 618-623.

Lappin-Scott, H. M. (1999). Claude E. Zobell – his life and contributions to biofilm microbiology. *In* "International Symposium on Microbial Ecology", Vol. Proceedings of the 8th International Symposium on Microbial Ecology, pp. 1-6.

Lasa, I., and Penades, J. R. (2006). Bap: a family of surface proteins involved in biofilm formation. *Research Microbiology* 157, 99-107.

Lawrence, J. R., and Snyder, R. A. (1998). Feeding behaviour and grazing impacts of a *Euplotes* sp. on attached bacteria. *Canadian Journal of Microbiology* 44, 623-629.

Legendre, G., Fay, F., Linossier, I., and Vallee-Rehel, K. (2011). Evaluation of antibacterial activity against *Salmonella enteritidis*. *Journal of Microbiology* 49, 349-354.

Lemon, K. P., Earl, A. M., Vlamakis, H. C., Aguilar, C., and Kolter, R. (2008). Biofilm development with an emphasis on *Bacillus subtilis*. *Current Topics in Microbiology and Immunology* 322, 1-16.

Lemon, K. P., Freitag, N. E., and Kolter, R. (2010). The virulence regulator PrfA promotes biofilm formation by *Listeria monocytogenes*. *Jouranl of Bacteriology* 192, 3969-3976.

Lequette, Y., Boels, G., Clarisse, M., and Faille, C. (2010). Using enzymes to remove biofilms of bacterial isolates sampled in the food-industry. *Biofouling* 26, 421-431.

Macfarlane, S., and Macfarlane, G. T. (2006). Composition and metabolic activities of bacterial biofilms colonizing food residues in the human gut. *Applied and Environmental Microbiology* 72, 6204-6211.

Majowicz, S. E., Musto, J., Scallan, E., Angulo, F. J., Kirk, M., O'Brien, S. J., Jones, T. F., Fazil, A., and Hoekstra, R. M. (2010). The global burden of nontyphoidal *Salmonella* gastroenteritis. *Clinical Infectious Diseases* 50, 882-889.

Mangalappalli-Illathu, A. K., Vidovic, S., and Korber, D. R. (2008). Differential adaptive response and survival of *Salmonella enterica* serovar Enteritidis planktonic and biofilm cells exposed to benzalkonium chloride. *Antimicrobial Agents and Chemotherapy* 52, 3669-3680.

Mara, D. D., and Horan, N. J. (2002). Sludge to land: Microbiological double standards. *Journal of the Chartered Institution of Water and Environmental Management* 16, 249-252.

Mathew, S., and Dudani, A. (1955). Lysis of human pathogenic bacteria by *Myxobacteria*. *Nature* 175, 125.

Mathews, F. (2010). Wild animal conservation and welfare in agricultural systems. *Animal Welfare* 19, 159-170.

Mikkelsen, H., Duck, Z., Lilley, K. S., and Welch, M. (2007). Interrelationships between colonies, biofilms, and planktonic cells of *Pseudomonas aeruginosa*. *Journal of Bacteriology* 139, 2411-2416.

Nadell, C. D., Xavier, J. B., Levin, S. A., and Foster, K. R. (2008). The evolution of quorum sensing in bacterial biofilms. *PLoS Biology* 6, e14.

Niemira, B. A. (2007). Irradiation sensitivity of planktonic and biofilm-associated *Escherichia coli* O157:H7 isolates is influenced by culture conditions. *Applied and Environmental Microbiology* 73, 3239-3244.

Niemira, B. A., and Solomon, E. B. (2005). Sensitivity of planktonic and biofilm-associated *Salmonella* spp. to ionizing radiation. *Applied and Environmental Microbiology* 71, 2732-2736.

Nunez, M. E., Martin, M. O., Chan, P. H., and Spain, E. M. (2005). Predation, death, and survival in a biofilm: *Bdellovibrio* investigated by atomic force microscopy. *Colloids and Surfaces B: Biointerfaces* 42, 263-271.

Papavasileiou, K., Papavasileiou, E., Tseleni-Kotsovili, A., Bersimis, S., Nicolaou, C., Ioannidis, A., and Chatzipanagiotou, S. (2010). Comparative antimicrobial susceptibility of biofilm versus planktonic forms of *Salmonella enterica* strains isolated from children with gastroenteritis. *European Journal of Clinical Microbiology & Infectious Diseases* 29, 1401-5.

Park, S. H., Jarquin, R., Hanning, I., Almeida, G., and Ricke, S. C. (2011). Detection of *Salmonella* spp. survival and virulence in poultry feed by targeting the hilA gene. *Journal of Applied Microbiology* 111, 426-432.

Parry, J. D. (2004). Protozoan grazing of freshwater biofilms. *Advances in Applied Microbiology* 54, 167-196.

Patel, J., and Sharma, M. (2010). Differences in attachment of *Salmonella enterica* serovars to cabbage and lettuce leaves. *International Journal of Food Microbiology* 139, 41-47.

Pe´ Rez-Conesa, D., Cao, J., Chen, L., McLandsborough, L., and Weiss, J. (2011). Inactivation of *Listeria monocytogenes* and *Escherichia coli* O157:H7 Biofilms by Micelle-Encapsulated Eugenol and Carvacrol. *Journal of Food Protection* 74, 55-62.

Penteado, A. L., and Leitao, M. F. (2004). Growth of *Listeria monocytogenes* in melon, watermelon and papaya pulps. *International Journal of Food Microbiology* 92, 89-94.

Petersen, F. C., Pecharki, D., and Scheie, A. A. (2004). Biofilm mode of growth of *Streptococcus intermedius* favored by a competence-stimulating signaling peptide. *Journal of Bacteriology* 186, 6327-6331.

Petersen, F. C., Tao, L., and Scheie, A. A. (2005). DNA binding-uptake system: a link between cell-to-cell communication and biofilm formation. *Journal of Bacteriology* 187, 4392-4400.

Petersilka, G. J. (2011). Subgingival air-polishing in the treatment of periodontal biofilm infections. *Periodontology* 55, 124-42.

Prouty, A. M., Schwesinger, W. H., and Gunn, J. S. (2002). Biofilm formation and interaction with the surfaces of gallstones by *Salmonella* spp. *Infection and Immunity* 70, 2640-2649.

Ramesh, N., Joseph, S. W., Carr, L. E., Douglass, L. W., and Wheaton, F. W. (2002). Evaluation of chemical disinfectants for the elimination of *Salmonella* biofilms from poultry transport containers. *Poultry Science* 81, 904-910.

Rochex, A., Masse, A., Escudie, R., Godon, J. J., and Bernet, N. (2009). Influence of abrasion on biofilm detachment: evidence for stratification of the biofilm. *Journal of Industrial Microbiology and Biotechnology* 36, 467-470.

Rodrigues, D., Cerca, N., Teixeira, P., Oliveira, R., Ceri, H., and Azeredo, J. (2011a). *Listeria monocytogenes* and *Salmonella enterica* Enteritidis biofilms susceptibility to different disinfectants and stress-response and virulence gene expression of surviving cells. *Microbial Drug Resistance* 17, 181-189.

Rodrigues, D., Teixeira, P., Oliveira, R., and Azeredo, J. (2011b). *Salmonella enterica* Enteritidis Biofilm Formation and Viability on Regular and Triclosan-Impregnated Bench Cover Materials. *Journal of Food Protection* 74, 32-37.

Rosenberg, L. E., Carbone, A. L., Romling, U., Uhrich, K. E., and Chikindas, M. L. (2008). Salicylic acid-based poly(anhydride esters) for control of biofilm formation in *Salmonella enterica* serovar Typhimurium. *Letters in Applied Microbiology* 46, 593-599.

Roy, V., Adams, B. L., and Bentley, W. E. (2011). Developing next generation antimicrobials by intercepting AI-2 mediated quorum sensing. *Enzyme and Microbial Technology* 49, 113-123.

Ryu, J. H., and Beuchat, L. R. (2005). Biofilm formation by *Escherichia coli* O157:H7 on stainless steel: effect of exopolysaccharide and Curli production on its resistance to chlorine. *Applied and Environmental Microbiology* 71, 247-254.

Sha, Q., Gunathilake, A., Forstner, M. R., and Hahn, D. (2011). Temporal analyses of the distribution and diversity of *Salmonella* in natural biofilms. *Systematic and Applied Microbiology* 34, 353-359.

Shah, D. H., Zhou, X., Addwebi, T., Davis, M. A., Orfe, L., Call, D. R., Guard, J., and Besser, T. E. (2011). Cell invasion of poultry-associated *Salmonella enterica* serovar Enteritidis isolates is associated with pathogenicity, motility and proteins secreted by the type III secretion system. *Microbiology* 157, 1428-1445.

Sheffield, C. L., Crippen, T. L., Andrews, K., Bongaerts, R. J., and Nisbet, D. J. (2009a). Characterization of planktonic and biofilm communities of day-of-hatch chicks cecal microflora and their resistance to *Salmonella* colonization. *Journal of Food Protection* 72, 959-965.

Sheffield, C. L., Crippen, T. L., Andrews, K., Bongaerts, R. J., and Nisbet, D. J. (2009b). Planktonic and biofilm communities from 7-day-old chicken cecal microflora cultures: characterization and resistance to *Salmonella* colonization. *Journal of Food Protection* 72, 1812-1820.

Sibille, I., Sime-Ngando, T., Mathieu, L., and Block, J. C. (1998). Protozoan bacterivory and *Escherichia coli* survival in drinking water distribution systems. *Applied and Environmental Microbiology* 64, 197-202.

Sivapalasingam, S., Friedman, C. R., Cohen, L., and Tauxe, R. V. (2004). Fresh produce: a growing cause of outbreaks of foodborne illness in the United States, 1973 through 1997. *Journal of Food Protection* 67, 2342-2353.

Smith, D. R., and Chapman, M. R. (2010). Economical evolution: microbes reduce the synthetic cost of extracellular proteins. *MBio* 1.

Soni, K. A., and Nannapaneni, R. (2010). Removal of *Listeria monocytogenes* biofilms with bacteriophage P100. *Journal of Food Protection* 73, 1519-1524.

Stepanović, S., Cirković, I. C., Ranin, L., and Svabić-Vlahović, M. (2004). Biofilm formation by Salmonella spp. and *Listeria monocytogenes* on plastic surface. *Letters in Applied Microbiology* 38, 428–432.

Stoodley, P., Wilson, S., Hall-Stoodley, L., Boyle, J. D., Lappin-Scott, H. M., and Costerton, J. W. (2001). Growth and detachment of cell clusters from mature mixed-species biofilms. *Applied and Environmental Microbiology* 67, 5608-5613.

Tamayo, R., Patimalla, B., and Camilli, A. (2010). Growth in a biofilm induces a hyperinfectious phenotype in *Vibrio cholerae. Infection and Immunity* 78, 3560-3569.

Telgmann, U., Horn, H., and Morgenroth, E. (2004). Influence of growth history on sloughing and erosion from biofilms. *Water Research* 38, 3671-3684.

Teplitski, M., Al-Agely, A., and Ahmer, B. M. (2006). Contribution of the SirA regulon to biofilm formation in *Salmonella enterica* serovar Typhimurium. *Microbiology* 152, 3411-3424.

Thormann, K. M., Duttler, S., Saville, R. M., Hyodo, M., Shukla, S., Hayakawa, Y., and Spormann, A. M. (2006). Control of formation and cellular detachment from *Shewanella oneidensis* MR-1 biofilms by cyclic di-GMP. *Journal of Bacteriology* 188, 2681-2691.

Trevors, J. T. (2011). Viable but non-culturable (VBNC) bacteria: Gene expression in planktonic and biofilm cells. *Journal of Microbiolical Methods* 86, 266-273.

Vestby, L. K., Moretro, T., Langsrud, S., Heir, E., and Nesse, L. L. (2009). Biofilm forming abilities of *Salmonella* are correlated with persistence in fish meal- and feed factories. *BMC Veterinary Research* 5, 20.

Vieira-Pinto, M., Morais, L., Caleja, C., Themudo, P., Torres, C., Igrejas, G., Poeta, P., and Martins, C. (2011). *Salmonella* sp. in Game (*Sus scrofa* and *Oryctolagus cuniculus*). *Foodborne Pathogens and Disease* 8, 739-740.

Wang, S., Phillippy, A. M., Deng, K., Rui, X., Li, Z., Tortorello, M. L., and Zhang, W. (2010). Transcriptomic responses of *Salmonella enterica* serovars Enteritidis and Typhimurium to chlorine-based oxidative stress. *Applied and Environmental Microbiology* 76, 5013-5024.

Ward, J. P., King, J. R., Koerber, A. J., Croft, J. M., Sockett, R. E., and Williams, P. (2004). Cell-signalling repression in bacterial quorum sensing. *Mathematical Medicine and Biology* 21, 169-204.

Waters, C. M., and Bassler, B. L. (2005). Quorum sensing: cell-to-cell communication in bacteria. *Annual Review of Cell and Developmental Biology* 21, 319-346.

Watnick, P., and Kolter, R. (2000). Biofilm, city of microbes. *Journal of Bacteriology* 182, 2675-2679.

Wilson, J. W., Ott, C. M., Honer zu Bentrup, K., Ramamurthy, R., Quick, L., Porwollik, S., Cheng, P., McClelland, M., Tsaprailis, G., Radabaugh, T., Hunt, A., Fernandez, D., Richter, E., Shah, M., Kilcoyne, M., Joshi, L., Nelman-Gonzalez, M., Hing, S., Parra, M., Dumars, P., Norwood, K., Bober, R., Devich, J., Ruggles, A., Goulart, C., Rupert, M., Stodieck, L., Stafford, P., Catella, L., Schurr, M. J., Buchanan, K., Morici, L., McCracken, J., Allen, P., Baker-Coleman, C., Hammond, T., Vogel, J., Nelson, R., Pierson, D. L., Stefanyshyn-Piper, H. M., and Nickerson, C. A. (2007). Space flight alters bacterial gene expression and virulence and reveals a role for global regulator Hfq. *Proceedings of the National Academy of Sciences of the United States of America* 104, 16299-16304.

Wong, P. C. F., Chai, L. C., Lee, H. Y., Tang, J. Y. H., Noorlis, A., Farinazleen, M. G., Cheah, Y. K., and Son, R. (2011). Biofilm formation by *Salmonella Typhi* and *Salmonella Typhimurium* on plastic cutting board and its transfer to dragon fruit. *International Food Research Journal* 18, 31-38.

Wu, Y. T., Zhu, H., Willcox, M., and Stapleton, F. (2011). The effectiveness of various cleaning regimens and current guidelines in contact lens case biofilm removal. *Investigative Ophthalmology & Visual Science* 52, 5287-5292.

Xia, X., Zhao, S., Smith, A., McEvoy, J., Meng, J., and Bhagwat, A. A. (2009). Characterization of Salmonella isolates from retail foods based on serotyping, pulse field gel electrophoresis, antibiotic resistance and other phenotypic properties. *International Journal of Food Microbiology* 129, 93-98.

Xu, H., Lee, H. Y., and Ahn, J. (2010). Growth and virulence properties of biofilm-forming *Salmonella enterica* serovar Typhimurium under different acidic conditions. *Applied and Environmental Microbiology* 76, 7910-7917.

Zobell, C. E., and Allen, E. C. (1935). The significance of marine bacteria in the fouling of submerged surfaces. *Journal of Bacteriology* 29, 239-251.

Influence of a Salt Water Regulator on the Survival Response of *Salmonella Paratyphi* in Vembanadu Lake: India

Chandran Abhirosh[1,2], Asit Mazumder[1], Sherin Varghese[2],
A.P Thomas[2] and A.A.M Hatha[3]

[1]*Water and Aquatic Sciences Research Lab*
Department of Biology, University of Victoria, Victoria
[2]*School of Environmental Sciences, Mahatma Gandhi University, Kottayam, Kerala*
[3]*School of Marine Sciences, Cochin University of Science*
and Technology, Cochin, Kerala
[1]*Canada*
[2,3]*India*

1. Introduction

Contamination of environmental water by pathogenic microorganisms and subsequent infections originated from such sources during different contact and non- contact recreational activities are a major public health problem worldwide particularly in developing countries. The main pathogen frequently associated with enteric infection in developing countries are *Salmonella* enterica serovar typhi and paratyphi. Although the natural habitat of *Salmonella* is the gastrointestinal tract of animals, it find its way into natural water through faecal contamination and are frequently identified from various aquatic environments (Baudart *et al.*, 2000; Dionisio *et al.*, 2000; Martinez -Urtaza *et al.*, 2004., Abhirosh *et al.*, 2008). Typhoid fever caused by *S. enterica* serotype typhi and paratyphi are a common infectious disease occurring in all the parts of the world with its highest endemicity in certain parts of Asia, Africa, Latin America and in the Indian subcontinent with an estimated incidence of 33 million cases each year with significant morbidity and mortality (Threlfall, 2002). In most cases the disease is transmitted by polluted water (Girard *et al.*, 2006) because of the poor hygienic conditions, inadequate clean water supplies and sewage treatment facilities. However in developed countries the disease is mainly associated with food (Bell *et al.*, 2002) especially shellfish (Heinitz *et al.*, 2000).

Salmonella, since being allochthonous to aquatic environments, the potential health hazard is dependent on their period of survival outside the host and retention of critical density levels in the receiving water in a given time frame during transmission via the water route. In general, the major environmental factors influencing the enteric bacterial survival following their exposure to aquatic environments are water temperature (Anderson *et al.*, 1983), adsorption and sedimentation processes (Auer & Niehaus, 1993), sunlight action (Sinton *et al.*, 1999), lack of nutrients (Sinclair & Alexander, 1984), predation by bacteria or protozoa (Hahn & Hofle, 2001), bacteriophage lysis (Ricca & Cooney, 1999), competition with autochthonous microbiota (McCambridge & McKeekin, 1981) and antibiosis (Colwell, 1978).

Although *Salmonella* spp. has been isolated from fresh, estuarine and marine waters, they showed differential survival response to those aquatic environments and the results were sometimes contradictory in relation to salinity. For instance, it has been reported that *Salmonella* showed very low survival in sea water (Lee *et al.*, 2010) on the contrary Sugumar & Mariappan (2003) found that they exhibited very long survival up to 16 to 48 week in sea water. But it is also documented that it survived for 54 days (Moore *et al.*, 2003) and 58 days in freshwater Sugumar & Mariappan (2003). However, when *Salmonella* suspended in stabilization ponds effluent and rapidly mixed with brackish water, survival time was particularly short, whereas it was prolonged when the bacteria was submitted to a gradual increase in salinity (Mezrioui *et al.*, 1995). Therefore the survival of pathogenic bacteria in estuarine environments in response to varying saline concentration due to the mixing of salt water with freshwater has of particular health significance especially in locations where contact and non recreation takes place.

Hence the present study has been carried out in Vembanadu Lake that lies 0.6-2.2 m below mean sea level (MSL) along the west coast of India ($9°35'N$ $76°25'E$) and has a permanent connection with the Arabian Sea (Fig.1). As the north-east monsoon recedes, the area is exposed to tidal incursion of saline water from the Arabian Sea. In order to prevent the saline incursion during certain periods of the year, a salt water regulator is constructed in the lake. It divides the lake into a freshwater region on the southern part and a saline lagoon on the northern part. As a result, during the closure and opening of the regulator the water quality on both regions of the regulator may change in terms of its salinity and a progressive saline gradient may occur throughout the lake when the regulator is open. On the other hand over 1.6 million people directly or indirectly depend on it for various purposes such as agriculture, fishing, transportation and recreation. As a result water related diseases are very common in this region particularly in young children but none of them were reported officially. Enteric fever caused by *Salmonella* enterica serovars paratyphi A, B and C and Newport have been reported in India (Misra *et al.*, 2005; Gupta *et al.*, 2009).

Since die-off of enteric bacteria in aquatic environment could be attributed to a variety of interacting physical, chemical and biological factors and processes (Rhoder & Kator, 1988), in our previous studies in the Vembanadu lake we have evaluated the effect of sunlight, chemical composition of the estuarine water (Abhirosh & Hatha, 2005) effect of biological factors such as protozoan predation, predation by bacteriophages, autochthonous bacterial competition (Abhirosh *et al.*, 2009) on the survival of *Salmonella* and other organisms. However, the effect of salinity, since being important on the survival of enteric bacteria has not been evaluated in Vembanadu lake. As we already reported the presence of different *Salmonella* serotypes such as *Salmonella paratyphi* A, B, C and *Salmonella Newport* in Vembanadu lake (Abhirosh *et al.*, 2008), in this study our aim was to evaluate the health risk associated with *S. paratyphi* when released into the water by studying the survival responses to the salinity changes (saline gradient) caused by the saltwater regulator in Vembanadu lake using microcosm experiments at 20°C and 30°C.

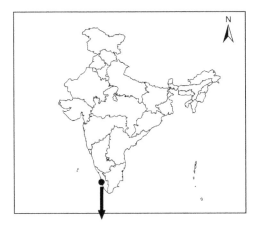

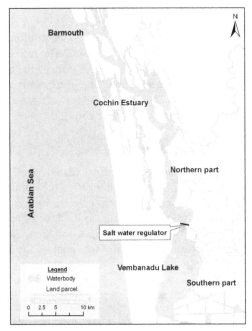

Fig. 1. Map showing Vembanadu Lake

2. Materials and methods

2.1 Test organism and water sample

A pure culture of *S. paratyphi* isolated from the Vembanadu lake was used for the survival experiments. All experiments were conducted in filter sterilized lake water in order to avoid the effect of predation. When saltwater regulator is closed the saline intrusion from northern part is prevented and the water on the southern part becomes freshwater. Therefore, to imitate the actual condition on the southern part of the lake, experiments were conducted in water collected from the lake when the salinity was 0 ppt (freshwater microcosm). To study the survival of the test organisms during mixing of water from northern and southern part of the Vembanadu lake, experiments were conducted in mixing water samples collected when the regulator was open (mixing water microcosm). Besides, in order to study the survival in all possible saline gradient throughout the year, survival experiments were conducted in lake water with salinity concentration ranged from 0-25 ppt. The test solutions of desired saline concentrations were prepared using fresh lake water with NaCl.

2.2 Preparation of inocula

The inocula were prepared as previously described by Abhirosh & Hatha (2005). *S. paratyphi* was grown in Tryptone Soy Broth (TSB) and incubated at 37°C for 24 h. After incubation, the cells were concentrated by centrifugation at 1400 × g for 15 min and washed twice with sterile isotonic saline. After the final wash, the cells were re-suspended in sterile isotonic saline for inoculation into the microcosms. Then 1 ml washed cell suspension of *S. paratyphi* was inoculated into each microcosm containing different test solution (250 ml Erlenmeyer flask with 100 ml) at a concentration of 10^{6-7} CFU/ mL. All inoculated microcosm were incubated at 20°C and 30°C. The microcosms were incubated at 20°C in order to find out the survival at low temperature as the temperature goes down to 20°C in winter as well as at a certain depth. The enumeration of culturable bacteria were done after 2, 4, 6, 8, 10, 13, 16, 22, 28 and 34 days using spread plate technique on TSA agar plates and the colony forming units were counted.

2.3 Decay rate and statistical calculation

The decay rates of culturable *S. paratyphi* cells were calculated as per first order decay model using the following equation Log Nt/N_0= $-kt$, where Nt is the number of bacteria at time t, N_0 is number of bacteria at time 0, and t is expressed in days; k is the first-order constant calculated by linear regression technique. T_{99} (time required for 2 log reduction) values were calculated using the decay constant (k) in the following equation, $T_{99}=-2/k$. The difference in the survival at different salinities and temperature was analysed using two way analysis of variance (ANOVA).

3. Results and discussion

The survival curves of *S. paratyphi* in freshwater and mixing water at 20°C and 30°C are given in Fig. 2 and the inactivation rates and T_{99} values are given in Table 1. The results revealed that *S. paratyphi* showed significantly (p<0.01) higher survival at 20°C (T_{99}= 25.99) compared to 30°C (T_{99}= 17.68) in freshwater water indicating their better survival capacity at low temperature. However *S. paratyphi* did not show much difference in the survival response in mixing water at both temperature and the T_{99} respectively was 16.37 days at 20°C and 15.12 days at 30°C. The results revealed that *S. paratyphi* cells remained viable until 34 days at a high density of 10^5 CFU/mL. The salinity of the mixing water when it was collected was 12.77 ppt and the average saline concentration of the lake water was 12.5ppt when it was monitored over 2hr interval in a day.

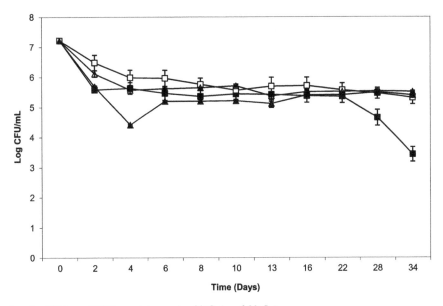

Freshwater 30°C■ and 20°C □; mixing water 30°C ▲ and 20°C Δ

Fig. 2. The survival curves of *S. paratyphi* in freshwater and mixing water at 20°C and 30°C

Days	Freshwater 30°C	Freshwater 20°C	Mixing water 30°C	Mixing water 20°C
0	0.00	0.00	0.00	0.00
2	-1.64	-0.74	-2.48	-2.05
4	-1.59	-1.23	-3.75	-2.59
6	-1.75	-1.25	-2.96	-2.54
8	-1.86	-1.46	-2.96	-2.51
10	-1.78	-1.66	-2.95	-2.45
13	-1.80	-1.53	-3.04	-2.79
16	-1.84	-1.50	-2.76	-2.65
22	-1.86	-1.65	-2.75	-2.64
28	-2.57	-1.75	-2.67	-2.63
34	-3.78	-1.90	-2.76	-2.65
K value	-0.11	-0.07	-0.13	-0.12
T99	17.68	25.99	15.20	16.37

Table 1. Inactivation rates of *S. paratyphi* in freshwater and mixing water at 30°C and 20°C

Even though the survival time was longer, in agreement with our results Sugumar & Mariyappan (2003) reported that *Salmonella* survived up to 24 weeks in sterile freshwater microcosm at 30°C but at low temperature it survived for 58 weeks. It is also documented that it survived for 54 days (Moore *et al.*, 2003) in freshwater. Since *S. paratyphi* did not show much difference in survival response in mixing water at both temperatures, similar to our results Rhodes and Kator (1988) reported that *Salmonella* populations exhibited significantly less die-off in filtered estuarine water at temperatures of <10°C. In sterile estuarine water virtually unaltered bacterial densities over a 10-day period have also been reported by McCambridge & McMeekin (1980a,b). It has been documented in other studies that low temperature is favorable for the survival of *Salmonella* in (Vasconcelos & Swartz, 1976; Hernroth *et al.*, 2010) and other enteric bacteria in aquatic environments (Craig *et al.*, 2004; Sampson *et al.*, 2006; Silhan *et al.*, 2006).

The aim of conducting this survival experiments in freshwater and mixing water was to evaluate the public health risk associated with *S. paratyphi* in Vembanadul lake during the closure and subsequent opening of the regulator. While addressing this issue it has been noticed that similar to other studies *S. paratyphi* could survive very long time in freshwater and mixing water until the end of the experimental period. Therefore the log term survival potential *S. paratyphi* in freshwater may pose health risk since people use this region for their freshwater needs and we have already recorded high abundance of indicator bacteria and enteric pathogens (*Salmonella* serotypes such as *S. paratyphi* A, B, C and S. Newport) on the southern part during the closure of the saltwater regulator (Abhirosh *et al.*, 2008). During the closure of the saltwater regulator the water on southern part of the lake become fresh and the natural flow is prevented which results in the accumulation of organic load in the southern part of the lake, giving proper environmental conditions for the multiplication of bacteria. Besides, the high survival capacity noticed at low temperature further increases the health risk during monsoon season because of the drop down of the water temperature to nearly 20°C and we already reported high prevalence of indicator and pathogenic bacteria in southern part of the lake during monsoon season (Abhirosh *et al.*, 2008) and every year waterborne disease outbreaks occur during monsoon season. Prolonged survival of *S. paratyphi* in mixing water suggests that it can remain viable in water at high concentration (10^5CFU/ml) when the saltwater is open. It was almost similar to the results we obtained for *S. typhimurium* in Cochin estuary where we found it remained viable at even higher density (10^6 CFU/mL) until the end of experiment (Abhirosh & Hatha . 2005) at 20°C and 30°C. Our results are also in agreement with other studies that better survival of enteric bacteria in estuarine and other aquatic environments (Rhodes and Kator, 1988; Placha *et al.*, 2001).

It has been reported that *Salmonella* may be of prolonged public health significance once it is introduced into tropical surface waters than *E. coli* (Jimenez *et al.*, 1989). Sporadic outbreaks of enteric fever due to *S. enterica* serovars paratyphi A, have been reported in India with an annual incidence of 3 million cases (Threlfall,2002; Misra *et al.*, 2005). *S. enterica* serovar paratyphi A has emerged as an important cause of enteric fever in India Gupta *et al.* (2009). These reports suggest that the high survival of *S. paratyphi* in Vembanadu lake could be a public health concern.

In order to assess the survival in all possible saline concentrations on both sides of the salt water regulator, survival experiment were conducted in lake water at 5, 10, 15, 20, and

25ppt at 20°C and 30°C and the results are represented in Fig 3-8 and the inactivation rates are given in Table 2 and 3. When the saltwater is closed the saline concentration on Northern part was reported to a maximum of 20ppt. Even though no significant variation in the survival response of *S. paratyphi* was noticed at 0, 5, 10, 15 and 20 ppt ($p>0.05$), they exhibited an extended survival for 34 days at 20°C and 30°C. They showed enhanced survival in water at 0 ppt at both temperatures as evident from T_{99} values and it was 25.99 days at 20°C and 17.68 days at 30°C (Table 2 and 3). However as time goes depending on the increasing saline concentration from 5to 25 ppt it showed gradual decrease in the T_{99} values at both temperatures. The lowest T_{99} was observed at 25 ppt (8.61 and 7.25) and showed a significant ($p<0.0001$) decline of cultural cells at both temperature indicating the deleterious effect of high saline concentration. However the most suitable condition for their growth was found to be at 0 and 5 ppt and suggests that they can survive well at low salinity levels in Vembanadu lake. The results indicate that *Salmonella* can survive well in water weakly diluted or with gradually increasing saline concentrations. In agreement with our results Mezrioui *et al.* (1995) reported that when *Salmonella* suspended in stabilization ponds effluent and rapidly mixed with brackish water survival time was particularly short as we found at 25 ppt where it showed a sudden decline at both temperature, whereas it was prolonged when the bacteria was submitted to a gradual increase in salinity.

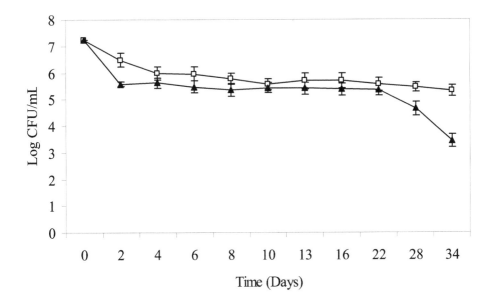

Fig. 3. Survival curves of *S. paratyphi* in fresh sterile water at 0 ppt at 20°C (□)and 30°C (▲) (Mean ±SD, n = 4).

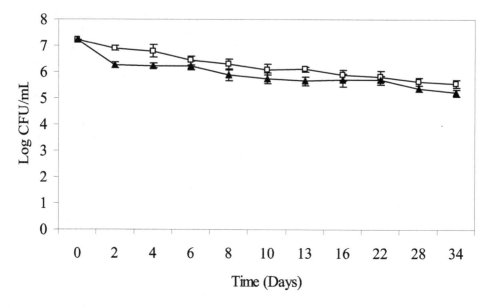

Fig. 4. Survival curves of *S. paratyphi* in sterile water at 5 ppt at 20ºC (□)and 30ºC (▲) (Mean ±SD, n = 4).

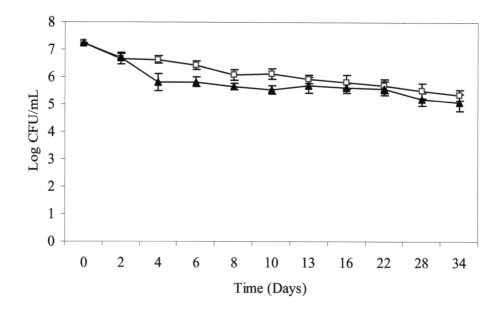

Fig. 5. Survival curves of *S. paratyphi* in sterile water at 10 ppt at 20ºC (□)and 30ºC (▲) (Mean ±SD, n = 4).

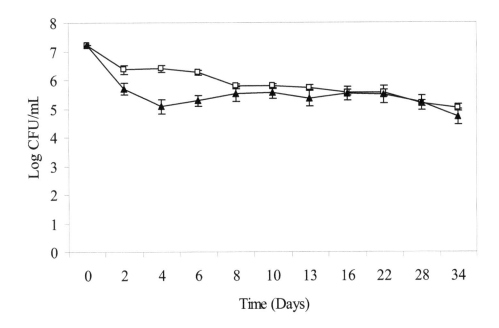

Fig. 6. Survival curves of *S. paratyphi* in sterile water at 15 ppt at 20°C (□)and 30°C (▲) (Mean ±SD, n = 4).

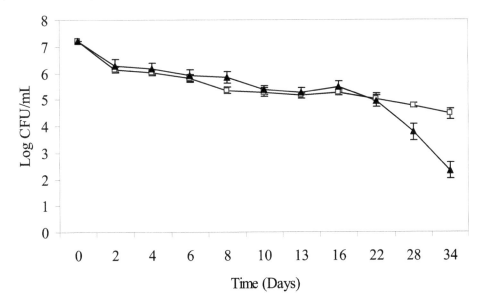

Fig. 7. Survival curves of *S. paratyphi* in sterile water at 20 ppt at 20°C (□)and 30°C (▲) (Mean ±SD, n = 4).

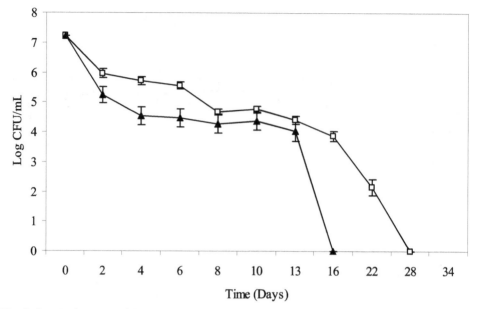

Fig. 8. Survival curves of *S. paratyphi* in sterile water at 25 ppt at at 20°C (□)and 30°C (▲) (Mean ±SD, n = 4).

Days	Saline concentration					
	0 ppt	5 ppt	10ppt	15ppt	20ppt	25 ppt
0	0.00	0.00	0.00	0.00	0.00	0.00
2	-0.74	-0.31	-0.58	-0.85	-1.10	-1.26
4	-1.23	-0.43	-0.59	-0.82	-1.19	-1.49
6	-1.25	-0.78	-0.79	-0.96	-1.40	-1.67
8	-1.46	-0.92	-1.14	-1.44	-1.87	-2.54
10	-1.66	-1.13	-1.10	-1.44	-1.95	-2.46
13	-1.53	-1.12	-1.28	-1.47	-2.07	-2.80
16	-1.50	-1.34	-1.39	-1.65	-1.96	-3.35
22	-1.65	-1.38	-1.53	-1.67	-2.20	-5.07
28	-1.75	-1.59	-1.72	-2.03	-2.44	-7.22
34	-1.90	-1.66	-1.87	-2.19	-2.76	-
k	-0.07	-0.06	-0.06	-0.08	-0.10	-0.23
T99	25.99	31.74	28.85	24.54	19.23	8.61

Table 2. Inactivation rates of *S. paratyphi* in water at different saline concentration at 20°C

Days	Saline concentration					
	0 ppt	5 ppt	10ppt	15ppt	20ppt	25 ppt
0	0.00	0.00	0.00	0.00	0.00	0.00
2	-1.64	-0.97	-0.53	-1.52	-0.94	-1.98
4	-1.59	-0.98	-1.40	-2.14	-1.07	-2.68
6	-1.75	-0.98	-1.39	-1.95	-1.32	-2.74
8	-1.86	-1.35	-1.56	-1.70	-1.37	-2.94
10	-1.78	-1.47	-1.68	-1.67	-1.85	-2.85
13	-1.80	-1.56	-1.51	-1.85	-1.97	-3.20
16	-1.84	-1.53	-1.61	-1.69	-1.74	-7.22
22	-1.86	-1.52	-1.63	-1.72	-2.28	-
28	-2.57	-1.83	-2.02	-2.01	-3.45	-
34	-3.78	-2.00	-2.15	-2.52	-4.90	-
K	-0.11	-0.07	-0.08	-0.09	-0.13	-0.27
$T99$	17.68	26.14	24.03	21.56	14.95	7.25

Table 3. Inactivation rates of S. paratyphi in water at different saline concentration at 30°C

We clearly observed that the decline of cell density with increasing saline concentration. Similar results were reported previously when a freshwater bacteria was exposed to brakish water, Painchaud et al. (1987, 1995; Painchaud and Therriault 1989). Similar gradients were reported in other estuaries (Albright, 1983) Rivers (Prieur, 1987). Painchaud et al.(1995) reported that no mortality resulted from exposure to water with a salinity of >10ppt and high bacterial count at saline concentration between 0-5. He also reported drastic decline of bacteria at higher salinity (20ppt). This is in agreement with our results that we observed high survival rate at 0 and 5 ppt which was found to be the most suitable condition for the growth whereas at 25 ppt a drastic decline was noticed indicates the deleterious effect of high saline concentration.

At higher saline concentration, for example in sea water, enteric bacteria are subjected to an immediate osmotic upshock, and their ability to overcome this by means of several osmoregulatory systems could largely influence their subsequent survival in the marine environment (Gauthier et al., 1987; Davies et al., 1995). This osmotic shock might be the reason for the sudden decline of cells at 25ppt. However there are contradicting reportes related to enteric bacterial survival in sea water. Lee et al (2010) Gerba and McLeod, (1976) reported that noh halophilic bacterial like Salmonella and E. coli do not survive well in seas water whereas Sugumar & Mariappan (2003) reported very long survival up to 16 to 48 week in sea water. Upon an osmotic upshift, bacterial cells accumulate or synthesize specific osmoprotectant molecules, in order to equalize osmotic pressure and avoid drastic loss of water from the cytoplasm (Csonka & Epstein, 1996). Although the accumulation or synthesis of such molecules (trehalose, glycine betaine, glutamic acid) has been reported in *Salmonella* spp. in estuarine waters, in the present study *S. paratyphi* might not overcome the stress caused by the high saline concentration at 25ppt whereas all other saline concentration tested were not found to be lethal.

The maximum saline concentration during the closure of the regulator on the Northern part of the Vembanadu lake is 20 ppt and minimum is 0ppt. Therefore in a year the possible seasonal salinity changes in Vembanadu lake could be between 0-20ppt. It has been generally assumed that when the regulator is closed the bacterial density on the Northern part would be very low because of the increasing saline concentration compared to Southern part. But it has been clearly observed that *S. paratyphi* exhibited high survival capacity in all possible saline gradients from 0 to 20ppt. The result indicates that *S. paratyphi* could survive very long time throughout Vembanadu lake irrespective of the saline concentration. Since the opening and closing of the regulator related to water quality and recreational activities has always been a topic of endless debate, the results indicates that the opening and closing of the salt water regulator does not have any significant impact on the survival (reduction in survival) of the enteric pathogens in relation to saline concentration in Vembanadu lake. However, if the saline concentration reaches 25 ppt it will negatively affect their survival ($p < 0.0001$) but the maximum salinity so far reported is 20ppt. Since the lake is being used for various recreational activities the long term survival of *S. paratyphi* in all season regardless of saline concentration in Vembanadu lake could be a public health concern.

4. Conclusions

The results of the microcosm experiment revealed that *S. paratyphi* has a better survival capacity over a wide range of saline concentration from 0 to 20 ppt in Vembanadu lake. It exhibited significantly higher survival at 20°C compared to 30°C. It also showed prolonged survival in all other saline concentration at a higher density at both the temperature and the most suitable saline concentration was found to be 5 ppt. The result indicates that *S. paratyphi* could survive very long time throughout Vembanadu lake irrespective of the saline concentration. The opening and closing of the salt water regulator does not have any significant impact on the survival (reduction in survival) of the enteric pathogens in relation to saline concentration in Vembanadu lake. However, if the saline concentration reaches 25 ppt it will negatively affect their survival but the maximum salinity so far reported is 20ppt. Since the lake is being used for various recreational activities the long term survival of *S. paratyphi* in all season regardless of saline concentration in Vembanadu lake could be a public health concern.

5. References

Abhirosh, C.; & Hatha, A.A.M. (2005). Relative survival of *Escherichia coli* and *Salmonella typhimarium* in a tropical estuary. *Water Research*, Vol. 39, No.7 (April), pp.. 1397-1403, ISSN 0043-1354

Abhirosh, C.; Hatha, A.A.M.; & Sherin, V. (2008). Increased prevalence of indicator and pathogenic bacteria in Vembanadu Lake: a function of salt water regulator, along south west coast of India. Journal *of Water and Health Water Health*, Vol.6, No. 4, (December), pp.539–546, ISSN 1477-8920

Abhirosh, C.; Sheeja, K.M.; Hatha, A.A.M.; Sherin, V.; & Thomas, A.P. (2009). Role of biological factors on the survival of Escherichia coli, *Salmonella* paratyphi and Vibrio parahaemolyticus in a tropical estuary, India. *Water Journal*, Vol.1,No.0, (October),pp. 76–84.

Albright, L. J. (1983). Influence of river-ocean plume upon bacterioplankton production of the Strait of Georgia, British Columbia. *Marine Ecology Progress Series*, Vo.12, No.2, (May), pp. 107–113, ISSN 0171-8630

Anderson, J.C.; Rhodes, M.W.; &Kator, H.I. (1983). Seasonal variation in the survival of *E. coli* exposed *in situ* in membrane diffusion chambers containing filtered and nonfiltered estuarine water. *Applied and Environmental Microbiology*, Vol. 45, No.6, (June), pp. 1877-1883.ISSN 0099-2240

Auer, M.; &Niehaus, J. (1993). Modelling faecal coliform bacteria I. Field and laboratory determination of loss kinetics. *Water Research*, Vol. 27, No. 4, (April), pp. 693-701, ISSN 0043-1354

Baudart, J.; Grabulos, J.; Barusseau, J.P. & Lebaron, P. (2000). *Salmonella* spp. and faecal coliform loads in coastal waters from a point vs. nonpoint source of pollution. *Journal of Environmental Quality*, Vol. 29, No.1, (January), pp. 241-250, ISSN 0047-2425

Bell, C. & Kyriakides, A. (2002). *Salmonella*. A practical app.roach to the organism and its control in foods. Practical Food Microbiology Series. Blackwell, ISBN 978-0-632-05519-7 Oxford, United Kingdom.

Colwell, R.R. (1978). Bacteria and viruses, indicators of environmental changes occurring in the estuaries. *Environment International*, Vol. 1, No.5, (May), pp. 223-231, ISSN 0160-4120.

Craig, D.L.; Fallowfield, H.J. & Cromar, N.J. (2004). Use of microcosms to determine persistence of Escherichia coli in recreational coastal water and sediment and validation with in situ measurements. *Journal of Applied Microbiology*, Vol. 96, No. 5, (March),pp. 922–930, ISSN 1365-2672

Csonka, L.N.; & Epstein, W. (1996). Osmoregulation. In: *Escherichia coli and Salmonella*: Cellular and Molecular Biology, Vol. 1 Neidhardt, F.C. Curtiss, R. Ingraham, J.L.Lin, E.C.C.Low, K.B.Magasanik, B. Rezniko.; W.S. Riley.; M. Schaechter.;& H.E. Umbarger, (Eds.), pp. 1210-1223. ASM Press, Washington, DC.

Davies, C.M.; Apte, S.C; &Peterson, S.M. (1995). L-D-Galactosidase activity of viable, non-culturable coliform bacteria in marine waters. *Letters in Applied Microbiology*, Vol. 21,No. 2, (August), pp. 99-102.

Dionisio, L.P.C.; Joao, M.; Ferreiro, V.S.; Fidalgo, M.L.; Garcia Rosado, M.E.; &Borrego, J.J.; 2000. Occurrence of*Salmonella* spp. in estuarine and coastal waters of Portugal. *Antonie van Leewenhoek*, Vol. 78, No.1, (July), pp. 99–106, ISSN 0003-6072

Gauthier, M.J.; Munro, P.M.; & Mohajer, S. (1987). Influence of salts and sodium chloride on the recovery of Escherichia coli from seawater. *Current Microbiology*, Vol.15, No.1, (January), pp. 5-10, ISSN 0343-8651

Gerba, C.P.; & McLeod, J.S. (1976). Effect of sediments on the survival of *Escherichia coli* in marine water. *Applied and Environmental Microbiology* , Vol.32, No.1, (July), pp. 114-120, ISSN 0099-2240

Girard, M.P.; Steele, D.; Chaignat, C.L.; & Kieny, M.P. (2006). A review of vaccine research and development: human enteric infections. *Vaccine*, Vol.24, No. 15, (April), pp. 2732-2750, ISSN 0264-410X.

Gupta, V.; Kaur, J.; & Chander, J. (2009). An increase in enteric fever cases due to *Salmonella* Paratyphi A in and around Chandigarh. *Indian Journal of Medical Research*,Vol. 129, No.1, (January),pp.95-98, ISSN 0971-5916

Hahn, M.W.; &Hofle, M.G. (2001). Grazing of protozoa and its effect on populations of aquatic bacteria. FEMS Microbiology Ecology, Vol. 35, No. 2, (April), pp. 113-121, ISSN 0168-6496

Heinitz, M.L.; Ruble, R.D.; &Wagner, D.E. *et al.* 2000. Incidence of *Salmonella* in fish and seafood. *Journal of Food Protection*, Vo.l 63, No. 5, (May), pp. 579-592. ISSN 0362-028X

Hernroth, B.; Lothigius, A.; & Bölin, I. (2010). Factors influencing survival of enterotoxigenic Escherichia coli, *Salmonella*enterica (serovar Typhimurium)and Vibrio parahaemolyticus in marineenvironments. *FEMS Microbiology Ecology*, Vo.71, No.2, (February), pp.272-80, ISSN 0168-6496

Jimenez, L.; Muniz, I.; Toranzos, G.A.; & Ilazen, T.C.(1989). Survival and activity of *Salmonella typhimurium* and *Escherichia coli* in tropical freshwater. *Journal of Applied Bacteriology*, Vol.67, No.1(July), 61 – 69, ISSN0021-8847

Lee, C.W.; Ng, A.Y.; Bong, C.W.; Narayanan, K.; Sim, E.U.; Ng, C.C. (2010) Investigating the decay rates of Escherichia coli relative to Vibrio parahemolyticus and *Salmonella* Typhi in tropical coastal waters. *Water Research*, Vol. 54, No. 4, (February), pp. 1561-70, ISSN 0043-1354

Martinez-Urtaza, J.; Liebana, E.; Garcia-Migura, L.; Perez-Pineiro, P.; & Saco, M. (2004). Characterization of *Salmonella enterica* Serovar Typhimurium from marine environments in coastal waters of Galicia (Spain). *Applied and Environmental Microbiology*, Vol.70, No. 7, (July),pp. 4030- 4034, ISSN 0099-2240

McCambridge, J.; & McMeekin T. A. (1980a). *Effect of temperature on activity of predators of Salmonella typhimurium and Escherichia coli in estuarine water, Australian Journal of Marine and Freshwater Research*, Vol. 31,No. 6, (June), pp. 851 - 855 ,ISSN 0067-1940

McCambridge, J.; & McMeekin, T. A. (1980b). Relative effects of bacterial and protozoan predators on survival of Escherichia coli in estuarine water samples. *Applied and Environmental Microbiolgy*, Vol. 40,No.5, (November), pp..907-911, ISSN 0099-2240

McCambridge, J.; &McMeekin, T.A. (1981). Effect of solar radiation and predacious microorganisms on faecal and other bacteria. *Applied and Environmental Microbiology*, Vol. 41, No. 5,(May), pp.1083-1087, ISSN 0099-2240

Mezrioui, N.; Baleux, B.; Trousselier, M. (1995). A microcosm study of the survival of Escherichia coli and *Salmonella* typhimurium in brackish water. *Water Research* Vol.29, No. 2, (February), pp. 459–465, ISSN 0043-1354

Misra, R.N.; Bawa, K.S.; Magu, S.K.; Bhandari, S.; Nagendra, A.; & Menon, P.K. (2005). Outbreak of multi-drug resistant *Salmonella* Typhienteric fever in Mumbai

Garrison. Medical Journal Armed Forces India, Vol.61, No.1,(January),pp. 48-50. ISSN: 0377-1237

Moore,B.C.; Edward Martinez,, E.; Gay, J.M.; & Rice, D.H. (2003). Survival of *Salmonella enterica* in Freshwater and Sediments and Transmission by the Aquatic Midge *Chironomus tentans* (Chironomidae: Diptera). *Applied and Environmental Microbiology*, Vol.69, No.8, (August)pp.4556–4560, ISSN 0099-2240

Painchaud, J.; & Therriault, J.C. (1989). Relationships between bacteria, phytoplankton and particulate organic carbon in the Upp.er St. Lawrence Estuary. *Marine Ecology Progress Series*, Vol. 56, No. (August)pp. 301–311, ISSN 0171-8630

Painchaud, J.; Lefaivre, D. ;&J.C.; Therriault. C. (1987). Box model analysis of bacterial fluxes in the St. Lawrence Estuary. *Marine Ecology Progress Series*, Vol. 41, No.3, (December), pp. 241– 252, ISSN0171-8630

Painchaud,J.; Therriault,J.C.; Legendre, L. (1995). Assessment of salinity-related mortality of freshwater bacteria in the saint Lawrence estuary. Applied and Environmental Microbiology, Vol. 61, No.1, (January)pp. 205–208, ISSN 0099-2240

Placha, I.; Venglovsky, J.; Sasakova, N.; &Svoboda, I.F. (2001). The effect of summer and winter seasons on the survival of *Salmonella typhimurium* and indicator microorganismsduring the storage of solid fractions of pig slurry. Journal of *Applied Microbiology*,Vol. 91, No.6, (December), pp. 1036 – 1043, ISSN 1365-2672

Prieur, D.; Troussellier, M.; Romana, A.; Chamroux, S.; Mevel, G.; & Baleux, B. (1987). Evolution of bacterial communities in the Gironde Estuary(France) according to a salinity gradient. Estuarine, Coastal and Shelf ScienceVol. 24, No.1, (January), pp.95– 108, ISSN 0272-7714

Rhodes, M.W.; & Kator, H.I. 1988. Survival of *Escherichia coli* and *Salmonella* spp.. in estuarine environments. *Applied and Environmental Microbiology*, Vol.54, No. 12,(December),pp. 2902-2907, ISSN 0099-2240

Ricca, D.M.; & Cooney, J.J. (1999). Coliphages and indicator bacteria in Bosten Harbour, Massachusetts. *Environmental Toxicology*, Vol. 14, No. 4, (August), pp. 404 – 408, ISSN 1520-4081

Sampson, R.W.; Swiatnicki, S.A.; Osinga, V.L.; Supita, J.L.; McDermott, C.M.; & Kleinheinz, G.T. (2006). Effects of temperature and sand on E. coli survival in a northern lake water microcosm. *Journal of Water and Health*, Vol. 4, No. 3, (September),pp. 389– 393, ISSN 0043-1354

Silhan, J.; Corfitzen, C.B. & Albrechtsen. H.J. (2006). Effect of temperature and pipe material on biofilm formation and survival of Escherichia coli in used drinking water pipes: a laboratory-based study. *Water Science and Technology*, Vol. 54, No.3, (March), pp. 49–56, ISSN 0273-1223

Sinclair, J.L; & Alexander, M. (1984). Role of resistance to starvation in bacterial survival in sewage and lake water. *Applied and Environmental Microbiology*, Vol. 48, No.2, (August), pp. 410- 415. ISSN 0099-2240

Sinton, L.W.; Finaly, R.K.; &Lynch, P.A. (1999). Sunlight inactivation of faecal bacteriophages and bacteria in sewage polluted seawater. *Applied and Environmental Microbiology*, Vol.65, No.8, (August), pp.3605-3613, ISSN 0099-2240

Sugumar, G.; & Mariappan, S. (2003) Survival of *Salmonella* sp. in Freshwater and Seawater
    Microcosms Under Starvation. *Asian Fisheries Science, Vol.* 16, No. 3,(March), pp.
    247-255, ISSN: 0116-6514

Threlfall, E.J. (2002). Antimicrobial drug resistance in *Salmonella*: problems and perspectives
    in food- and water-borne infections. FEMS Microbiology Reviews, Vol. 26, No.
    2,(June), pp. 141-148, ISSN1574-6976

Vasconcelos, G.J.; & Swartz, R.G. (1976). Survival of bacteria in seawater using a diffusion
    chamber app.aratus in situ. *Applied and Environmental Microbiology* , Vo. 31,No. 6,
    (June), pp. 913-920, ISSN 0099-2240

3

Motility and Energy Taxis of *Salmonella* spp.

Andreas E. Zautner
Universitätsmedizin Göttingen
Germany

1. Introduction

1.1 Flagellar motility

The essential morphological prerequisite for active bacterial motility is the flagellum. Besides that, only flotation with the help of self-produced gas vesicles in some cyanobacteria and gliding of filamentous rod-shaped bacteria are known mechanisms of flagella-independent active motion (Madigan & Martinko, 2006). The flagellum is a tail-like protrusion that can be found in many bacterial species. Besides its main function – locomotion – it is also involved in passing the mucosa barrier (Erdem et al., 2007), regulation of auto-aggregation (Ulett et al., 2006), aggregation on solid surfaces, which precedes biofilm formation (O'Toole & Kolter, 1998), and in the export of virulence factors and other proteins (Samudrala et al., 2009). Some bacteria own even more than one flagellum. According to number and arrangement of flagella, different schemes are distinguished (Hahne et al. 2004). Monotrichous bacteria, like *Vibrio cholerae* have only a single flagellum at one cell pole. The amphitrichous flagella arrangement scheme is characterized by single flagella on each of both cell poles, as observable for most *Campylobacter* spp. Lophotrichous flagellated bacteria, e.g. *Pseudomonas aeruginosa*, have multiple flagella on one cell pole and peritrichous bacteria, like *Salmonella* spp. have multiple flagella randomly distributed over the whole cellular surface.

The flagellum acts in principle like a marine screw propeller. Its rotational direction is by definition described by an external observer looking down the flagellar filament toward the bacterial cell (Adler, 1975). The flagellar mechanics is the only known real-rotating joint in the biological world. Its rotation frequency is around 100 Hz (Lauga et al., 2006). The direction of the flagellar motor and in consequence of the flagellar filament determines whether there is a thrust or drag impulse acting on the bacterium. The rotational direction can be reversed in a very short time, thus thrust and drag impulse momentum can switch suddenly. In general, the flagellum pushes the bacterium by providing a pressure gradient, which is relatively high near the filament and acts as a centrifugal force (Gebremichael et al., 2006). According to the physical law of the conservation of the angular momentum, the bacterial body rotates slowly in the counter direction at a rotation frequency of about 10 Hz (Lauga et al., 2006). A counter-clockwise rotation of the flagella causes a bacterial cell to move straight forwards, whereas a clockwise rotation causes the bacterium to tumble. The bacterial movement is controlled by conformational transitions in the flagellar filament between left- and right-handed supercoils (Kitao et al. 2006). These transitions are realized by a high flexible structure of the flagellar filament, due to "sliding"-interactions and "switch"-interactions", which stabilize inter- and intrasubunit interactions (Kitao et al. 2006). In case of a counter-clockwise flagellar rotation, several filaments of a left-handed helical structure form a bundle and act as propeller. If the flagellar motor rotates clockwise a transition into a right-handed helix of the filament structure is induced and the bundle is feazed (Larsen et al. 1974). The flagella of peritrichous bacteria are synchronized some way that they all rotate in the same orientation. They unite to form a rear-facing bundle that pushes the bacterium forward (Adler, 1975). In amphitrichous bacteria, the flagella of both poles rotate in opposite directions. Thus, the flagellum of the rear-end rotates comparable to monotrichous bacteria in order to provoke a thrust impulse, whereas the flagellum of the bow-end is bent backwards and turns around the front end of the bacterium. Thereby, the thrust impulse is increased. If the direction of the flagellar rotation is reversed, the filaments are fold over. The rear-end of the bacterium becomes the bow-end and the bow-end becomes the rear-end. In consequence, the bacterium swims in the opposite direction. In case of Gram-negative bacteria like *Salmonella* sp., the process of active bacterial movement is divided into continuously alternating phases of slow, non-directed movement called "tumbles" and phases of fast, straight-lined movement called "runs" (Adler, 1975). During a "tumble", the bacterium stops and turns in a more or less randomly chosen direction. It is a passive phase of re-orientation due to a rotational motion, where the non-spherical shape of the bacterial cell affects the way that it is rotated by the shear flow of the surrounding medium. Then the bacterium starts a fast, rectilinear "run", driven by the rotation of the flagella until it stops again and the next motion cycle begins. When the rotational direction of the flagella of peritrichous-flagellated bacteria is to be inverted, the individual flagellum is directed radially from the bacterial cell body in a way that it is sticking out. The dragging effects on the bacterial body outweigh each other to the mean positions in which the bacterium tumbles in a random motion in one place. The reversal of the flagellar rotation and the associated change in the direction of motion plays an important role in (chemo)-tactic movements (Adler, 1975).

1.2 Chemotaxis

Chemotaxis is the process in which bacteria direct their locomotion dependent on the concentration of certain substances in their environment. Compounds affecting chemotaxis are called chemotaxins or chemoeffectors. Chemotaxis in the direction of a higher concentration of the chemoeffector is defined as positive and these kind of compounds are called chemoattractors. On the contrary, chemotaxis away from the higher concentration is defined as negative and these chemotaxins are called chemorepellents. Energy sources usually attract motile bacteria whereas bacteriotoxic agents act as repellents (Fig. 1). The finding, that bacteria move actively towards or away from certain substances, was already made at the end of the 19th century by Engelmann (Engelmann, 1881) and Pfeffer (Pfeffer, 1884 & 1888). Thus, with the help of chemotaxis bacteria direct their movement to find favourable niches with high chemoattracor and low chemorepellent concentrations. This decision-making is based on temporal sensing. As indicated above the overall motion of a bacterium is composed of alternating phases of straight swimming and thumbling. In the presence of a chemical gradient the straight swimming phases last longer, and if the bacterium is moving nat along this gradient, it starts sooner to tumble and tries to reorientate depending on the chemotaxins concentration (Adler, 1975). The essential prerequisites for chemotaxis are, as already mentioned, a flagella mediated motility, a variety of individual chemoreceptors and a highly conserved chemosensory signal-transduction system.

2. Flagellar motility and chemotaxis

2.1 Experimental approaches

Before the mechanisms of flagellar motility and chemotaxis will be discussed, the most common tools or experimental approaches to study and record bacterial motility and taxis will be presented: microscopy and chemotaxis assay.

2.1.1 Microscopy

Conventional light microscopy is not sufficient to visualize flagellar filaments because of their thinness and the swiftness. One very early approach to visualize flagella of living bacterial cells is dark field microscopy (Macnab, 1976). Since light is scattered by dirt particles reducing the contrast, it has to be considered that the medium and the specimen slides must be remarkably clean. A great advance in this field is video-enhanced differential interference-contrast microscopy (Block et al., 1991). Video microscopy combined with computer based image processing made it possible to detect very small objects like particular microtubules of ≈ 25 nm in diameter. Computerized image analysis offers the option to estimate values like mean cell run speed and average tumbling frequency and their variation in the presence or absence of attractants or repellents (Staropoli & Alon, 2000). Phase-contrast video microscopy combined with the analysis of superimposed image series is a very useful tool, especially for the study of the taxis to and the motion near solid surfaces (Lauga et al., 2006). A further helpful method, although not specifically associated with flagellar motility and chemotaxis, is fluorescence microscopy, which can be used to visualize protein-protein-interactions in the chemoreceptor signaltransduction pathway and the fagellar motor, in combination with green fluorescence fusion proteins (Pierce et al., 1999; Khan et al., 2000).

2.1.2 Chemotaxis assays

Another easy to handle experimental set of tools is composed from different kinds of chemotaxis assays (Miller et al. 2009). One semiquantitative variant is based on changes of the opalescence of a semi-solid agar due to the concentration of bacterial cells (Hugdahl et al., 1988). In a first step, a phosphate buffered saline-agar solution is mixed with a bacterial suspension of a specifiy optical density and poured into a petri dish. After solidification, paper discs with the chemotaxins are placed onto the agar surface following incubation of three to four hours. A more opaque zone can be seen in the surrounding of chemoattractants (see Fig. 1A), whereas chemorepellents are girdled by a more transparent halo (see Fig. 1 B). Other versions of the agar based chemotaxis assay deal with pure – bacteria free - agar plates. After solidification of the agar small recessions are cut into the agar and are filled with either a bacterial suspension or the test solution (Köhidai, 1995). A variation of this assay uses parallel channels (PP-technique) cut from each of both recesses connected by a third perpendicular channel between these two to facilitate diffusion of bacterial suspension and test solution (Köhidai, 1995).

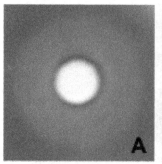 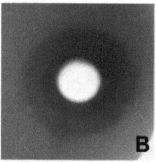 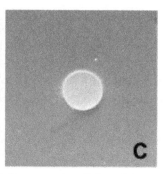

Fig. 1. Examples for chemotaxis: A: Attraction towards towards L-asparagine; B: Repulsion from deoxycholic acid; C: Control (PBS)

A second method is a capillary assay. In this assay, heparinized glass capillaries bridge between a bacterial suspension and a test solution (Koppelhus et al., 1994; Leick & Helle, 1983). Two-phase assays in spectrophotometer plastic cuvettes are suitable to monitor the chemotaxis-mediated migration between the two fluid phases (Koppelhus et al., 1994). The T-maze assay allows the quantification of the chemoresponse of two substances in direct comparison, using a T-shaped experimental arrangement of three containers (Van Houten et al., 1982).

The use of a so-called Boyden chamber is a third variant to study chemotaxis. Chambers divided by filters are a third variant of chemotaxis assays. The suspension of motile cells is placed into the upper vessel of the so-called *Boyden chamber* (Boyden, 1962). The test solution with the chemoeffector is filled into the lower vessel. A filter membrane separates both parts of the Boyden chamber. The pore diameter must be chosen according to the size of the organism allowing its transmigration. To simulate *in vivo* conditions, the filter membranes can be optionally covered with extracellular matrix proteins like collagen, elastin or fibrin. Modifications of this technique connect the vessels either horizontally (Zigmond, 1988) or as concentric rings (Zicha et al., 1991). Multiwell chambers make the parallel testing of different substances in one occasion feasible.

2.2 Molecular structure of the flagella motor and chemoreceptors

2.2.1 Molecular structure and synthesis of the flagellar apparatus

Non-flagellar Type III secretory systems and the flagellar apparatus share a common basic architecture. Thus, it seems apparently that both go back to a common evolutionary origin (Toft & Fares, 2008).

Basically, the flagellum is comprised of three parts: a helical filament, the hook and the basal body (see Fig. 2). The filament of bacterial flagella is built up of multiple subunits of the flagellin protein FliC (Samatey et al., 2004), which form so called protofilaments. Eleven circular arranged protofilaments stacked into a left- or right-handed helix, according to the direction of rotation (see above) comprise the filament (O'Brian & Bennet, 1972). It has a length of about 10 μm and a diameter of about 20 nm. If a flagellum is virtually flattened, it shows a constant interspace between adjacent turns, corresponding to the wavelength of a "sine wave" (Madigan & Martinko, 2006). This wavelength is specific for any bacterial species and is determined by the structure of the flagellin protein and the rotational direction of the filament (Madigan & Martinko, 2006).

The hook tethers the filament to the basal body. Electron microscopic studies demonstrated that the hook of *S. typhimurium* has length of 55 ± 6 nm (Hirano et al., 1994). The interdigitated hook-subunits make up a bended tube with a 2 to 3 nm wide central channel, which continues in the rod as well as in the filament (Shaik et al., 2005). The hook connects the filament to the motor portion of the flagellar apparatus. The hooks flexibility permits the transmission of torque from the motor to the helical propeller when both are not in a coaxial orientation to each other (Berg & Anderson, 1973). A so-called gap compression/extension mechanism and mutual sliding of the hook-subunits allows continuous structural change of the hook during flagellar rotation at low energy cost (Furuta et al., 2006). The hook facilitates the synchronization of several filaments bundled together at one cell pole (Macnab, 1977; Berg & Anderson, 1973). The two hook-associated proteins (HAP 1 and 3) form a small hook-filament junction, which acts as an adaptor for transition between the hook that is flexible in bending but rigid against twisting and the much more stiff filament (Samatey et al., 2004).

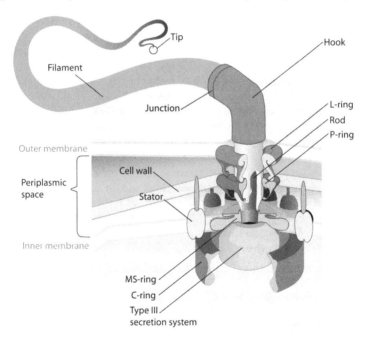

Fig. 2. The flagellum of Gram-negative bacteria, like *Salmonella* sp., is a complex structure consisting of the propeller like acting filament, the flexiple hook and the flagellar ATP-driven motor, which is comprised of four ring-structures and the static motor complexes. An integrated flagellar type III secretion system exports actively the proteins for flagellar assembly but also some virulence associated factors. (Copyright: Wikimedia Commons, public domain aviable from http://upload.wikimedia.org/wikipedia/commons/1/15/ Flagellum _base_diagram_en.svg)

The third part of the flagella – the basal body is comprised of a rod sticking in four ring-like structures: the L-ring (associated with lipopolysaccharides) and the P-ring (associated with peptidoglycanes) forming an outer cylinder embedded in the plasma membrane, the MS-ring, building up a motor mounting plate; and the versatile C-ring (Macnab, 1999). Gram-positive bacteria lacking an outer membrane lack consequentely the outer ring structures.

Overall more than 40 genes are involved in flagellar synthesis and subsequent motility in *S. typhimurium* (Shaik et al. 2005). The MS- and the C-ring, the export apparatus as well as the motor and switch are the first synthesized compounds of the flagella (Katayama et al., 1996; Macnab, 2003). The following assembly steps utilize the type III export apparatus, while the export substrates are supplied via an delivery apparatus located in a patch of membrane near the center of the MS-ring to the channel (Suzuki et al., 1998; Macnab, 2003). It follows the formation of the rod and the other two rings. The proteins of which the rod is comprised are FlgB, FlgC, FlgF and FlgG (Homma et al., 1990). The bifunctional FlgJ protein, which has a muraminidase activity to open the peptidoglycan layer for penetration of the sprouting rod, is also able to bind other rod constituents (Hirano et al., 2001), and thus may act as a rod capping protein promoting the assembly of the rod out of its four components (Nambu et al.; 1999, Macnab, 2003). The assembly of the basal body is finished by the synthesis of the periplasmic P-ring made out of FlgI (Homma et al., 1987) and the outer membrane L-ring consisting of FlgH-subunits (Jones et al., 1990). In the following step, the rod cap is dislodged, while the hook-cap, consisting of about 5 FlgD-subunits, is mounted (Macnab, 2003) and the hook is attached on the basal body. It consists, comparable to the flagellar filament, of about 120 copies of a single kind of protein, FlgE – the so-called hook protein (Samatey et al., 2004). Parallel, the L-ring is assembled (Kubori et al., 1992; Ohnishi et al., 1987). After this, two junction protein zones, made either of FlgK or FlgL, are attached, and a so-called filament-cap out of FliD-proteins is mounted on the hook (Homma et al., 1985; Ikeda et al., 1987 & 1989). Cap proteins assist the organization of the flagellin proteins to form a new filament (Ikeda et al., 1985). Between hook and cap, a junction zone is synthesized before the protofilaments are assembled (Macnab, 2003). The flagellin molecules pass the channel inside the hook and the filament and add on successively at the lower end. The flagellar assembly starts beneath the cap and grows from its tip to its base. A mature flagellum is composed of approximately 20 000 copies of flagellin protein. A broken flagella can be repaired with newly synthesized flagellin units from the cytoplasm passing through the filament channel (Homma & Iino, 1985). The proteins, which built up the flagellum, are translocated to the distal part of the growing flagellum through the central channel by a flagellar type III homologous protein secretion system (Ibuki et al., 2011). This secretion system is comprised of two classes of proteins: soluble and membrane associated ones. The essential soluble compounds of this ATP-consuming process are the soluble FliI-ATPase, its regulator FliH and the FliJ-protein, that promotes the hexamerization of FliI-ATPases (Ibuki et al., 2011). The remaining soluble components are specific chaperones: FlgN for the hook-filament junction proteins, FliT for the filament cap protein, and FliS for flagellin (Macnab, 2003). The six membrane associated components FlhA, FlhB, FliO, FliP, FliQ, and FliR form a complex within the MS-ring (Macnab, 2003).

A rotary motor is in principle built up of two functional components: the rotor and the stator. The flagellar motor consists of the static Mot-complexes, which were affixed in the inner cytoplasmatic membrane and the rotating C-ring. The Mot-complexes are transmembrane structures made from two proteins MotA and MotB (Macnab, 2003). The cytoplasmic C-ring contains the motor/switch proteins – the Fli-proteins. The FliG-proteins generate the torsional moment, while working against the Mot-complexes. The switch-proteins, in *Salmonella* sp.: FliG, FliM and FliN can reverse the flagellar rotational direction in response to intracellular signals (Francis et al., 1994; Yamaguchi et al. 1986). The FliM-protein is the final effector of a sensory transduction chain (Bren et al., 1998; Sockett et al., 1992). Thus, the stator is formed by the Mot-proteins, which mantle the C- and the MS-rings. C- and MS-rings, as well as the rod, collectively form the rotor.

Driven by a transmembrane proton gradient, the flagellar motor is able to perform the clockwise and a counterclockwise rotation of the filament, which is reversed stochastically in the absence of any stimulus. The protons are pumped from outside across the cytoplasmic membrane through the Mot-complexes. Calculations showed that about 1000 protons must be translocated to perform a single rotation of the flagellar filament. In the proposed proton pump model, the protons flowing through the stator channels exert electrostatic forces on helically arranged charge clusters on the rotor rings. Most probably, the protons bind temporary to a specific aspartate residue of MotB, causing a change of the stators conformation that drives the rotor through an elementary rotational step (Kojima & Blair, 2001; Macnab, 2003). In the next step, the aspartate residue gets deprotonated and the stator returns to its original conformation. These resulting series of interactions between positive and negative charges generate a torsional moment as protons flow through the Mot-complexes.

2.3 Chemoreceptor structure and signal transduction

Presence and concentration of chemotaxins are detected by a family of chemoreceptors sharing a common two-component system architecture (Miller et al. 2009). Such two-component systems are generally comprised of a membrane associated histidine autokinase (CheA) and a cytoplasmic response regulator (CheY; Lux et al. 2004). Methyl-accepting chemotaxis proteins (MCPs), embedded in the cytoplasmic membrane, sense the environmental signals via their N-terminal periplasmic sensory domain to their C-terminal cytoplasmic signaling domain. The MCP-monomers have a molecular mass of about 60 kDa and form constitutively very stable homodimers, which are arranged in groups of three (Lux et al. 2004). CheW linker proteins tether the CheA histidine kinase to the MCPs (Miller et al. 2009). This inhibits autophosphorylation of CheA, which in turn reduces the phosphorylation of CheY response regulator at a conserved histidine residue. Hypophosphorylated CheY can diffuse freely in the cytoplasm and can interact with FliM, the switch protein of the flagellar motor (Mot), which is the final effector of sensory transduction chain (Bren et al., 1998; Sockett et al., 1992). Thus, it triggers counterclockwise rotation of the flagella, which leads to bacterial "running". If a bacterium moves along a gradient of a chemoattractant, the intracellular concentration of phosphorylated CheY decreases. Consequently, the frequency of flagella switching decreases, and the number of site directed "runs" along the gradient increases. Thus, addition of an attractant triggers a counterclockwise rotation of the filament (Bren & Eisenbach, 2000).

In the opposite case, decreasing ligand occupancy of the MCPs leads to increased autophoshorylation of CheA and in consequence to an amplified phosphorylation of CheY and CheB. Phosphorylated CheY binds as well to the motor switch but triggers a clockwise flagellar rotation resulting in bacterial "tumbling".

CheB is a receptor-demethylating enzyme, which is also activated by phosphorylation. The phosphatase CheZ is responsible for the dephosphorylation of phosphate-activated CheY (Bourret & Stock, 2002).

The result of this chemosensing in three-dimensional spatial gradients of different chemoattractors and chemorepellents is a stereoscopic "zigzag" path of motion (Berg & Brown, 1972), until the bacterium reaches a niche with an equilibrium between the varying chemoeffectors (Miller et al., 2009).

2.4 Sensory adaptation

Sensory adaption means reestablishment of the prestimulus state in the perpetual presence of the stimulus. Adaptation to chemotactical stimuli is mostly due to modulation of the methylation of certain sites of the MCP receptors. The central players in the process of sensory adaption are the methyltransferase CheR, the methylesterase CheB, and the cytoplasmic domains of the MCP-receptors that have adjacent to the CheA and CheW binding sites, sites for methylation and demethylation of glutamyl side chains (Macnab, 2003).

CheR catalyzes in a S-adenosylmethionine consuming reaction the methylation of the specific glutamate residues on the cytoplasmic domains of the MCPs (Bren & Eisenbach, 2000). This reaction enhances the CheA autophosphorylation favouring clockwise flagallar rotation and is triggered by attracting stimuli (Borkovich et al., 1992; Ninfa et al., 1991). The cytoplasmatic domains of the MCPs have a specific domain, which is methylated by CheR, and a distinct CheR-binding site consisting of a pentapepetide that is only present in high-abundance receptors. It was shown that CheR bound to the binding sites onto the high-abundance receptors methylates the designated sites of the low-abundance receptors (Le Moual et al., 1997; Li et al., 1997).

Its antagonist is the methylesterase CheB, which demethylates the MCPs during adaptation to repelling stimuli. Additionally CheB has an amidase activity catalyzing the conversion of glutamamine residues into glutamate on the MCPs (Djordjevic et al., 1998). The liberation of glutamate residues inhibits the autophosphorylation of CheA favouring a counterclockwise rotation of the flagellum. In addition CheB itself is regulated by CheA-mediated phosphorylation (Hess et al., 1988; Lupas & Stock, 1989). Phosphorylation inhibits the methylesterase activity, while the unphosphorylated enzyme has less methylesterase activity. The binding sites on CheA for CheY and CheB are identical. Thus CheB competes with CheY (Li et al., 1995).

Furthermore a high methylation rate decreases the receptors affinity to chemoattractants (Bornhorst et al., 2000; Li et al., 2000). It was also suggested that a deferred activation of CheZ, which is responsible for an enhanced CheY dephosphorylation, is involved in the process of sensory adaptation (Blat et al., 1998).

These regulatory effects occur only after the initial chemotactic response and the steady state of all these parallel-acting adaptational processes determines the extent of reaction to a certain chemoeffector (Alon et al., 1999).

2.5 Specific Salmonella chemoreceptors

Altogether, it is difficult to identify chemoreceptors specific for a certain taxin, because the different MCPs can compensate each other in many cases, Thus, knockout mutants of chemoreceptor genes show often no defects in their phenotype (Vegge et al., 2009; Tareen et al. 2010). Up to now, four chemoreceptor specificities are identified for *Salmonella* spp.

The Tar chemoreceptor is specific for aspartate and initiates attractant signalling (Foster et al. 1985; Milburn et al., 1991). The same receptor molecule interacts also with the periplasmic maltose-binding protein, and senses in this way chemoattraction towards maltose (Mowbray & Koshland 1987; Gardina et al., 1992). It was also demonstrated that this receptor mediates attractant responses to phenol and repellent responses to glycerol and nickel or cobalt ions in *Escherichia coli* as well as thermoresponses (Lee & Imae, 1990).

The ligand serine mediates positive taxis via the Tsr receptor, whereas Tsr sensing due to leucine and glycerol results in a repulsion of the bacteria. (Lee & Imae, 1990; Jeffrey & Koshland, 1993; Oosawa & Imae, 1984; Springer et al., 1977). Tsr functions also as thermoreceptor. Temperature increase leads to smooth swimming of bacterial cells, whereas temperature decrease induces tumbling (Lee et al., 1988).

S. typhimurium demonstrates attraction towards citrate and metal-citrate complexes, but repulsion from phenol. This behavior gives the name to the third chemoreceptor in this schedule – Tcp, that stands for taxis to citrate and away from phenol (Yamamoto & Imae, 1993).

The *trg* gene encodes a fourth chemoreceptor of the MCP family specific for ribose/galactose (Blat & Eisenberg, 1995, Kasinkas et al. 2007).

2.6 Virulence factors secreted via the flagellar type III secretion system

As mentioned above, the flagellar apparatus is a homologue of a type III secretion system that is able to secrete specific peptides and proteins in an ATP dependent mechanism into the environment (Collazo & Galán, 1996; Eichelberg et al., 1994). Among these secreted proteins are mostly structural components of the flagella, for example flagellin monomers, and the hook protein, but also several virulence factors. It functions as a molecular syringe – the so-called injectisome – that is used by bacteria to inject effector proteins directly into the interior of host cells (Mota et al., 2005a+b; Arnold et al. 2009). Thus, these proteins play an important role for host cell invasion and the pathogenesis of salmonellosis. It was shown, that the N-terminal 30 residues of these effector proteins form a taxonomically universal, type III specific secretion signal (Arnold et al. 2009; Samurdrala et al. 2009). About 65 type III secretion system substrates are known for *S. typhimurium* (Samurdrala et al. 2009). Five well described proteins involved in host cell invasion, typically the M-cells of the ileal Peyers' Patches, are InvE, Sipa, Sipb, SipC, and SipD. InvE plays a pivotal role for triggering cellular mechanisms, which lead to bacterial entry. It is required for translocation of other effector proteins into the cytosol of host cells and forms complexes with SipA, SipB, and SipC. (Kubori & Galán, 2002). Comparable to InvE the effector protein SipD, which has been shown to be important for liver and ileum colonization, is suggested to modulate the secretion of SipA, SipB, and SipC (Gong et al., 2010). Cell invasion occurs via a ruffle-mediated mechanism, which is initiated by the activation of specific signal transduction cascades and rearrangement of the actin cytoskeleton. The actin rearrangements are realized by SipA interworking with SopE, a guanine-nucleotide exchange factor for Rho GTPases, and SptP, a protein tyrosine phosphatase, (Brumell et al., 1999) as well as SipC that binds and bundles F-actin (Myeni & Zhou, 2010).

SipB interacts after entering the cytosol of macrophages with cell signalling pathways to induce apoptosis (Hersh et al. 1999). It associates with caspase-1 and promotes the proteolytic activation of this protease.

Two further proteins entering macrophages are SrfN and PagK2, which were shown to be essential for full virulence and are suggested to interact with host cellular components (Yoon et al. 2011). These two effector proteins are translocated independently of the injectisome. Thus, the flagellar type III secretion system is the only protein export mechanism in *Salmonella* sp.

2.7 Role of chemotaxis and flagellar motility for the pathoegenesis of salmonelosis

The ability for directed movement and taxis towards and away from chemoeffectors plays a crucial role for the pathogenesis of salmonellosis. Amongst others, *Salmonella* bacteria are able to persist inside the inner leaf tissue of plants (Kroupitski et al., 2009; Goldberg et al., 2011). It was shown that flagellar motility and chemotaxis towards nutrients produced by photosynthetically active cells are crucial for entry into iceberg lettuce leaves via open stomata and invasion into the plant tissue (Kroupitski et al., 2009). Enteropathogens have the ability to adapt to the phyllosphere environment. They obviously interact with epiphytic bacteria (Beuchat, 2002; Brandl, 2006; Heaton & Jones 2008) and become part of phylloplane biofilms, where they gain protection from environmental stressors (Fett, 2000). Plants that might become contaminated by the use of germ-containing water for irrigation or *Salmonella*-containing liquid manure for fertilization might function as source of infection (Beuchat, & Ryu, 1997; Brandl, 2006; Horby et al., 2003). Internal persistence after entering the plant tissue explains the failure of lavation and sanitizers to eradicate *Salmonella* in leafy greens.

Furthermore, flagellar movement and chemotaxis are also pivotal for the intestinal colonization of the different *Salmonella* hosts, especially for the competition for nutrients with other bacteria of human microbiome (Stecher et al., 2008). Even the induction of colitis depends on a functioning flagellar movement and chemotaxis (Stecher et al., 2004).

3. Conclusion

The flagellar apparatus is an evolutionary ancient multifunctional tool involved in motility, bacterial cell aggregation, biofilm formation, protein export, and a virulence factor injection via the injectisome. It is also the prototype of a sensing system, coupling energy taxis and motility. The research on chemotaxis and flagellar motility is almost as old as bacteriology itself, starting at the end of the 19th century. The research on *Salmonella* sp. plays here a special role, as most of the knowledge about thermo- and chemotaxis, MCP-receptor signal transduction, MCP-receptor sensory adaptation, structure, synthesis, and function of the flagellar apparatus as well as effector protein secretion via a flagellar type III homologue secretion system was made using *Salmonella* sp. and *E. coli* as model organism.

Thus, the flagellar apparatus regulated by energy taxis may be the most important structure for intestinal colonization and pathogenesis of salmonellosis.

4. Acknowledgement

The work of the author is funded by the Forschungsförderungsprogramm of the Universitätsmedizin Göttingen, Germany and by the Deutsche Forschungsgemeinschaft (PAK 400).

5. References

Adler, J. (1975) Chemotaxis in Bacteria. *Annual Review of Biochemistry*, Vol. 44, (1975), pp. 341-356, ISSN 0066-4154

Alon, U., Surette, M.G., Barkai, N. & Leibler, S. (1999). Robustness in bacterial chemotaxis. *Nature*, Vol. 397, No. 6715, (January 1999), pp. 168-171, ISSN 0369-3392

Arnold, R., Brandmaier, S., Kleine, F., Tischler, P., Heinz, E., Behrens, S, Niinikoski, A., Mewes, H.W., Horn, M. & Rattei, T. (2009) Sequence-based prediction of type III secreted proteins. PLoS Pathogens. Vol. 5, No. 4, (April 2009), pp. e1000376, ISSN 1553-7366

Barnakov, A.N., Barnakova, L.A. & Hazelbauer, G.L. (1999). Efficient adaptational demethylation of chemoreceptors requires the same enzyme-docking site as efficient methylation. Proceedings of the National Academy of Sciences of the United States of America. No. 96, Vol. 19, (September 1999), pp. 10667-10672, ISSN 0027-8424

Berg, H.C. & Anderson, R.A. (1973). Bacteria swim by rotating their flagellar filaments. *Nature*. Vol. 245, No. 5425, (October 1973), pp. 380-282, ISSN 0028-0836

Berg, H.C. & Brown D.A. (1972). Chemotaxis *in Escherichia coli* analysed by three-dimensional tracking. *Nature*. Vol. 239, No. 5374, (October 1972), pp 500- 504, ISSN 0028-0836

Beuchat L.R. (2002). Ecological factors influencing survival and growth of human pathogens on raw fruits and vegetables. *Microbes and infection / Institut Pasteur*. Vol. 4, No. 4), (April 2002), pp. 413-423, ISSN 1286-4579

Beuchat, L.R. & Ryu, J.H. (1997). Produce handling and processing practices. *Emerging infectious diseases*. Vol. 3, No. 4, (October-December 1997), pp. 459-465., ISSN 1080-6040

Blat, Y. & Eisenbach, M. (1995). Tar-dependent and -independent pattern formation by *Salmonella typhimurium*. *Journal of bacteriology*. Vol. 177, No. 7, (April 1995), pp. 1683-1691, ISSN 0021-9193

Blat, Y., Gillespie, B., Bren, A., Dahlquist, F.W. & Eisenbach, M. (1998). Regulation of phosphatase activity in bacterial chemotaxis. *Journal of molecular biology*. Vol. 284, No, 4, (December 1998), pp. 1191-1199, ISSN 0022-2836

Block, S.M., Fahrner, K.A. & Berg, H.C. (1991). Visualization of bacterial flagella by video-enhanced light microscopy. Journal of Bacteriology, Vol. 173, No. 2, (January 1991), pp. 933-936, ISSN 0021-9193

Borkovich, K.A., Alex, L.A. & Simon, M.I. (1992). Attenuation of sensory receptor signaling by covalent modification. *Proceedings of the National Academy of Sciences of the United States of America*. Vol. 89, No. 15, (August 1992), pp. 6756-6760, ISSN 0027-8424

Bornhorst, J.A. & Falke, J.J. (2000). Attractant regulation of the aspartate receptor-kinase complex: limited cooperative interactions between receptors and effects of the receptor modification state. *Biochemistry*. Vol. 39, No. 31, (August 2000), pp. 9486-9493, ISSN 0006-2960

Bourret, R.B. & Stock, A.M. (2002). Molecular information processing: lessons from bacterial chemotaxis. *The Journal of biological chemistry*. Vol. 277, No. 12, (March 2002), pp. 9625-9628, ISSN 0021-9258

Boyden, S.V. (1962). The chemotactic effect of mixtures of antibody and antigen on polymorphonuclear leucocytes. *The Journal of Experimental Medicine*, (February 1962), Vol. 115, No. 3, pp. 453-466, ISSN 0022-1007

Brandl, M.T. (2006). Fitness of human enteric pathogens on plants and implications for food safety. *Annual review of phytopathology*. Vol. 44, (2006), pp. 367-392, ISSN 0066-4286

Bren, A. & Eisenbach, M. (1998). The N terminus of the flagellar switch protein, FliM, is the binding domain for the chemotactic response regulator, CheY. Journal of molecular biology. Vol. 278, No. 3, (May 1998), pp. 507-514. ISSN 0022-2836

Bren, A. & Eisenbach, M. (2000). How signals are heard during bacterial chemotaxis: protein-protein interactions in sensory signal propagation. *Journal of bacteriology*. Vol. 182, No. 24, (December 2000), pp. 6865-6873, ISSN 0021-9193

Brumell, J.H., Steele-Mortimer, O. & Finlay, B.B. (1999). Bacterial invasion: Force feeding by *Salmonella*. *Current biology*. Vol. 9, No. 8, (April 1999), pp. R277-80, ISSN 0960-9822

Collazo, C.M. & Galán, J.E. (1996). Requirement for exported proteins in secretion through the invasion-associated type III system of *Salmonella typhimurium*. *Infection and immunity*. Vol. 64, No. 9. (September 1996) pp. 3524-3531, ISSN 0019-9567

Djordjevic, S. & Stock, A.M. (1998). Structural analysis of bacterial chemotaxis proteins: components of a dynamic signaling system. *Journal of structural biology*. Vol. 124, No. 2-3, (December 1998), pp. 189-200, ISSN 1047-8477

Eichelberg, K., Ginocchio, C.C. & Galán, J.E. (1994). Molecular and functional characterization of the *Salmonella typhimurium* invasion genes invB and invC: homology of InvC to the F0F1 ATPase family of proteins. *Journal of bacteriology*. Vol. 176, No. 15, (August 1994), pp. 4501-4510, ISSN 0021-9193

Engelmann, T.W. (1881). Neue Methode zur Untersuchung der Sauerstoffausscheidung pflanzlicher und thierischer Organismen. *Pflügers Archiv für die Gesamte Physiolgie der Menschen und Tiere*, Vol. 25, (1881) pp. 285-292, ISSN 0365-267X

Erdem, A.L., Avelino, F., Xicohtencatl-Cortes. J. & Girón J.A. (2007). Host protein binding and adhesive properties of H6 and H7 flagella of attaching and effacing Escherichia coli. *Journal of Bacteriology*. Vol. 189, No. 20, (October 2007), pp. 7426-7435, ISSN 0021-9193

Fett W.F. (2000). Naturally occurring biofilms on alfalfa and other types of sprouts. Journal of food protection. Vol. 63, No. 5, (May 2000), pp. 625-632, ISSN 0362-028X

Foster, D.L., Mowbray, S.L., Jap, B.K. & Koshland D.E. Jr. (1985). Purification and characterization of the aspartate chemoreceptor. *The Journal of biological chemistry*. Vol. 260, No. 21, (September 1985), pp. 11706-11710, ISSN 0021-9258

Francis, N.R., Sosinsky, G.E., Thomas, D. & DeRosier, D.J. (1994). Isolation, characterization and structure of bacterial flagellar motors containing the switch complex. *Journal of molecular biology*. Vol. 235, No. 4, (January 1994), pp. 1261-1270; ISSN 0022-2836

Frye, J., Karlinsey, J.E., Felise, H.R., Marzolf, B., Dowidar, N., McClelland, M. & Hughes, K.T. (2006). Identification of new flagellar genes of Salmonella enterica serovar Typhimurium. *Journal of Bacteriology*, Vol. 188, No. 6, (March 2006), pp. 2233-2243, ISSN 0021-9193

Furuta, T., Samatey, F.A., Matsunami, H., Imada, K., Namba, K. & Kitao, A. (2007). Gap compression/extension mechanism of bacterial flagellar hook as the molecular universal joint. *Journal of Structural Biology*, Vol. 157, No. 3, (March 2007), pp. 481-490, ISSN 1047-8477

Gardina, P., Conway, C., Kossman, M. & Manson, M. (1992). Aspartate and maltose-binding protein interact with adjacent sites in the Tar chemotactic signal transducer of *Escherichia coli*. *Journal of bacteriology*. Vol. 174, No. 5, (March 1992), pp.1528-1536, ISSN 0021-9193

Gebremichael Y, Ayton GS, Voth GA. (2006). Mesoscopic modeling of bacterial flagellar microhydrodynamics. *Biophysical Journal*. Vol. 91, No. 10, (November 2006), pp. 3640-3652, ISSN 0006-3495

Golberg, D., Kroupitski, Y., Belausov, E., Pinto, R. & Sela, S. (2011). *Salmonella Typhimurium* internalization is variable in leafy vegetables and fresh herbs. *International journal of food microbiology*. Vol. 145, No. 1, (January 2011), pp. 250-257, ISSN 0168-1605

Gong, H., Vu, G.P., Bai, Y., Yang, E., Liu, F. & Lu, S. (2010). Differential expression of *Salmonella* type III secretion system factors InvJ, PrgJ, SipC, SipD, SopA and SopB in cultures and in mice. *Microbiology*. Vol. 156, No. Pt 1, (January 2010), pp. 116-127, ISSN 1350-0872

Hahn, H., Klein, P., Giesbrecht, R.E. & Streeck, R.E. (2004). Bakterien: Definition und Aufbau, In: *Medizinische Mikrobiologie und Infektologie*, Hahn, H., Falke, D., Kaufmann, S.H.E. & Ullmann, U., pp. 169-179, Springer Medizin Verlag Heidelberg, ISBN 3-540-21971-4, Berlin, Germany

Heaton, J.C. & Jones, K. (2008). Microbial contamination of fruit and vegetables and the behaviour of enteropathogens in the phyllosphere: a review. *Journal of applied microbiology*. Vol. 104, No. 3, (March 2008), pp. 613-626, ISSN 1364-5072

Hersh, D., Monack, D.M., Smith, M.R., Ghori, N., Falkow, S. & Zychlinsky, A. (1999). The *Salmonella* invasin SipB induces macrophage apoptosis by binding to caspase-1. *Proceedings of the National Academy of Sciences of the United States of America*. Vol. 96, No. 5, (March 1999), pp. 2396-2401, ISSN 0027-8424

Hess, J.F., Oosawa, K., Kaplan, N. & Simon, M.I. (1988). Phosphorylation of three proteins in the signaling pathway of bacterial chemotaxis. *Cell*. Vol. 53, No. 1, (April 1988), pp. 79-87, ISSN 0092-8674

Hirano, T., Minamino, T. & Macnab, R.M. (2001). The role in flagellar rod assembly of the N-terminal domain of *Salmonella* FlgJ, a flagellum-specific muramidase. *Journal of molecular biology*. Vol. 312, No. 2, (September 2001), pp. 359-369, ISSN 0022-2836

Hirano, T., Yamaguchi, S., Oosawa, K. & Aizawa, S. (1994). Roles of FliK and FlhB in determination of flagellar hook length in *Salmonella typhimurium*. *Journal of bacteriology*. Vol. 176, No. 17, (September 1994) pp. 5439-5449, ISSN 0021-9193

Homma, M. & Iino. T. Excretion of unassembled hook-associated proteins by *Salmonella typhimurium*. (1985). *Journal of bacteriology*. Vol. 164, No. 3, (December 1985), pp. 1370-1372, ISSN 0021-9193

Homma, M., Komeda, Y., Iino, T. & Macnab, R.M. (1987). The *flaFIX* gene product of *Salmonella typhimurium* is a flagellar basal body component with a signal peptide for export. *Journal of bacteriology*. Vol. 169, No. 4, (April 1987), pp. 1493-1498, ISSN 0021-9193

Homma, M., Kutsukake, K. & Iino, T. (1985). Structural genes for flagellar hook-associated proteins in *Salmonella typhimurium*. *Journal of bacteriology*. Vol. 163, No. 2, (August 1985), pp. 464-471, ISSN 0021-9193

Homma, M., Kutsukake, K., Hasebe, M., Iino, T, & Macnab R.M. (1990). FlgB, FlgC, FlgF and FlgG. A family of structurally related proteins in the flagellar basal body of *Salmonella typhimurium*. *Journal of molecular biology*. Vol. 211, No. 2, (January 1990), pp. 465-477, ISSN 0022-2836

Horby, P.W., O'Brien, S.J., Adak, G.K., Graham, C., Hawker, J.I., Hunter, P., Lane, C., Lawson, A.J., Mitchell, R.T., Reacher, M.H., Threlfall, E.J., Ward, L.R. & PHLS Outbreak Investigation Team (2003). A national outbreak of multi-resistant *Salmonella enterica* serovar *Typhimurium* definitive phage type (DT) 104 associated with consumption of lettuce. *Epidemiology and infection*. Vol. 130, No. 2, (April 2003), pp. 169-178, ISSN 0950-2688

Hugdahl, M.B., Beery, J.T. & Doyle MP. (1988). Chemotactic behavior of *Campylobacter jejuni*, *Infection and Immunity*, Vol. 56, No. 6, (June 1988), pp.1560-1566, ISSN 0019-9567

Ibuki, T., Imada, K., Minamino, T., Kato, T., Miyata, T. & Namba, K. (2011) Common architecture of the flagellar type III protein export apparatus and F- and V-type ATPases. *Nature structural & molecular biology*. Vol. 18, No. 3, (March 2011), pp. 277-282 , ISSN 1545-9993

Ikeda, T., Asakura, S. & Kamiya, R. (1985). "Cap" on the tip of *Salmonella* flagella. *Journal of molecular biology*. Vol. 184, No. 4, (August 1985), pp. 735-737, ISSN 0022-2836

Ikeda, T., Asakura, S. & Kamiya, R. (1989). Total reconstitution of *Salmonella* flagellar filaments from hook and purified flagellin and hook-associated proteins *in vitro*. *Journal of molecular biology*. Vol. 209, No. 1, (Sep 1989), pp. 109-114, ISSN 0022-2836

Ikeda, T., Homma, M., Iino, T., Asakura, S. & Kamiya, R. (1987). Localization and stoichiometry of hook-associated proteins within *Salmonella typhimurium* flagella. *Journal of bacteriology*. Vol. 169, No. 3, (March 1987), pp.1168-1173, ISSN 0021-9193

Iwama, T., Ito, Y., Aoki, H., Sakamoto, H., Yamagata, S., Kawai, K. & Kawagishi I. (2006). Differential recognition of citrate and a metal-citrate complex by the bacterial chemoreceptor Tcp. *The Journal of Biological Chemistry*, Vol. 281, No. 26, (June 2006), pp. 17727-17735, ISSN 0021-9258

Jeffery, C.J. & Koshland, D.E. Jr. Three-dimensional structural model of the serine receptor ligand-binding domain. *Protein science : a publication of the Protein Society*. Vol. 2, No. 4, (April 1993), pp. 559-566, ISSN 0961-8368

Jones, C.J. & Aizawa S. (1991). Genetic control of the bacterial flagellar regulon. *Current opinion in genetics & development*, Vol. 1, No. 3, (October 1991), 319-923, ISSN 0959-437X

Jones, C.J., Macnab, R.M., Okino, H. & Aizawa, S. (1990). Stoichiometric analysis of the flagellar hook-(basal-body) complex of *Salmonella typhimurium*. *Journal of molecular biology*. Vol. 212, No. 2, (March 1990), pp. 377-387, ISSN 0022-2836

Kasinskas, R.W. & Forbes, N.S. (2007). *Salmonella typhimurium* lacking ribose chemoreceptors localize in tumor quiescence and induce apoptosis. *Cancer Research*, Vol. 67, No. 7, (April 2007), pp. 3201-3209, ISSN 0008-5472

Katayama, E., Shiraishi, T., Oosawa, K., Baba, N. & Aizawa, S. (1996). Geometry of the flagellar motor in the cytoplasmic membrane of *Salmonella typhimurium* as determined by stereo-photogrammetry of quick-freeze deep-etch replica images. *Journal of molecular biology*. Vol. 255, No. 3, (January 1996), pp. 458-475, ISSN 0022-2836

Khan, S., Pierce, D. & Vale, R.D. (2000). Interactions of the chemotaxis signal protein CheY with bacterial flagellar motors visualized by evanescent wave microscopy. *Current Biology*, Vol. 10, No. 15, (July – August 2000), pp. 927-930, ISSN 0960-9822

Kitao, A., Yonekura, K., Maki-Yonekura, S., Samatey, F.A., Imada, K., Namba, K. & Go, N. (2006). Switch interactions control energy frustration and multiple flagellar filament structures. *Proceedings of the National Academy of Sciences of the United States of America.* Vol. 103, No. 13 (March 2006), pp. 4894-4899, ISSN 0027-8424

Köhidai L. (1995). Method for determination of chemoattraction in Tetrahymena pyriformis. *Current Microbiology.* Vol. 30, No. 4 (April 1995), pp. 251-253, ISSN 0343-8651

Kojima, S. & Blair, D.F. (2001). Conformational change in the stator of the bacterial flagellar motor. *Biochemistry.* Vol. 40, No. 43, (October 2001), pp. 13041-13050, ISSN 0006-2960

Koppelhus, U., Hellung-Larsen, P. & Leick V. (1994). An improved quantitative assay for chemokinesis in Tetrahymena. *The Biological Bulletin (Marine Biological Laboratory).* Vol. 187, No. 1, (August 1994), pp. 8–15, ISSN 0006-3185

Kroupitski, Y., Golberg, D., Belausov, E., Pinto, R., Swartzberg, D., Granot, D. & Sela, S. (2009). Internalization of *Salmonella enterica* in leaves is induced by light and involves chemotaxis and penetration through open stomata. *Applied and environmental microbiology.* Vol. 75, No. 19, (October 2009), pp. 6076-6086, ISSN 0099-2240

Kubori, T. & Galán, J.E (2002). *Salmonella* type III secretion-associated protein InvE controls translocation of effector proteins into host cells. *Journal of bacteriology.* Vol. 184, No. 17, (September 2002), pp. 4699-4708, ISSN 0021-9193

Kubori, T., Matsushima, Y., Nakamura, D., Uralil, J., Lara-Tejero, M., Sukhan, A., Galán, J.E. & Aizawa, S.I. (1998). Supramolecular structure of the Salmonella typhimurium type III protein secretion system. Science. Vol. 280, No. 5363, (April 1998), pp. 602-605, ISSN 0036-8075

Larsen, S.H., Reader, R.W., Kort, E.N., Tso, W.W. & Adler J. (1974). Change in direction of flagellar rotation is the basis of the chemotactic response in *Escherichia coli. Nature.* Vol. 249, No. 452, (May 1974), pp. 74-77, ISSN 0028-0836

Lauga, E., DiLuzio, W.R., Whitesides, G.M. & Stone, H.A. (2006). Swimming in circles: motion of bacteria near solid boundaries. *Biophysical Journal,* Vol. 90, No. 2, (January 2006), pp. 400-412, ISSN 0006-3495

Le Moual, H., Quang, T. & Koshland, D.E. Jr. (1997). Methylation of the *Escherichia coli* chemotaxis receptors: intra- and interdimer mechanisms. *Biochemistry.* Vol. 36, No. 43, (October 1997), pp. 13441-13448, ISSN 0006-2960

Lee, L. & Imae, Y. (1990). Role of threonine residue 154 in ligand recognition of the tar chemoreceptor in *Escherichia coli. Journal of bacteriology.* Vol. 172, No. 1, (January 1990), pp. 377-382, ISSN 0021-9193

Lee, L., Mizuno, T. & Imae, Y. (1988). Thermosensing properties of *Escherichia coli* tsr mutants defective in serine chemoreception. Journal of bacteriology. Vol. 170, No. 10, (October 1988), pp. 4769-4774, ISSN 0021-9193

Leick, V. & Helle, J. (1983). A quantitative assay for ciliate chemotaxis. *Analytical Biochemistry,* Vol. 135, No. 2, (December 1983), pp. 466–469, ISSN: 0003-2697

Li, G. & Weis, R.M. (2000). Covalent modification regulates ligand binding to receptor complexes in the chemosensory system of *Escherichia coli.* (2000) *Cell.* Vol. 100, No. 3, (February 2000), pp. 357-365, ISSN 0092-8674

Li, J., Li, G. & Weis, R.M. (1997). The serine chemoreceptor from *Escherichia coli* is methylated through an inter-dimer process. *Biochemistry.* Vol. 36, No. 39, (September 1997), pp. 11851-11857, ISSN 0006-2960

Li, J., Swanson, R.V., Simon, M.I. & Weis, R.M. (1995). The response regulators CheB and CheY exhibit competitive binding to the kinase CheA. *Biochemistry.* Vol. 34, No. 45, (November 1995), pp. 14626-14636, ISSN 0006-2960

Lupas, A. & Stock, J. (1989). Phosphorylation of an *N*-terminal regulatory domain activates the CheB methylesterase in bacterial chemotaxis. *The Journal of biological chemistry.* Vol. 264, No. 29, (October 1989), pp. 17337-17342, ISSN 0021-9258

Lux, R. & Shi, W. (2004). Chemotaxis-guided movements in bacteria. *Critical Reviews in Oral Biology & Medicine,* Vol. 15, No. 4, (July 2004), pp. 207-220, ISSN 1045-4411

Macnab, R.M. (1977). Bacterial flagella rotating in bundles: a study in helical geometry. *Proceedings of the National Academy of Sciences of the United States of America.* Vol. 74, No. 1, (January 1977) pp. 221-225, ISSN 0027-8424

Macnab, R.M. (1999). The bacterial flagellum: reversible rotary propellor and type III export apparatus. *Journal of Bacteriology,* Vol. 181, No. 23, (December 1999), pp.7149-7153, ISSN 0021-9193

Macnab, R.M. (2003). How bacteria assemble flagella. *Annual review of microbiology.* Vol. 57, (May 2003), pp. 77-100, ISSN 0066-4227

Macnab, R.M. (2004). Type III flagellar protein export and flagellar assembly. *Biochimica et biophysica acta.* Vol. 1694, No. 1-3, (November 2004), pp. 207-217, ISSN 0006-3002

Macnab, R.M. Examination of bacterial flagellation by dark-field microscopy. (1976). *Journal of Clinical Microbiology.* Vol. 4, No. 3, (September 1976), pp. 2582-65, ISSN 0095-1137

Madigan, M.T. & Martinko, J. M. (2006). Mirobial locomation, In: *Brock's Biology of Microorganisms 11th Edition,* pp. 91-100, ISBN 0-13-196893-9, Upper Saddle River, New Jersey, USA

Mariconda, S., Wang. Q. & Harshey, R.M. (2006). A mechanical role for the chemotaxis system in swarming motility. *Molecular Microbiology,* Vol. 60, No. 6, (June 2006), pp. 1590-1602, ISSN 0950-382X

Milburn, M.V., Privé, G.G., Milligan, D.L., Scott, W.G., Yeh, J., Jancarik, J., Koshland, D.E. Jr. & Kim, S.H. Three-dimensional structures of the ligand-binding domain of the bacterial aspartate receptor with and without a ligand. *Science.* Vol. 254, No. 5036, (November 1991), pp. 1342-1347, ISSN 0036-8075

Miller, L.D., Russell, M.H. & Alexandre, G. (2009). Diversity in bacterial chemotactic responses and niche adaptation. Advances in Applied Microbiology, Vol. 66, (2009), pp. 53-57, ISSN 0065-2164

Mota LJ, Journet L, Sorg I, Agrain C, Cornelis GR. (2005a). Bacterial injectisomes: needle length does matter. *Science.* Vol. 307, No. 5713, (February 2005), pp. 1278, ISSN 0036-8075

Mota, L.J., Sorg, I. & Cornelis, G.R. (2005b). Type III secretion: the bacteria-eukaryotic cell express. *FEMS Microbiol Letters.* Vol. 252, No. 1, (November 2005), pp. 1-10, ISSN 0378-1097

Mowbray, S.L. & Koshland, D.E. Jr. (1987). Additive and independent responses in a single receptor: aspartate and maltose stimuli on the tar protein. *Cell.* Vol. 50, No. 2, (July 1987), pp. 171-180, ISSN 0092-8674

Myeni, S.K. & Zhou D. (2010). The C terminus of SipC binds and bundles F-actin to promote *Salmonella* invasion. *The Journal of biological chemistry.* Vol. 285, No. 18, (April 2010), pp. 13357-13363, ISSN 0021-9258

Nambu, T., Minamino, T., Macnab, R.M. & Kutsukake, K. (1999). Peptidoglycan-hydrolyzing activity of the FlgJ protein, essential for flagellar rod formation in

Salmonella typhimurium. Journal of bacteriology. Vol. 181, No. 5, (March 1999), pp. 1555-1561, ISSN 0021-9193

Ninfa, E.G., Stock, A., Mowbray, S. & Stock, J. (1991). Reconstitution of the bacterial chemotaxis signal transduction system from purified components. *The Journal of biological chemistry.* Vol. 266, No. 15, (May 1991), pp. 9764-9770, ISSN 0021-9258

O'Brien E.J. & Bennett, P.M. (1972). Structure of straight flagella from a mutant *Salmonella. Journal of molecular biology.* Vol. 70, No. 1, (September 1972), pp. 133-152, ISSN 0022-2836

Ohnishi, K., Homma, M., Kutsukake, K. & Iino, T. (1987). Formation of flagella lacking outer rings by flaM, flaU, and flaY mutants of *Escherichia coli. Journal of bacteriology.* Vol. 169, No. 4, (April 1987), pp.1485-1488, ISSN 0021-9193

Oosawa, K. & Imae, Y. Demethylation of methyl-accepting chemotaxis proteins in *Escherichia coli* induced by the repellents glycerol and ethylene glycol. *Journal of bacteriology.* Vol. 157, No. 2, (February 1984), pp. 576-581, ISSN 0021-9193

O'Toole, G.A. & Kolter, R. (1998). Flagellar and twitching motility are necessary for *Pseudomonas aeruginosa* biofilm development. *Molecular Microbiology*, Vol. 30, No. 2, (October 1998), pp. 295-304, ISSN 0950-382X

Pfeffer, W. (1884). Locomotorische Richtungsbewegungen durch chemische Reize. *Untersuchungen aus dem Botanischen Institut zu Tübingen*, Vol. 1, No. III, (1884), pp. 363-482, Aviable from
http://www.archive.org/stream/untersuchungena01pfefgoog#page/n4/mode/2up

Pfeffer, W. (1888). Über chemotaktische Bewegungen Von Bakterien, Flagellaten und Volvocineen. *Untersuchungen aus dem Botanischen Institut zu Tübingen* Vol. 2, No. III, (1888), pp. 582-661, Aviable from
http://www.archive.org/stream/untersuchungena00pfefgoog#page/n3/mode/2up

Pierce, D.W. & Vale, R.D. (1999). Single-molecule fluorescence detection of green fluorescence protein and application to single-protein dynamics. *Methods in Cell Biology.* Vol. 58, (1999), pp. 49-73, ISSN 0091-679X

Samatey, F.A., Matsunami, H., Imada K, Nagashima, S., Shaikh, T.R., Thomas, D.R., Chen, J.Z., Derosier, D.J., Kitao, A. & Namba, K. (2004). Structure of the bacterial flagellar hook and implication for the molecular universal joint mechanism. *Nature.* Vol. 431, No. 7012, (October 2004), pp. 1062-1068, ISSN 0028-0836

Samudrala, R., Heffron, F. & McDermott, J.E. (2009). Accurate prediction of secreted substrates and identification of a conserved putative secretion signal for type III secretion systems. *PLoS Pathogens.* Vol. 5, No. 4, (April 2009), pp. e1000375, ISSN 1553-7366

Shaikh, T.R., Thomas, D.R., Chen, J.Z., Samatey, F.A., Matsunami, H., Imada, K., Namba, K. & Derosier, D.J. (2005). A partial atomic structure for the flagellar hook of *Salmonella typhimurium. Proceedings of the National Academy of Sciences of the United States of America.* Vol. 102, No. 4 (January 2005), pp. 1023-1028, , ISSN 0027-8424

Sockett, H., Yamaguchi, S., Kihara, M., Irikura, V.M. & Macnab, R.M. (1992). Molecular analysis of the flagellar switch protein FliM of *Salmonella typhimurium. Journal of bacteriology.* Vol. 174, No. 3, (February 1992) pp. 793-806, ISSN 0021-9193

Springer, M.S., Goy, M.F. & Adler, J. Sensory transduction in *Escherichia coli*: two complementary pathways of information processing that involve methylated proteins. *Proceedings of the National Academy of Sciences of the United States of America.* Vol. 74, No. 8, (August 1977), pp. 3312-3316, ISSN 0027-8424

Staropoli, J.F. & Alon, U. (2000). Computerized analysis of chemotaxis at different stages of bacterial growth. *Biophysical Journal*, Vol. 78, No. 1, (January 2000), pp. 513-519, ISSN 0006-3495

Stecher, B., Barthel, M., Schlumberger, M.C., Haberli, L., Rabsch, W., Kremer, M. & Hardt, W.D. (2008). Motility allows *S. Typhimurium* to benefit from the mucosal defence. *Cellular microbiology*. Vol. 10, No. 5, (May 2008), pp. 1166-1180, ISSN 1462-5814

Stecher, B., Hapfelmeier, S., Müller, C., Kremer, M., Stallmach, T. & Hardt W.D. (2004). Flagella and chemotaxis are required for efficient induction of Salmonella enterica serovar Typhimurium colitis in streptomycin-pretreated mice. *Infection and Immunity*, Vol. 72, No. 7, (July 2004), pp. 4138-4150, ISSN 0019-9567

Suzuki, H., Yonekura, K., Murata, K., Hirai, T., Oosawa, K. & Namba, K. (1998). A structural feature in the central channel of the bacterial flagellar FliF ring complex is implicated in type III protein export. *Journal of structural biology*. Vol. 124, No. 2-3, (December 1998), pp. 104-114, ISSN 1047-8477

Tareen, A.M., Dasti, J.I., Zautner, A.E., Groß, U. & Lugert R. *Campylobacter jejuni* proteins Cj0952c and Cj0951c affect chemotactic behaviour towards formic acid and are important for invasion of host cells. Microbiology. Vol. 156, No. Pt 10, (October 2010), pp. 3123-3135, ISSN 1350-0872

Toft, C. & Fares, M.A. (2008). The evolution of the flagellar assembly pathway in endosymbiotic bacterial genomes. *Molecular biology and evolution*. Vol. 25, No. 9, (September 2008), pp. 2069-2076, ISSN 0737-4038

Ulett, G.C., Webb, R.I. & Schembri, M.A. (2006). Antigen-43-mediated autoaggregation impairs motility in Escherichia coli. Microbiology. Vol. 152, No. PT 7, (July 2006), pp. 2101-2110, ISSN 1350-0872

Van Houten, J., Martel, E. & Kasch T. (1982). Kinetic analysis of chemokinesis of Paramecium. *The Journal of Protozoology*, Vol. 29, No. 2, (May 1982), pp. 226–230, ISSN 0022-3921

Vegge, C.S., Brøndsted, L., Li, Y.P., Bang, D.D. & Ingmer, H. (2009). Energy taxis drives *Campylobacter jejuni* toward the most favorable conditions for growth. *Applied and environmental microbiology*. Vol. 75, No. 16, (August 2009), pp. 5308-5314, ISSN 0099-2240

Yamaguchi, S., Fujita, H., Ishihara, A., Aizawa, S. & Macnab, R.M. (1986). Subdivision of flagellar genes of *Salmonella typhimurium* into regions responsible for assembly, rotation, and switching. *Journal of bacteriology*, Vol. 166, No. 1, (April 1986), pp. 187-193, ISSN 0021-9193

Yamamoto, K. & Imae, Y. (1993). Cloning and characterization of the *Salmonella typhimurium*-specific chemoreceptor Tcp for taxis to citrate and from phenol. *Proceedings of the National Academy of Sciences of the United States of America*. Vol. 90, No. 1, (January 1993), pp. 217-221, ISSN 0027-8424

Yoon, H., Ansong, C., McDermott, J.E., Gritsenko, M., Smith, R.D., Heffron, F. & Adkins J.N. (2011). Systems analysis of multiple regulator perturbations allows discovery of virulence factors in Salmonella. *BMC systems biology*. Vol. 5, No. 1, (June 2011) pp. 100, ISSN 1752-0509

Zicha, D., Dunn, G.A. & Brown A.F. (1991). A new direct-viewing chemotaxis chamber. *Journal of Cell Science*, (August 1991), Vol. 99, No. 4, pp. 769–775, ISSN: 0021-9533

Zigmond, S.H. (1988). Orientation chamber in chemotaxis. *Methods in Enzymology*, Vol 162, (November 2003), pp. 65–72, ISSN 0076-6879

Part 2

Antimicrobial Chemotherapy

4

Antibiotic Resistance and the Prospects of Medicinal Plants in the Treatment of Salmonellosis

A.A. Aliero[1] and A.D. Ibrahim[2]

[1]*Department of Biological Sciences, Usmanu Danfodiyo University, Sokoto*
[2]*Department of Microbiology, Usmanu Danfodiyo University, Sokoto*
Nigeria

1. Introduction

Salmonella enterica serotype typhi is the aetiological agent of typhoid fever, a multisystemic disease with protean manifestations and initial lesions in the bowel. Typhoid fever still remains a major public health problem in developing countries even in the twenty first century (Lin et al., 2000; Otegbayo et al., 2003). This was also the case in America and Europe three centuries ago, until measures for sanitary disposal and supply of potable water were put in place. Unacceptable morbidity and mortality are still recorded in developing countries in spite of availability of several drugs over the years for the treatment of typhoid fever. There is enough evidence to show that the prevalence of typhoid fever in any community is an index of communal hygiene and effectiveness of sanitary disposal. In Nigeria, as in other developing countries of the world, studies have estimated over 33 million cases and 500,000 deaths due to typhoid fever per year (Institute of Medicine, 1986). Otegbayo (2005) enumerated several factors responsible for the failure of public health measures to tame the tide of the continuing rise in the incidence, prevalence, morbidity and mortality of typhoid fever.

Salmonellosis is an infection with Salmonella bacteria, often restricted to the gastro-intestinal tract and is often a self limiting disease. Most individuals infected with *Salmonella typhimurium* experience mild gastrointestinal illness involving diarrhoea, chills, abdominal cramps, fever, head and body aches, nausea, and vomiting (Honish, 1999). Infections are usually self-limiting, and antimicrobial treatment is not recommended for uncomplicated illnesses (Aserkoff and Bennet, 1969; Gill and Hammer, 2001). However, extraintestinal infection can occur, particularly in very young, elderly, and immunocompromised patients (Angulo and Swerdlow, 1995; Thuluvath and McKendrick, 1998). In these cases, effective antimicrobial treatment is essential (Cruchaga et al., 2001). Every year, approximately 40,000 cases of salmonellosis are reported in the United States. The actual number of infections may be thirty or more times greater (CDC, 2006). In many parts of the world, such cases are either not documented or because many milder cases are not diagnosed or reported. Cases, however, of systemic disease due to *Salmonella typhimurium* and other salmonellae have been reported (Panhot and Agarwal, 1982; Varma et al., 2005). Salmonelloses have been reported to be season dependent and occur more in the winter than summer and often referred to as gastroenteritis or diarrheoa. Likewise more cases of diarrhoea caused by enterobacteriacea especially *E. coli*, occurring more during wet season than dry season (Olowe et al., 2003). Children are the most likely to get salmonellosis, however the elderly, and the immunocompromised are the most likely to have severe infections. It is estimated that approximately 600 persons die each year with acute salmonellosis as reported by Centre for disease control (CDC, 2006).

2. Multidrug-resistant (MDR) strains of *Salmonella*

Multidrug-resistant (MDR) strains of *Salmonella* are now encountered frequently and the rates of multidrug-resistance have increased considerably in recent years (CDC, 2006). Even worse, some variants of *Salmonella* have developed multidrug-resistance as an integral part of the genetic material of the organism, and are therefore likely to retain their drug-resistant genes even when antimicrobial drugs are no longer used, a situation where other resistant strains would typically lose their resistance (CDC, 2006). Most of the strains of *Salmonella typhimurium* isolated in a study in western part on Nigeria were resistant to drugs like streptomycin, amoxicilin, tetracycline, ampicillin, kanamycin and chloramphenicol (Olowe et al., 2007). This data is alarming since the isolates were already showing high resistance to drugs that are meant as alternate therapy to salmonellosis treatment; especially isolates from blood were resistant to the commonly used antibiotics. Drug-resistant *Salmonella* emerged in response to antimicrobial usage in food animals, which has also contributed or resulted in major outbreaks of salmonellosis (Olowe et al., 2007). Selective pressure from the use of antimicrobials is a major driving force behind the emergence of resistance, but other factors also need to be taken into consideration. Four types of species namely *S. typhi* (55.5%) *S. paratyphi* A (48.1%), *S. paratyphi* B (25.9%) and *S. typhimurium* (22.2%) were isolated from food samples in Namakkal and the isolates showed multiple drug resistance, and 100 % resistant to Vancomycin, Novobiocin, Nitrofurantoin, Ciproflaxacin, and Methicillin (Jegadeeshkumar et al., 2010). Jegadeeshkumar et al. (2010) also observed increasing resistance for Amoxiclave (92.55 %) and Bacitracin (78.57 %). Akinyemi et al. (2000) reported that out of the total blood samples cultured, 101 (15.9%) isolates of *Salmonella* species were isolated of which 68 (67.3%) were *S. typhi*, 17 (16.8%) and 16 (15.8%) were *S. paratyphi* A and *S. arizonae* respectively. All the *S. typhi* and *S. paratyphi* isolates showed resistance to two or more of the 10 of 12 antibiotics tested particularly the 3-first-line antibiotics commonly used (chloramphenicol, ampicillin and cotrimoxazole) in the treatment of typhoid fever in Nigeria.

Since the discovery of antibiotics and their uses as chemotherapeutic agents, there was a belief in the medical fraternity that this would lead to the eradication of infectious diseases. However, diseases and disease agents that were once thought to have been controlled by antibiotics are returning in new forms resistant to antibiotic therapies (Levy and Marshall, 2004). Incidents of epidemics due to such drug resistant microorganisms are now a common global problem posing enormous public health concerns (Iwu et al., 1999). The global emergence of multi-drug resistant bacterial strains is increasingly limiting the effectiveness of current drugs and significantly causing treatment failure of infections (Hancock, 2005). Examples include methicillin-resistant staphylococci, pneumococci resistant to penicillin and macrolides, vancomycin-resistant enterococci as well as multidrug resistant gram-negative organisms (Norrby et al., 2005). As resistance to old antibiotics spreads, the development of new antimicrobial agents has to be expedited if the problem is to be contained. However, the past record of rapid, widespread and emergence of resistance to newly introduced antimicrobial agents indicates that even new families of antimicrobial agents will have a short life expectancy (Coates et al., 2002). Confronted with a possible shortage of new antimicrobials, there is need to ensure a careful use of our available drugs. This has led to calls for controlled use of antibiotics through the reduction of dosage used per regime of treatment or by regulating prescriptions in areas such as animal husbandry and aquaculture (Hernandez, 2005). While, reduced use could lead to delayed resistance development, the emergence of resistant strains is from an evolutionary viewpoint inevitable. It becomes imperative therefore that alternative approaches are explored. Targeting and blocking resistance processes could be an attractive approach. The presence of efflux pumps and multidrug resistance (MDR) proteins in antibiotic resistant organisms contribute significantly to the intrinsic and acquired resistance in these pathogens. The discovery and development of new compounds that either block or circumvent resistance mechanisms could improve the containment, treatment, and eradication of these strains (Oluwatuyi et al., 2004).

3. Problems of antibiotic resistance

The origin of antibiotic resistance extends much further back in evolutionary terms and reflects the attack and counter-attack of complex microbial flora in order to establish ecological niches and survive. Early treatment failures with antibiotics did represent a significant clinical problem because other classes of agents, with different cellular targets, were available. The emergence of multiple resistances is causing major problems in the treatment options today. Several factors drove this situation in the 1970s and 1980s, including the introduction of extended-spectrum agents and advances in medical techniques such as organ transplantation and cancer chemotherapy. The net result has been a huge selective pressure in favour of multiply resistant species. Coupled with this, there has been a sharp decline in the introduction of agents acting on new cellular targets over the last 30 years compared with the 20-year period following World War II. Smith (2004) reported that the resistant organisms causing concern among Gram-positive organisms at present are methicillin resistant *Staphylococcus aureus* (MRSA) and coagulase-negative staphylococci, glycopeptides intermediate sensitivity *S. aureus* (GISA), vancomycin-resistant *Enterococcus* (VRE) species and penicillin-resistant *Streptococcus pneumoniae*. Concerns among the Gram-negative organisms include multidrug-resistant *Pseudomonas aeruginosa*, *Stenotrophomonas maltophilia* and *Acinetobacter baumannii* and members of the Enterobacteriaceae with extended-spectrum beta- lactamases.

When infections become resistant to first choice or first line antimicrobials, treatment has to be switched to second-or third-line drugs, which are nearly always expensive (Sibanda and Okoh, 2007). In many poor countries in developing nations, the high cost of such replacement drugs is not easy to come-by, with the result that some diseases can no longer be treated in areas where resistance to first-line drugs is widespread (WHO, 2002). The alarming challenges facing physicians and pharmacist now, is the need to develop alternative approaches in addition to the search for new antimicrobial compounds (Sibanda and Okoh, 2007). Plants might hold a promise for combating the problem of antibiotic resistance.

4. Mechanisms of antibiotic resistance in pathogenic bacteria

Resistance to antimicrobial agents typically occurs as a result of four main mechanisms namely enzymatic inactivation of the drug (Davies, 1994), alteration of target sites (Spratt, 1994), reduced cellular uptake (Smith, 2004) and extrusion by efflux (Nakaido, 1994). It has been reported that chemical modifications could be significant in antibiotic resistance, though exclusion from the cell of unaltered antibiotic represents the primary means in denying the antibiotic access to its targets and this is believed to enhance resistance even in cases where modification is the main mechanism (Li et al., 1994b).

4.1 Alteration of target site

Chemical modifications in the antibiotic target may result in reduced affinity of the antibiotic to its binding site (Lambert, 2005). These mechanisms have been reported to be employed by a number of pathogenic bacteria in over-powering the effect of antibiotics and are usually mediated by constitutive and inducible enzymes. Resistance to macrolides, lincosamide and streptogramin B antibiotics (MLSB resistance) in pathogenic *Streptococcus* species is a result of methylation of the N6 amino group of an adenine residue in 23S rRNA. This is presumed to cause conformational changes in the ribosome leading to reduced binding affinity of these antibiotics to their binding sites in the 50S ribosomal subunit (Seppala et al., 1998; Kataja et al., 1998). Beta-lactams antibiotics function by binding to and inhibiting the biosynthetic activity of Penicillin Binding Proteins (PBPs), thereby blocking cell wall synthesis.

4.2 Enzymatic inactivation

The production of hydrolytic enzymes and group transferases is a strategy employed by a number of pathogens in evading the effect of antibiotics (Wright, 2005). Genes that code for antibiotic degrading enzymes are often carried on plasmids and other mobile genetic elements. The resistance to-lactam antibiotics by both gram negative and gram positive bacteria has long been attributed to -lactamases (Frere, 1995). These enzymes confer significant antibiotic resistance to their bacterial hosts by hydrolysis of the amide bond of the four membered-lactam ring (Wilke et al., 2005). Resistance to aminoglycosides in gram-negative bacteria is most often mediated by a variety of enzymes that modify the antibiotic molecule by acetylation, adenylation or phosphorylation (Over et al., 2001).

4.3 Antibiotic efflux pump

It is now widely recognized that constitutive expression of efflux pump proteins encoded by house-keeping genes that are widespread in bacterial genomes are largely responsible for the phenomenon of intrinsic antibiotic resistance (Lomovskaya and Bostian, 2006). Several studies have shown that active efflux can be a mechanism of resistance for almost all antibiotics (Li et al., 1994a; Gill et al., 1999; Lin et al., 2002). The majority of the efflux systems in bacteria are non-drug-specific proteins that can recognize and pump out a broad range of chemically and structurally unrelated compounds from bacteria in an energy-dependent manner, without drug alteration or degradation (Kumar and Schweizer, 2005).

The consequence of this drug extrusion is that, it leads to a reduced intracellular concentration of the antimicrobial such that the bacterium can survive under conditions of elevated antimicrobial concentration (Marquez, 2005). The MIC of the drug against such organisms will be higher than predicted. Multi-drug resistance efflux pumps are ubiquitous proteins present in both gram-positive and gram-negative bacteria as either chromosomally encoded or plasmid encoded (Akama et al., 2005). Although, such proteins are present constitutively in bacteria, the continued presence of the substrate induces over-expression (Teran et al., 2003). This increased transcription is responsible for the acquired resistance. In gram negatives bacteria, the effect of the efflux pumps in combination with the reduced drug uptake due to the double membrane barrier is responsible for the high inherent and acquired antibiotic resistance often associated with this group of organisms (Lomovskaya and Bostian, 2006).

The NorA protein which is the best studied chromosomally encoded pump in pathogenic gram positive bacteria (Hooper, 2005) has been reported to be present in S. epidermidis but appears to be absent in Enterococcus faecalis or in gram-negative organisms, such as E. coli and K. pneumoniae (Kaatz et al., 1993). Over expression of the NorA gene in S. aureus confers resistance to chloramphenicol and hydrophilic fluoroquinolone antimicrobials (Hooper, 2005; Kaatz and Seo, 1995; Hooper, 2005 instead of Hooper, 2005; Kaatz and Seo, 1995). QacA is a member of the major facilitator super-family of transport proteins, which are involved in the uniport, symport, and antiport of a wide range of substances across the cell membrane (Mitchell et al., 1998). The QacA multidrug exporter from S. aureus mediates resistance to a wide array of monovalent or divalent cationic, lipophilic, antimicrobial compounds. QacA provides resistance to these various compounds via a proton motive force-dependent antiport mechanism (Brown and Skurray, 2001).

Baucheron et al. (2004) reported the resistance of Salmonella enteric serovar Typhimurium to fluororoquinolones, chloramphenicol-florfenicol and tetracycline is highly dependent on the presence of AcrAB-TolC efflux pump. S. enterica includes nontyphoidal Salmonella belonging to different serotypes on the basis of the flagellar and somatic antigens, and represents one of the most important food-borne pathogens causing gastroenteritis in humans (Neidhardt, 1996). Nontyphoidal S. enterica strains are easily passed from animals to humans and are thus classified as zoonotic pathogens. They can colonize or infect humans as well as a variety of domesticated and wild animals ranging from mammals to birds and reptiles. Most infections are related to ingestion of contaminated food products rather than person-to-person transmission or direct fecal-oral transmission (Mead et al., 1999). Several S. enterica isolates are characterized by the presence of host-adapted virulence plasmids encoding genes contributing to colonization and resistance to complement killing, such as the spvA, spvB and spvC (salmonella plasmid virulence) and the rck (resistance to complement killing) genes (Guiney et al., 1994).

In the last few years, Salmonella shows increasing antimicrobial resistance rates in isolates obtained from both food animals and humans. S. enterica strains belonging to different serotypes and showing multiple antibiotic resistance (to four or more antimicrobials) are now widespread in both developed and developing countries, most of these strains are zoonotic in origin, acquire their resistance in the food-animal host and cause human infections through the food chain (Threlfall, 2002). Salmonella infections have been associated with the ingestion of poultry, meat, milk and dairy products (Bean et al., 1996).

Most infections result in self-limiting diarrhea and do not require antimicrobial treatment. However, severe life-threatening bacteremias and other deep-seated infections do occur, particularly in children and immunocompromised hosts and in these cases an antimicrobial therapy is recommended (Blaser and Feldman, 1981). Good drugs for Salmonella infections include fluoroquinolones, ampicillin, trimethoprimsulfamethoxazole or third-generation cephalosporins. Rising rates of resistance to ampicillin and trimethoprimsulfamethoxazole have significantly reduced their efficacy and fluoroquinolones are not approved for the use in children. Consequently, extended-spectrum cephalosporins have become the current drugs of choice for the treatment for invasive infections in children (Hohmann, 2001). The emergence of Salmonella species that are resistant to extended-spectrum cephalosporins (Herikstad et al., 1997; Threlfall et al., 1997) is cause of worldwide concern.

4.4 Importance of plasmids in the dissemination of resistance in *S. enterica*

Since the aminoglycoside antibiotic apramycin was licensed for veterinary use in the 1980s, resistance to apramycin and the related antibiotic gentamicin, one the most frequently used aminoglycoside in human therapy (Sibanda and Okoh, 2007). In the United Kindom, during the period 1982-84, the incidence of resistance to apramycin in salmonellas increased from 0.1% in 1982 to 1.4% in 1984 (Wray et al., 1986). Resistance to both apramycin and gentamicin was detected in different Salmonella serotypes, as well in *Escherichia coli*. In particular, the incidence of *S. enterica* (S.) *typhimurium* definitive type 204c (DT204c) from calves showed a dramatically increase (Wray et al., 1986). In *S. typhimurium* DT204c the gentamicin resistance was specified by three types of plasmids of the I1 incompatibility group, which also conferred resistance to apramycin (Threlfall et al., 1986). Most of these plasmids produced the enzyme aminoglycoside 3-N-acetyltransferase IV and the resistance was transferable by conjugation in most of the strains examined. The increasing incidence of the gentamicin-resistant *S. typhimurium* DT204c was also observed in humans, providing the first evidence that the use of apramycin in animal husbandry gave rise to resistance to gentamicin, an antimicrobial used for human therapy (Threlfall et al., 1986).

In the 1990s, the increasing frequency of *E. coli* and Salmonella with plasmids conferring resistance to gentamicin and apramycin was reported in other European countries (Chaslus-Dancla et al., 1991; Pohl et al., 1993). Gentamicin- and apramycin- resistant strains were isolated from both humans and cattle in France and Belgium and six different types of replicons were identified (Pohl et al., 1993). During the 1990-1997 alarming reports pointed out the rapid development in several countries of resistance to β-lactam antibiotics in Salmonella (Threlfall et al., 1997). A survey conducted between 1987 and 1994 in France, demonstrated a dramatic increase (from 0 to 42.5%) in the prevalence of β-lactam resistance among Salmonella isolates. Several types of β-lactamases were found on plasmids belonging to different incompatibility groups Q, P, F and HI (Llanes et al., 1999).

Resistance to β-lactams in Gram-negative bacteria is mediated predominantly by two major types of β- lactamases: the chromosomally-encoded enzymes of the Amber class C (e.g. AmpC β-lactamase in *Citrobacter, Enterobacter, Serratia spp, Morganella morganii* and *Pseudomonas aeruginosa*) or by plasmid-encoded enzymes of the Amber class A, in species that do not produce AmpC β-lactamases, such as *E. coli, Salmonella spp.*, and *Shigella spp.* (Bauernfeind et al., 1998a).

Extended-spectrum β-lactamase (ESBL), evolved from the blaTEM-1, blaTEM-2, and blaSHV-1 genes, extending resistance to new third-generation cephalosporins. During the last decade, infections caused by *S. enterica* carrying ESBLs have been reported, and most of the ESBL-producing strains were found to carry plasmids encoding the blaTEM- 1, and blaSHV-1 gene derivatives (Hammani et al., 1991; Morosini et al., 1995; Tassios et al., 1997; Villa et al., 2000; Mulvey et al., 2003).

In Southern Italy, during the period 1990 to 1998, several epidemiologically unrelated *S. enteriditis* isolates showing resistance to expanded-spectrum cephalosporins were recurrently isolated from ill patients. Most of these strains carried the blaSHV-12 gene located on conjugative plasmids (Villa et al., 2002a). Notably, this was the first case of acquisition of the blaSHV-12 gene by Salmonella in Italy and worldwide. However, SHV-12-encoding plasmids were previously encountered in *K. pneumoniae* isolated from hospitals throughout Italy (Laksai et al., 2000; Pagani et al., 2000). Therefore, it is plausible that SHV-12-encoding plasmids originated from nosocomial bacterial pathogens and were horizontally transmitted to *S. enteriditis* strains.

In 1999, the spread of a *S. typhimurium* clone resistant to third generation cephalosporins has been reported in Russia, Hungary and Greece. In this case non distinguishable institutional and community outbreak of *S. typhimurium* isolates harboured a transferable plasmid containing the blaCTX-M gene (Tassios et al., 1999). The relatedness of resistance plasmids harboured by strains of various origins can be demonstrated by incompatibility grouping, restriction fragmentation pattern analysis and identification of specific resistance determinants located within the plasmids. These analyses may allow a better understanding of how resistance plasmids propagate, helping to trace their evolution.

The *S. typhimurium* pSEM plasmid has a very similar restriction pattern to that of the *K. oxytoca* plasmid, pACM1. The pSEM plasmid was identified in 1997 in *S. typhimurium* strains isolated from children in Albania (Villa et al., 2000), while pACM1 was isolated from *K. oxytoca* strains responsible of a nosocomial outbreak in the USA (Preston et al., 1997; Preston et al., 1999). Both plasmids belonged to the same IncL/M group and conferred resistance to expanded-spectrum cephalosporins by the blaSHV-5 gene. Both plasmids carried a class 1 integron conferring aminoglycoside resistance by the aacA4, aacA1 and aadA1 resistance gene cassettes (Villa et al., 2000; Preston et al., 1997). Thus, these plasmids could be members of a family of broad-host-range replicons widely spreading among Gram-negative pathogens. Other plasmids of the IncL/M group showing similar restriction profiles, and carrying the blaSHV-5 gene and a class 1 integron, were previously described in several countries in Europe from clinical isolates of *P. aeruginosa* and *K. pneumoniae* (de Champs et al., 1991; Petit et al., 1990; de Champs et al., 1991; Prodinger et al., 1996; Preston et al., 1997 instaed of de Champs et al., 1991; Petit et al., 1990; Preston et al., 1997, Prodinger et al., 1996). The identification of a family of related plasmids has serious public health implications, since it demonstrates that broad-host-range plasmids carrying resistance to clinically relevant antibiotics can spread worldwide among bacteria responsible of both nosocomial and community-acquired infections. The fact that conserved plasmids can be identified in a wide variety of pathogens isolated in different countries, illustrates the important role of plasmids in the dissemination of antimicrobial resistance among Gram-negative bacteria.

Recently, a case of treatment failure due to ceftriaxone resistant *S. anatum* has been reported in Taiwan (Su et al., 2003). In this study, ceftriaxone-susceptible *S. anatum* was initially isolated from the urine of a 70-year-old diabetic patient hospitalized for the treatment of a large pressure sore in the sacral area and urinary tract infection. The unexpected emergence of the resistance during the treatment with ceftriaxone led to systemic bacteraemia by *S. anatum* and to the fatal outcome in the patient. The emergence of the resistance has been linked to the *in vivo* acquisition of a resistance plasmid carrying the CTX-M3 β-lactamase by the susceptible *S. anatum* strain. In the same hospital this β-lactamase has been previously identified in clinical isolates of *E. coli, K. pneumoniae* and *Enterobacter cloacae*, suggesting that such bacteria may have acted as reservoirs of the resistance plasmid (Su et al., 2003).

4.5 Plasmid-mediated resistance to expanded-spectrum cephalosporins encoded by the CMY-2 AmpC β- lactamase

The ampC genes were regarded as exclusively chromosomal until 1989, when an AmpC-type β-lactamase was found for the first time on transmissible plasmids (Bauernfeind et al., 1998a). Plasmid-mediated AmpC β-lactamases belong to the homogeneous group of genes related to the chromosomal ampC gene of *Citrobacter freundii* (cmy-2, bil-1 and lat genes), the cmy-1, fox and mox family, or originate from the *Morganella morganii* AmpC β-lactamase (DHA-1) (Bauernfeind et al., 1998a). The latter was identified on a plasmid in *S. enteriditis* (Barnaud et al., 1998; Verdet et al., 2000). The first case of plasmid-mediated CMY-2 in Salmonella was reported on a conjugative plasmid of *S. senftenberg* recovered in 1994 from stool of an Algerian child (Koeck et al., 1997). Review of 1996 data from the National Antimicrobial Resistance Monitoring System (NARMS) in the United States identified only 1 (0.1%) Salmonella isolate among 1272 human Salmonella isolates showing expanded spectrum cephalosporin resistance (Dunne et al., 2000). However, in 1999, NARMS reported the emergence of domestically acquired broad-spectrum cephalosporin resistant Salmonella, most of them producing the CMY-2 AmpC β-lactamase (Dunne et al., 2000).

S. typhimurium strains carrying indistinguishable CMY-2-encoding plasmids were isolated in Nebraska from a patient and cattle during a local outbreak of salmonellosis, demonstrating that the Salmonella-resistant strain evolved primarily in livestock (Fey et al., 2000). Since 1998, *Salmonella* of human and animal origin were reported to show resistance to expanded-spectrum cephalosporins, in Iowa (Winokur et al., 2000; Winokur et al., 2001). Molecular studies demonstrate the emergence of plasmid encoded CMY-2 β-lactamase in most cephalosporin resistant isolates. During the 1998-1999 periods, nearly 16% of *E. coli* isolates and 5.1% of *Salmonella* isolates from clinically ill animals in Iowa produced CMY-2 (Winokur et al., 2000; Winokur et al., 2001).

Similar plasmids, carrying the cmy-2 gene were also reported in Salmonella isolated from animals in Illinois (Odeh et al., 2002). The spreading of the CMY-2-carrying plasmids in the USA was confirmed by molecular analysis of domestically acquired Salmonella strains of human origin isolated in nine different States, representing the 87% of the total expanded-spectrum cephalosporin resistant Salmonella collected by the Center for Disease Control and Prevention (CDC) during the 1996-1998 surveillance periods. The isolates were distinguishable by their chromosomal DNA patterns, thus demonstrating that they did not represent the epidemic spread of a clonal strain (Dunne et al., 2000).

4.6 The use of resistance modifying agents in combination with antibiotics to overcome resistance

The selection pressure exerted by the continued presence of bactericidal or bacteriostatic agents facilitates the emergence and dissemination of antibiotic resistance genes. Over generations, the genotypic makeup of bacterial populations is altered (Taylor et al., 2002). The clinical implications of this are that many infections become untreatable resulting in serious morbidity and mortality. Although the introduction of new compounds into clinical use has helped to curtail the spread of resistant pathogens, resistance to such new drugs, has developed in some cases. It has been observed by several studies that antibiotic combinations can have synergistic benefits and interactions between existing antibiotics (Bayer et al., 1980; Hallander et al., 1982; Hooton et al., 1984; Cottagnoud et al., 2000; instaed of Bayer et al., 1980; Hooton et al., 1984; Cottagnoud et al., 2000; Hallander et al., 1982.). Several current therapeutic regimes are based on synergistic interactions between antibiotics with different target sites. As new antimicrobial compounds are discovered, there is need to assess their potentials in combination therapies with old antibiotics that have been rendered ineffective by the development of resistant strains, even when such compounds are not directly evidently inhibitory. Taylor et al. (2002) suggested that the use of agents that do not kill pathogenic bacteria but modify them to produce a phenotype that is susceptible to the antibiotic could be an alternative approach to the treatment of infectious disease. Such agents could render the pathogen susceptible to a previously ineffective antibiotic, and because the modifying agent applies little or no direct selective pressure, this concept could slow down or prevent the emergence of resistant genotypes. The inhibition of resistance expression approach was successfully used in the production of Augmentin, a combination of amoxycillin and clavulanic acid (Reading and Cole, 1977). In this case, clavulanic acid is an inhibitor of class-A-lactamases which is coadministered with amoxicillin. The combination has been used clinically since the late 1970s (Neu et al., 1993). A similar approach can be used for target-modifying enzymes and for efflux systems. A number of *in vitro* studies have reported the use of plant extracts in combination with antibiotics, with significant reduction in the MICs of the antibiotics against some resistant strains (Al-hebshi et al., 2006; Darwish et al., 2002; Al-hebshi et al., 2006; Betoni et al., 2006 instead of Al-hebshi et al., 2006; Darwish et al., 2002; Betoni et al., 2006). The curative effect of plant extracts in this combination study has been variably referred to as resistance modifying/modulating activity (Gibbons, 2004). This ability of plant extracts to potentiate antibiotics has not been well explained. It is speculated that inhibition of drug efflux, and alternative mechanisms of action could be responsible for the synergistic interactions between plant extracts and antibiotics (Lewis and Ausubel, 2006; Zhao et al., 2001; Lewis and Ausubel, 2006; instead of Lewis and Ausubel, 2006; Zhao et al., 2001).

4.7 Efflux pump inhibition in combination with antibiotics as a strategy for overcoming resistance

The discovery and development of clinically useful Efflux Pump Inhibitors (EPIs) that decrease the effectiveness of efflux pumps represents a significant advance in the development of therapeutic regimes for the treatment of MDR-related conditions. This approach termed the EPI strategy (Lomovskaya and Bostian, 2006), is based on blocking the activity of the pumps, resulting in the accumulation of the antibiotic inside the bacterial cell, consequently increasing access to its target sites. In addition, this will lead to increased susceptibility of the bacterium, thus implying that the therapeutic effect of the drug is achieved with low concentrations. Combining broad spectrum efflux pump inhibitors with current drugs that are pump substrates can recover clinically relevant activity of those compounds and thus may provide new dimensions to the ever increasing need for development of new antimicrobial agents. This approach will in addition lead to the preservation and improvement of the usefulness of old and cheap antibacterial agents. Ultimately this could reduce the appearance and spread of resistant mutants (Kaatz, 2002).

4.8 Role of Ethnopharmacology in the treatment of Salmonellosis

Herbal medicine is used globally and has a rapidly growing economic importance. In developing countries, traditional medicine is often the only accessible and affordable treatment available. In Africa, 80% of the population uses traditional medicine as the primary health care system (Fisher and Ward, 1994). Traditional medicine is also gaining more respect by national governments and health providers. Peru's national program in complementary medicine and the Pan American health organization recently compared complementary medicine in clinics and hospitals within the Peruvian social security system (Lima, 2000). Plants have been used in traditional medicine for several thousand years (Abu – Rabia, 2005). The knowledge of medicinal plants has been accumulated in the course of many centuries based on different medicinal system. During the last few decades there is an increasing interest in the study of medicinal plants and their traditional use in different parts of the world (Rossato et al., 1999). There are considerable economic benefits in the use of medicinal plants for the treatments of various diseases (Azaizeh, 2003). Due to less communication means, poverty, ignorance and unavailability of modern health facilities, most of the rural people are forced to practice traditional medicines for their common day ailments. Most of these people form the poorest link in the trade of medicinal plants (Khan, 2002). A vast knowledge of how to use the plants against different illness may be expected to have accumulated in areas were the use of plants is still of great importance (Diallo et al., 1999). In the developed countries, 25% of the medical drugs are based on plants and their derivatives (Principe, 1991). A group of World Health Organization experts, who met in Congo, Brazzaville in 1976, sought to define traditional African medicine as the sum total of practices, measures, ingredients and procedures of all kinds whether material or not, which from time immemorial has enabled the African to guard against diseases to alleviate his / her suffering and cure him / herself (Busia, 2005). Traditional medical knowledge of medicinal plants and their use by indigenous cultures are not only useful for conservation of community health care and biodiversity but also for community health care and drug development in the present and future (Pei, 2001).

A number of workers have evaluated the anti typhoidal activities of medicinal plants and some of them proved to be promising. For example: Aliero and Wara (2009) evaluated the efficacy of *Leptadenia hastata* (Pers.) Decne extracts against five selected bacterial species and two fungal species. Aqueous extract markedly inhibited the growth of *Salmonella paratyphi* and *Escherichia coli* and *Pseudomonas aeruginosa*. The result obtained in this study has provided a scientific support for the claimed ethnomedical uses of aqueous extracts of *L. hastata* in the treatment of bacterial diseases and suggest the potential of methanol extract as a source of antifungal agent. Evans et al. (2002) evaluated the efficacy of *Euphobia hirta; Citrus aurantifolia, Cassia occidentalis,* and *Cassia eucalyptus* claimed by the Nupes tribe of Nigeria to be effective in the treatment of typhoid fever. The result of *invitro* antimicrobial analysis showed that, only *Cassia eucalyptus* showed inhibition of *Salmonella typhi* growth and concluded that the plant is efficacious and contains natural compounds that could be used in the treatment of typoid fever. Similarly, In Camerron, Nkuo-Akenji et al. (2001) evaluated the effects of herbal extracts derived from plants commonly prescribed by traditional practitioners for the treatment of typhoid fever against *Salmonella typhi*, *S. paratyphi* and *S. typhimurium* using formulations often prescribed by the traditional healers which includes; 1) Formulation A comprising *Cymbogogon citratus* leaves, *Carica papaya* leaves, and *Zea mays* silk. 2) Formulation B comprising *C. papaya* roots, *Mangifera indica* leaves, *Citrus limon* fruit and *C. citratus* leaves. 3) *C. papaya* leaves. 4) *Emilia coccinea* whole plant. 5) *Comelina bengalensis* leaves. 6) *Telfaria occidentalis* leaves. 7) *Gossypium arboreum* whole plant. The result obtained in this study, showed that Formulation A elicited inhibitory activity at a lower range of 0.02 to 0.06 mg/ml. Similarly, Formulation B elicited bacterial activity at the lowest range of 0.06 to 0.25mg/ml. *C. bengalensis* leaves on the other hand, showed the lowest activity with a concentration range of 0.132 to 2.0 mg/ml and 1 to 4 mg/ml in MIC and MBC assays respectively. The result demonstrated that *S. paratyphi* was most sensitive to the formulations (concentration range of 0.02 to 1 mg/ml in both MIC and MBC assays) while *S. typhimurium* was the least sensitive and concentrations of up to 4 mg/ml were required to be bactericidal.

Iroha et al. (2010) evaluated the anti *Salmonella typhi* activity of ethanol, hot and cold crude water extracts of *Vitex doniana* (root), *Cassia tora* (Leaf), *Alstonia boonei* (bark), *Stachytarpheta jamaicensis* (leaf), and *Carica papaya* (leaf) used as traditionally medicine in Ebonyi state, Nigeria. Ethanol extracts of *Vitex doniana* exhibited anti-typhoid activity against 9(90%) of the test organisms, *A. boonei* exhibited activity against 8(80%) of the test organisms, *C. papaya* against 2(20%), *C. tora* against 6(60%), and *S. jamaicensis* against 6(60%). Hot water extract of *Vitex doniana* showed anti-typhoid activity against 7(70%) of the test organisms, *A. boonei* against 9(90%), *C. papaya* against 1(10%), *C. tora* against 8(80%) and *S. jamaicensis* against 7(70%). Cold water extract of *V. doniana*, had anti-typhoid activity against 6(60%) of the test organisms, *A. boonei*, against 6(60%), *C. papaya* against 0(0%), *C. tora* against 6(60%) and *S. jamaicensis* against 4(40%). MIC of ethanol, hot and cold water extracts of *V. doniana*, *A. boonei*, *C. papaya*, *C. tora* and *S. jamaicensis,* fall within 0.4 -128, 0.8 -128, 64 -128, 32 – 128 and 32 – 128. MIC of hot water extracts were within 16 -128, 0.8 – 128, 128 -512, 0.8 – 512 and 0.8 – 128 while MIC of cold water extract are within 64 – 128, 64 – 512, 64 – 512, 64 – 512 and 128-512 respectively. The results of these findings showed that ethanol and hot water extracts of *V. doniana* and *A. boonei* had the best antityphoid activity.

The work of Oluduro and Omoboye (2010) investigated the antibacterial potency and synergistic effect of crude aqueous and methanolic extracts of nine plant parts against multi-drug resistant *S. typhi* tested against nine plant parts: unripe *Carica papaya* fruit, *Citrus aurantifoliia*, *Anana sativus*, *Citrus paradisi*, *Cymbopogon citratus*, *Cocos nucifera* leaves of *Carica papaya,* leaves of *Euphorbia heterophylla* and *Gossypium* spp. Both the aqueous and methanol extracts of each plant material and mixture showed appreciable antimicrobial activities on *S. typhi*. Antimicrobial activity increased with increasing concentration of the extracts. Synergistic activity of crude aqueous and crude methanolic extracts of the plant parts, in various combinations of two to nine against the test organism ranged from 10-33 mm zone of growth inhibition. The antibacterial efficacy of the mixture of extracts from plant parts increased considerably compared to the low activities recorded with the extract of individual plant parts (P>0.05). Methanolic extracts of each plant material and mixture produced greater antimicrobial activity than the aqueous extracts at all concentrations. The minimum inhibitory concentration (MIC) of the individual plant parts ranged between 0.1 and 1.0 mg/ml in aqueous extracts and 0.01 and 0.1 mg/ml in methanol extracts while the MICs of the combined extracts ranged between 0.1 and 0.01 mg/ml in aqueous extracts and 0.01 and 0.0001 mg/ml in methanolic extracts. The combined or synergistic activity of the plant parts compared favourably with the standard antibiotics of choice for salmonella-infections therapy, and contained two or more phytochemicals responsible for their antimicrobial activities. There is the need therefore to develop effective combination of antimicrobial agents in purified form from higher plants and their parts for clinical trials. Frey and Meyers (2010) reported that *Achillea millefolium, Ipomoea pandurata, Hieracium pilosella,* and *Solidago canadensis* exhibited antimicrobial properties as expected, with particularly strong effectiveness against *S. typhimurium.* In addition, extracts from *Hesperis matronalis* and *Rosa multiflora* also exhibited effectiveness against this pathogen.

4.9 Plants as sources of new antimicrobials and resistance modifying agents

Plants have traditionally provided a source of hope for novel drug compounds, as plant herbal mixtures have made large contributions to human health and well-being (Iwu et al., 1999). Owing to their popular use as remedies for many infectious diseases, searches for substances with antimicrobial activity in plants are frequent (Shibata et al., 2005; Betoni et al., 2006). Plants are rich in a wide variety of secondary metabolites, such as tannins, terpenoids, alkaloids, and flavonoids, which have been found *in vitro* to have antimicrobial properties (Cowan, 1999; Lewis and Ausubel, 2006). Examples of some of these compounds are shown in Table 1. Literature is awash with compounds that have been isolated from a variety of medicinal plants. Despite this abundant literature on the antimicrobial properties of plant extracts, none of the plant derived chemicals have successfully been exploited for clinical use as antibiotics (Gibbons, 2004). A significant part of the chemical diversity produced by plants is thought to protect plants against microbial pathogens. Gibbons (2004) reported that a number of plant compounds often classified as antimicrobial produce MIC greater than 1,000 µg/ml, which are of no relevance from a clinical perspective. Tegos et al. (2002) suggested that a vast majority of plant compounds showing little *in vitro* antibacterial activity are not antimicrobial but are regulatory compounds playing an indirect role in the plant defence against microbial infections. The observation that plant derived compounds are generally weak compared to bacterial or fungal produced antibiotics and that these compounds often show considerable activity against gram-positive bacteria than gram-negative species has been made by many (Nostro et al., 2000; Gibbons, 2004). This observations led Tegos et al. (2002) hypothesizing that; Plants produce compounds that can be effective antimicrobials if they find their way into the cell of the pathogen especially across the double membrane barrier of Gram negative bacteria. Production of efflux pump inhibitors by the plant would be one way to ensure delivery of the antimicrobial compound. This hypothesis has been supported by the findings of Stermitz et al. (2000 a,b), who observed that *Berberis* plants which produce the antimicrobial compound, berberine, also make the MDR inhibitors 5-methoxyhydnocarpin D (5-MHC-D) and pheophorbide A. The MDR inhibitors facilitated the penetration of berberine into a model gram-positive bacterium, *S. aureus.* In testing their hypothesis, Tegos et al. (2002), showed that two MDR inhibitors (INF271 and MC207110) dramatically increased the effectiveness of thirteen putative plant antimicrobial compounds against gram-negative and gram positive bacteria including isolates known to express efflux pumps.

Class of compound	Examples	Plant sources	Reference
Coumarins and their derivatives	Asphodelin A 4'-O-D-glucoside Asphodelin A	*Asphodelus microcarpus*	El-Seedi (2007)
Simple phenols	Epicatechin Epigallocatechin Epigallocatechin gallate Epicatechin gallate	*Calophyllum brasiliense* *Camellia sinensis*	Pretto et al. (2004) Mabe et al. (1999) Hamilton-Miller (1995)
Flavonoids	Isocytisoside Eucalyptin	*Aquilegia vulgaris* L. *Eucalyptus maculate*	Bylka et al. (2004) Takahashi et al. (2004)
Flavones	Luteolin GB1(hydroxybiflavanonol)	*Senna petersiana* *Garcinia kola*	Tshikalange et al. (2005) Madubunyi (1995) Han et al. (2005)
Tannins	Ellagitannin	*Punica granatum*	Machado et al. (2002)
Alkaloids	Berberine	*Mahonia aquifolium*	Cernakova and Kostalova (2002)
Terpenes	Ferruginol, (Diterpene) Epipisiferol (Diterpene) 1-Oxoferruginol	*Chamaecyparis lawsoniana* *Salvia viridis*	Smith et al. (2007) Ulubelen et al. (2000)

Sibanda and Okoh, 2007

Table 1. Examples of some plants with active antimicrobial derived compounds.

These studies have provided the bases for understanding the action of plant antimicrobials, namely that vast majority of such compounds are agents with weak or narrow-spectrum activities that act in synergy with intrinsically produced efflux pump inhibitors. There is reason therefore to believe that, plants could be a source of compounds that can increase the sensitivity of bacterial cells to antibiotics. Such compounds could be useful particularly against antibiotic resistant strains of pathogenic bacteria. The rich chemical diversity in plants promises to be a potential source of antibiotic resistance modifying compounds and has yet to be adequately explored.

4.10 Resistance modifying activities of plants crude extracts: The basis for isolation of potentially useful compounds

If the isolation of resistance modifying compounds from plants is to be realistic, screening for such activities in crude extracts is the first step in identifying leads for isolation of such compounds, and some plants have provided good indications of these potentials for use in combination with antimicrobial therapy. Typical examples are as follows: Aqueous extracts of tea (*Camellia sinensis*) have been shown to reverse methicillin resistance in MRSA and also, to some extent, penicillin resistance in beta lactamase-producing *Staphylococcus aureus* (Stapleton et al., 2004). Forty to one hundred fold dilutions of tea extracts was able to reduce the MICs of high- level resistant MRSA (256 µg/ml) to less than 0.12 µg/ml for methicillin and penicillin (Yam et al., 1998; Stapleton et al., 2004). Aqueous crude khat (*Catha edulis*) extracts of Yemen showed varying antibacterial activities with a range of 5-20 mg/ml-1 against periodontal bacteria when tested in isolation. Addition of the extracts at a sub- MIC (5 mg/ml) resulted in a 2 to 4-folds potentiation of tetracycline against resistant strains *Streptococcus sanguis* TH-13, *Streptococcus oralis* SH-2, and *Fusobacterium nucleatum* (Al-hebshi et al., 2006). Betoni et al. (2006), observed synergistic interactions between extracts of guaco (*Mikania glomerata*), guava (*Psidium guajava*), clove (*Syzygium aromaticum*), garlic (*Allium sativum*), lemongrass (*Cymbopogon citratus*), ginger (*Zingiber officinale*), carqueja (*Baccharis trimera*), and mint (*Mentha pieria*) from Brazil and some antibiotics which represented inhibitors of protein synthesis, cell wall synthesis, nucleic acid synthesis and folic acid synthesis against *Staphylococcus aureus*. Darwish et al. (2002) reported that sub-inhibitory levels (200 µgml−1) of methanolic extracts of some Jordanian plants showed synergistic interactions in combination with chloramphenicol, gentamicin, erythromycin and penicillin G against resistant and sensitive *S. aureus*. The methanolic extract of *Punica granatum* (PGME) showed synergistic interactions with chloramphenicol, gentamicin, ampicillin, tetracycline, and oxacillin. The bactericidal activity of the combination of PGME (0.1×MIC) with ampicillin (0.5×MIC) by time-kill assays, reduced cell viability b 99.9 and 72.5% in MSSA and MRSA populations, respectively (Braga et al., 2005). The ethanol extracts of the Chinese plants, *Isatis tinctoria* and *Scutellaria baicalensis* in combination with ciprofloxacin had synergistic activities against antibiotic resistant *S. aureus* (Yang et al., 2005). The combinations of pencillin with ethanolic extracts of *Paederia scandens* and *Taraxacun monlicum* showed a strong bactericidal activity on two strains of *S. aureus* (Yang et al., 2005). When Ciprofloxacin was incorporated at sub-inhibitory concentrations (1/8MIC) to the crude chloroform extracts of *Jatropha elliptica* and the mixture assayed against NorA expressing *S. aureus*, the activity of the extract was enhanced. This suggests the presence of an inhibitor of the pump which could restore the activity of Ciprofloxacin (Marquez et al., 2005). In another study, Ahmad and Aqil (2006) observed that crude extracts of Indian medicinal plants, *Acorus calamus, Hemidesmus indicus, Holarrhena antidysenterica* and *Plumbago zeylanica* showed synergistic interactions with tetracycline and ciprofloxacin against extended Spectrum beta-lactamase (ESBL), producing multidrug-resistant enteric bacteria with ciprofloxacin showing more synergy with the extracts than tetracycline.

4.11 Plant compounds as resistance modifying agents

Some isolated pure compounds of plant origin have been reported to have resistance modifying activities *in vitro*. Examples of some of the compounds are given in Table 2. This has prompted the search for such compounds from a variety of medicinal plants. Some of the compounds which have been observed to have direct antimicrobial activity have been

Compound	Plant source	Antibiotics potentiated	Reference
Ferruginol 5-Epipisiferol	*Chamaecyparis lawsoniana*	Oxacillin, Tetracycline, Norfloxacin Tetracycline	Smith et al. (2007)
2,6-dimethyl-4-phenylpyridine-3,5-dicarboxylic acid diethyl ester	*Jatropha elliptica*	Ciprofloxacin, Norfloxacin, Pefloxacin, Acriflavine and Ethidium bromide	Marquez et al. (2005)
Carnosic acid carnosol	*Rosmarinus officinalis*	Erythromycin	Oluwatuyi et al. (2004)
Ethyl gallate	*Caesalpinia spinosa*	Bate-lactams	Shibata et al. (2005)
Methyl-1-_-acetoxy-7-_-14 _-dihydroxy-8,15-isopimaradien-18-oate Methyl-1-_-14-_-diacetoxy-7-_-hydroxy-8,15-isopimaradien-18-oate	*Lycopus europaeus*	Tetracycline and Erythromycin	Gibbons et al. (2003)
Epicatechin gallate Epigallocatechin gallate	*Camellia sinensis*	Norfloxacin Imipenem Panipenem beta-Lactams	Gibbons et al. (2004) Hu et al. (2002) Zhao et al. (2001)

Sibanda and Okoh, 2007

Table 2. Some antibiotic resistance modifying compounds from plants.

shown to be potentiating against the activity of antibiotics when used at low MIC levels. The antimicrobial properties of tea (*Camellia sinensis*) have been found to be a result of the presence of polyphenols (Yam et al., 1998; Stapleton et al., 2004; Si et al., 2006). Bioassay directed fractionation of the extracts revealed that epicatechin gallate (ECG), epigallocatechin gallate (EGCG), epicatechin (EC), and caffeine (CN) are the bioactive components. ECG and CG reduced MIC values for oxacillin from 256 and 512 to 1 and 4 mgl-1 against MRSA (Shibata et al., 2005). Ethyl gallate, a conginer of alkyl gallates purified from a dried pod of tara (*Caesalpinia spinosa*) native to South America, intensified lactam susceptibility in MRSA an MSSA strains (Shibata et al., 2005). The abietane diterpenes, (carnosic acid carnosol) isolated from the aerial parts of *Rosmarinus officinalis* by fractionation of its chloroform extract at 10 µgml-1, potentiated the activity of erythromycin (16 - 32 fold) against strains of *S. aureus* that express the two efflux proteins MsrA and TetK. Additionally, carnosic acid was shown to inhibit ethidium bromide efflux in a NorA expressing *S. aureus* strain (Oluwatuyi et al., 2004). A penta-substituted pyridine, 2, 6-dimethyl-4-phenylpyridine-3, 5-dicarboxylic acid diethyl ester and proparcine have been isolated from an ethanol extract of rhizome of *Jatropha elliptica* by bioassay guided fractionation. The pyridine at a concentration of 75 µgml-1 was shown to increase by 4-fold, the activity of ciprofloxacin and norfloxacin against NorA expressing *S. aureus* when tested at sub-inhibitory concentrations (Marquez et al., 2005). Smith et al. (2007) screened active compounds from the cones of *Chamaecyparis Lawsoniana* for resistance modifying activities and observed that Ferruginol and 5-Epipisiferol were effective in increasing the efficacy of tetracycline, norfloxacin, erythromycin and Oxacillin against resistant *S. aureus*. The majority of researches on the combinations between plant extracts and antibiotics have been focused on the identification and isolation of potential resistance modifiers from such natural sources which are considered to be positive results. However, it is likely that such combinations could produce antagonistic interactions that most studies have considered irrelevant and therefore ignored (Sibanda and Okoh, 2007).

5. Suggested solutions to challenges in management

There are still loopfuls of challenges in many developing countries for the management of typhoid fever. Otegbayo (2005) gave the following suggestions as solution for typhoid fever management. This include among others, the improvement in personal and communal hygiene, effective waste disposal system and provision of potable water. Effective treatment of index cases, health education both for the populace and physicians are other important measures. Determination of drug sensitivity patterns and aggressive policy will be quite helpful. The difficulty in diagnosis could also be overcome by making laboratory facilities such as culture media available. Parry et al. (2002) recently suggested the use of conjugate Vi vaccine as part of the Expanded Programme of Immunization. The cost-effectiveness of this latter measure may however be negative for resource – poor countries, where preventive measures by way of improved sanitation and provision of potable water would be more beneficial. Above all, resources should be made available, accessible and affordable to the common man; National Health Insurance appears to be the answer to this as well as economic empowerment of the people in emerging economies.

6. References

Abu – Rabia, A. (2005). Urinary diseases and ethno botany among pastoral nomads in the Middle East. *Journal of Ethno biology and Ethno medicine*. 1:4.

Ahmad, I and Aqil, F. (2006). *In vitro* efficacy of bioactive extracts of 15 Medicinal plants against ESbL- producing multidrug-resistant enteric bacteria. *Microbio. Res.* 1-12.

Akama, H., Kanemaki, M., Tsukihara, T., Nakagawa, A. and Nakae, T. (2005). Preliminary crystallographic analysis of the antibiotic discharge outer membrane lipoprotein OprM of *Pseudomonas aeruginosa* with an exceptionally long unit cell and complex lattice structure. *Acta Cryst.* F61: 131-133.

Akinyemi, K. O., Coker, A. O. Olukoya, D. K. Oyefolu, A. O. Amorighoye, E. P. and among Clinically Diagnosed Typhoid Fever Patients in Lagos, Nigeria. *Z.Natureforsch.* 55:489-493.

Al-hebshi, N., Al-haroni, M. and Skaug, N. (2006). *In vitro* antimicrobial and resistance-modifying activities of aqueous crude khat extracts against oral microorganisms. *Arch. Oral Biol.* 51:183-188.

Aliero, A. A. and Wara, S. H. (2009). Validating the medicinal potential of *Leptadenia hastata*. *African Journal of Pharmacy and Pharmacology.* 3(6): 335-338.

Angulo, F.J. and Swerdlow, D.L. (1995). Bacterial enteric infections in persons infected with human immunodeficiency virus. *Clin Infect Dis*; 21(Suppl 1): S8493.

Aserkoff, B. and Bennett, J.V. (1969). Effect of antimicrobial therapy in acute salmonellosis on the fecal excretion of salmonellae. *N Engl J Med*; 281:63640.

Azaizeh, H., Fulder, S. and Khalil, K. and Said, O. (2003). Ethno medicinal knowledge of local Arab practioners in the Middle East Region. *Fitoterapia*,74: 98 – 108.

Barnaud, G., Arlet, G., Verdet, C., Gaillot, O., Lagrange, P.H. and Philippon, A. (1998). *Salmonella enteritidis*: AmpC plasmid-mediated inducible beta-lactamase (DHA-1) with an ampR gene from *Morganella morganii*. *Antimicrob Agents Chemother.* 42: 2352-2358.

Baucheron, S., Tyler, S., Boyd, D., Mulvey, MR, Chaslus-Dancla, E. and Cloeckaert, A. (2004). AcrAB -TolC Directs Efflux-Mediated Multidrug Resistance in *Salmonella enterica Serovar Typhimurium* DT104. *Antimicro. Agents Chemother.* 48(10): 3729-3735.

Bauernfeind, A., Chong, Y. and Lee, K. (1998a). Plasmidencoded AmpC β-lactamases: how far have we gone 10 years after the discovery? *Yonsei Medical J.* 6: 520-525.

Bauernfeind, A., Hohl, P., Schneider, I., Jungwirth, R. and Frei, R. (1998b). *Escherichia coli* producing a cephamycinase (CMY-2) from a patient from the Libyan- Tunisian border region. *Clin. Microbiol. Infect.* 4:168-170.

Bayer, A.S, Chow, A.W., Morrison, J.O. and Guze, L.B. (1980). Bactericidal synergy between penicillin or ampicillin and aminoglycosides against antibiotic-tolerant lactobacilli. *Antimicrob.Agents Chemother.* 17(3): 359-363.

Bean, N.H., Goulding, J.S., Lao, C. and Angulo, F.J. (1996). Surveillance for foodborne-disease outbreaks—United States, 1988-1992. MMWR CDC Surveill. *Summ.* 45: 1-66.

Betoni, J.E.C., Mantovani, R.P., Barbosa, L.N., Di-Stasi, L.C. and Fernandes, A. (2006). Synergism between plant extract and antimicrobial drugs used on *Staphylococcus aureus* diseases. Mem. Inst. Oswaldo Cruz. 101 No. 4.

Blaser, M.J. and Feldman, R.A. (1981). From the centers for disease control. Salmonella bacteremia: reports to the Centers for Disease Control, 1968-1979. *J. Infect. Dis.* 143: 743- 746.

Braga, L.C., Leite, A.A.M., Xavier, K.G.S., Takahashi, J.A., Bemquerer, M.P., Chartone-Souza, E. and Nascimento, A.M.A. (2005). Synergic interaction between pomegranate extract and antibiotics against *Staphylococcus aureus*. Can. *J. Microbio.* 51 (7): 541-547.

Bush, K. (2004). Antibacterial drug discovery in the 21st century. *Clin. Microbiol. Inf.*, 10 (s4):10-17.

Bylka, W., Szaufer-Hajdrych, M., Matlawska, I. and Goslinska, O. (2004). Antimicrobial activity of isocytisoside and extracts of *Aquilegia vulgaris* L. *Lett. Appl. Microbiol.* 39(1): 93-97.

Breevort, P. (1998). The Booming US. Botanical market. *Herbal Gram.* 44:34 - 46.

Brown, M.H. and Skurray, R.A. (2001). Staphylococcal multidrug efflux protein QacA. *J. Mol. Microbiol. Biotechnol.* 3(2): 163-170.

Busia, K. (2005). Medicinal provision in Africa past and present. *Phytotherapy Research.*19: 919-923.

CDC. (2006). Coordinating Centre for Infectious Diseases / Division of Bacteria and Mycotic Diseases. Bulletin; Nov 4, 2006.

Cernakova, M. and Kostalova, D. (2002). Antimicrobial activity of berberine, a constituent of *Mahonia aquifolium*. *Folia Microbiol.* (Praha). 47(4):375-378.

Chaslus-Dancla, E., Pohl, P., Meurisse, M., Marin, M. and Lafont, J.P. (1991). High genetic homology between plasmids of human and animal origins conferring resistance to the aminoglycosides gentamicin and apramycin. *Antimicrob. Agents Chemother.* 35:590-593.

Coates, A., Hu, Y., Bax, R. and Page, C. (2002). The future challenges facing the development of new antimicrobial drugs. *Nat. Rev. Drug Discov.* 1:895-910.

Cottagnoud, P., Acosta, F., Cottagnoud, M., Neftel, K. and Tauber, M.G. (2000). Synergy between Trovafloxacin and Ceftriaxone against Penicillin- Resistant Pneumococci in the Rabbit Meningitis Model and *In Vitro*. *Antimicrob. Agents Chemother.*44(8):2179-2181.

Cowan, M.M. (1999). Plant Products as Antimicrobial Agents. *Clin. Microbiol Rev.*12(4): 564 - 582.

Cruchaga, S., Echeita, A., Aladuena, A., Garcia-Pena, J., Frias, N. and Usera, M.A. (2001). Antimicrobial resistance in salmonellae from humans, food and animals in Spain in 1998. *J Antimicrob Chemother*; 47:31521.

Darwish, R.M., Aburjai, T., Al-Khalil. S. and Mahafzah, A. (2002). Screening of antibiotic resistant inhibitors from local plant materials against two different strains of *S. aureus*. *J. Ethnopharm.* 79: 359- 364.

Davies, J. (1994). Inactivation of antibiotics and the dissemination of resistance genes. *Science.* 264: 375-382.

De Champs, C., Sirot, D., Chanal, C., Poupart, M.C.,Dumas, M.P. and Sirot, J. (1991). Concomitant dissemination of three extended-spectrum β-lactamasesamong different Enterobacteriaceae isolated in French hospital. *J. Antimicrob. Chemother.* 27:441-457.

Diallo, D., Hveem, B., Mahmoud, M.A., Berge, G., Paulsen, B.S. and Maiga, A. (1999). An ethno botanical survey of herbal drugs of Gourma District, Mali. *Pharmaceutical Biology.* 37:80 -91.

Dryselius, R., Nekhotiaeva, N. and Good, L. (2005). Antimicrobial synergy between mRNA- and protein-level inhibitors. *J. Antimicrob. Chemother.* 56 (1): 97-103.

Dunne, E.F., Fey, P.D., Kludt, P., Reporter, R., Mostashari, F., Shillam, P., Wicklund, J., Miller, C., Holland, B., Stamey, K., Barret, T.J., Rasheed, J.K., Tenover, F.C., Ribot, E.M. and Angulo F.J. (2000). Emergence of domestically acquired ceftraxoneresistant Salmonella infections associated with AmpC beta-lactamase. *JAMA.* 284:3151-3156.

Elkins, R. and Hawaiian, N. (1997). Pleasant Groove, UT. *Woodland Publishing.*

El-Seedi, H.R. (2007). Antimicrobial Arylcoumarins from *Asphodelus microcarpus. J. Nat. Prod.,* 1:118 -120.

Evans, C.E., Banso, A. and Samuel, O. A. (2002). Efficacy of some Nupe medicinal plants against *Salmonella typhi*: an in vitro study. *J Ethnopharmacol.* 80 (1): 21-4.

Fey, P. D., Safranek, T. J., Rupp, M. E., Dunne, E. F., Ribot, E., Iwen, P. C., Bradford, P.A., Angulo, F.J., and Hinrichs, S. H. (2000). Ceftriaxone-resistant Salmonella infections acquired by a child from cattle. *N. Engl. J. Med.* 342:1242-1249.

Fisher, P. and Ward, A. (1994). Complementary medicine in Europe. *British Medical Journal.* 309: 107-111.

Frere, J. M. (1995). Beta-lactamases and bacterial resistance to antibiotics. *Mol. Microbiol.* 16(3): 385-395.

Frey, F. M. and Meyers, R. (2010). Antibacterial activity of traditional medicinal plants used by Haudenosaunee peoples of New York State. *BMC Complementary and Alternative Medicine,* 10(64):1-10

Gibbons, S. (2004). Anti-staphylococcal plant natural products. *Nat. Prod. Rep.,* 21:263-277.

Gibbons, S., Moser, E. and Kaatz, G.W. (2004). Catechin gallates inhibit multidrug resistance (MDR) in *Staphylococcus aureus. Planta Med.* 70(12): 1240-1242.

Gibbons, S., Oluwatuyi, M., Veitch, N.C. and Gray, A.I. (2003). Bacterial resistance modifying A gents from *Lycopus europaeus. Phytochem.* 62(1): 83-87.

Gill, C.J. and Hamer, D.H. (2001). Foodborne illnesses. *Curr Treat Options Gastroenterol*; 4:23-38.

Gill, M.J., Brenwald, N.P. and Wise, R. (1999). Identification of an efflux pump gene *pmrA,* associated with fluoroquinolone resistance in *Streptococcus pneumoniae. Antimicrob. Agents Chemother.* 43: 187-189.

Guiney, D.G., Fang, F.C., Krause, M. and Libby, S. (1994). Plasmid-mediated virulence genes in non-typhoidal Salmonella serovars. *FEMS Microbiol. Lett.* 124: 1-9.

Hallander, H.O., Dornbusch, K., Gezelius, L., Jacobson, K. and Karlsson, I. (1982). Synergism between aminoglycosides and cephalosporins with antipseudomonal activity: interaction index and killing curve method. *Antimicrob. Agents Chemother.* 22(5): 743-752.

Hamilton-Miller, J. M. (1995). Antimicrobial properties of tea (*Camellia sinensis* L.) *Antimicrob. Agents Chemother.* 39(11): 2375-2377.

Hammami, A., Arlet, G., Ben Redjeb, S., Grimont, F., BenHassen, A., Rekik, A. and Philippon, A. (1991). Nosocomial outbreak of acute gastroenteritis in a neonatal intensive care unit in Tunisia caused by multiply drug resistant Salmonella Wien producing SHV-2 ?-lactamase. *Eur. J.Clin. Microbiol. Infect. Dis.* 10: 641-646.

Han, Q.B., Lee, S.F., Qiao, C.F., He, Z.D., Song, J.Z., Sun, H.D. and Xu, H.X. (2005). Complete NMR Assignments of the Antibacterial Biflavonoid GB1 from *Garcinia kola. Chem. Pharm. Bull.* 53(8):1034-1036.

Hancock, E.W. (2005). Mechanisms of action of newer antibiotics for Gram-positive pathogens. *Lancet Infect. Dis.* 5(4): 209-218.

Herikstad, H., Hayes, P.S., Hogan, J., Floyd, P., Snyder, L. and Angulo, F.J. (1997). Ceftriaxone-Resistant *Salmonella* in the United States. *Pediatr. Infect. Dis. J.* 9: 904-905.

Hernandez, S.P. (2005). Responsible use of antibiotics in aquaculture. FAO Fisheries Technical paper 469.

Hirazumi, A., Furusawa, E., Chou, S.C. and Hokama, Y. (1996). Immunolomodulation contributes to the anticancer activity of Noni (noni) fruit juice. Proc West Pharmacol. Soc. 39:7-9.

Hohmann, E.L. (2001). Nontyphoidal salmonellosis. Clin.Infect. Dis. 15: 263-269.

Honish, L. (2000). Restaurant-associated outbreak of Salmonella typhimurium phage type 1 gastroenteritis- Edmonton,. Can Commun Dis Rep; 26:258.

Hooper, D.C. (2005). Efflux Pumps and Nosocomial Antibiotic Resistance: A Primer for Hospital Epidemiologists. Healthcare Epidemiol. 40:1811-1817.

Hooton, T. M., Blair, A.D., Turck, M. and Counts, G.W. (1984). Synergism at clinically attainable concentrations of aminoglycoside and betalactam antibiotics. Antimicrob. Agents Chemother. 26 (4): 535-538.

Hu, Z.Q., Zhao, W.H., Asano, N., Yoda, Y., Hara, Y. and Shimamura, T. (2002). Epigallocatechin gallate synergistically enhances the activity of carbapenems against methicillin-resistant Staphylococcus aureus. Antimicrob. Agents Chemother. 46(2): 558-560.

Institute of Medicine (1986). New vaccine development: establishing priorities, DC. National Academy Press.

Iroha, I. R., ILang, D. C., Ayogu, T. E., Oji, A. E. and Ugbo, E. C. (2010). Screening for anti-typhoid activity of some medicinal plants used in traditional medicine in Ebonyi state, Nigeria. African Journal of Pharmacy and Pharmacology 4(12): 860 – 864.

Iwu, M.W., Duncan, A.R. and Okunji, C.O. (1999). New antimicrobials of plant origin. Janick J (ed.), Perspectives on new crops and new uses, pp. 457-462.

Jerry, D. and Smilack, M.D. (1999). Trimethoprim-Sulfamethoxazole. Mayo Clin. Proc. 74: 730-734.

Jesudason, M.V., Sridharan, G., Arulsevan, R., Babu, P.G. and John, T.J. (1998). Diagnosis of typhoid fever by the detection of anti-LPS and anti-flagellin antibodies by ELISA. Indian J. Med Res; 107: 204-207.

Kaatz, G.W. (2002). Inhibition of bacterial efflux pumps: a new strategy to combat increasing antimicrobial agent resistance. Expert. Opin. Emerg. Drugs. 7(2): 223-233.

Kaatz, G.W and Seo, S.M. (1995). Inducible NorA-mediated multidrug resistance in S. aureus. Antimicrob. Agents Chemother. 39(12): 2650-2655.

Kaatz, G.W.., Seo SM, Ruble CA (1993). Efflux-Mediated Fluoroquinolone Resistance in Staphylococcus aureus. Antimicrob. Agents Chemother. 37(5): 1086-1094.

Kataja, J., Seppala, H., Skurnik, M., Sarkkinen, H. and Huovinen, P. (1998). Erythromycin Resistance Mechanisms in Group C and Group G Streptococci. Antimicrob. Agents Chemother. 42(6): 1493- 1494.

Khan, A.U. (2002). History of decline and present status of natural tropical thorn forest in Punjab. Pakistan biological conservation. 63:210 – 250.

Khan, M., Coovadia, Y. and Sturm, A.W. (1998). Typhoid fever complicated by acute renal failure and hepatitis: case reports and review. Am J Gastroenterol. 93: 1001-1003.

Kohler, T., Pechere, J.C. and Plesiat, P. (1999). Bacterial antibiotic efflux systems of medical importance. Cell. Mol. Life Sci. 56: 771-778.

Kumar, A. and Schweizer, H.P. (2005). Bacterial resistance to antibiotics: Active efflux and reduced uptake. Adv. Drug Deliv. Rev. 57: 1486-1513.

Laksai, Y., Severino, M., Perilli, M., Amicosante, G., Bonfiglio, G. and Stefani, S. (2000). First identification of an SHV-12 extended-spectrum beta-lactamase in *Klebsiella pneumoniae* isolated in Italy. *J. Antimicrob. Chemother.* 45: 349-351.

Lambert, P.A. (2005). Bacterial resistance to antibiotics: Modified target sites. *Adv. Drug Deliv. Rev.*57 (10): 1471-1485.

Lawrence, W.T. and Diegelmann, R.F. (1994). Growth factors in wound healing. *Clin Dermatol.* 12: 157-69.

Levy, S.B. and Marshall, B. (2004). Antibacterial resistance worldwide: causes, challenges and responses. *Nat. Med.* 10: S122-S129.

Lewis, K. and Ausubel, F.M. (2006). Prospects for plant-derived antibacterials. *Nat. Biotechnol.* 24 (12): 1504-1507.

Li, X. Z., Livermore, D. M. and Nikaido, H. (1994a). Role of efflux pump(s) in intrinsic resistance of *Pseudomonas aeruginosa*: resistance to tetracycline, chloramphenicol and norfloxacin. *Antimicrob. Agents Chemother.* 38 (8): 1732-1741.

Li, X. Z., Ma, D., Livermore, D.M. and Nikaido, H. (1994b). Role of efflux pump(s) in intrinsic resistance of *Pseudomonas aeruginosa*: active efflux as a contributing factor to beta- lactam resistance. *Antimicrob. Agents Chemother.* 38 (8): 1742-1752.

Lima, E. D. (2000). Traditional medicinal plants of Davangere District, Karnataka with reference to cure skin diseases. *Environmental and Ecology.* 18: 441-446.

Lin, J., Michel, L.O. and Zhang, Q. (2002). Cme ABC functions as a multidrug efflux system in *Campylobacter jejuni. Antimicrob. Agents Chemother.* 46: 2124-2131.

Lin, F.Y.C., Ho, V.A., Bay, P.V., Thuy, N.T.T., Bryla, D., Thanh, T.C., Ha, B.K. *et al* (2000).The epidemiology of typhoid fever in the Dong Thap Province, Mekong Delta region of Vietnam. *Am.J. Trop. Med. Hyg.* 62:644-648.

Llanes, C., Kirchgesner, V. and Plesiat, P. (1999). Propagation of TEM- and PSE-type β-lactamases among amoxicillin-resistanct Salmonella spp. isolated in France. *Antimicrob. Agents Chemother.* 43: 2430-2436.

Lomovskaya, O. and Bostian, K.A. (2006). Practical applications and feasibility of efflux pump inhibitors in the clinic - A vision for applied use. *Biochem Pharmacol.* 7(1): 910-918.

Luby, S.P., Faizan, M.K., Fisher-Hoch, S.P., Syed, A., Mintz, E.D., Bhutta, Z.A. *et al.* (1998). Risk factors for typhoid fever in an endemic setting, Karachi, Pakistan. *Epidemiol Infect.* 120:129-138.

Mabe, K., Yamada, M., Oguni, I. and Takahashi, T. (1999). *In Vitro* and *In Vivo* Activities of Tea Catechins against *Helicobacter pylori. Antimicrob. Agents Chemother.* 43(7): 1788-1791.

Machado, T.B., Leal, I.C.R., Amaral, A.C.F., Santos, K.R.N., Silva, M.G. and Kuster, R.M. (2002). Antimicrobial Ellagitannin of *Punica granatum* Fruits. *J. Braz. Chem. Soc.* 13(5): 606-610.

Madubunyi, I. I. (1995). Antimicrobial activities of the constituents of *Garcinia Kola* Seeds. *Int. J. Pharmacog.* 33(3): 232-237.

Marquez, B. (2005). Bacterial efflux systems and efflux pumps inhibitors. Biochimie 87(12): 1137-1147.

Marquez, B., Neuville, L., Moreau, N. J., Genet, J. P., Santos, A. F., Andrade, M. C. C. and Sant Ana, A.E.G. (2005). Multidrug resistance reversal agent from *Jatropha elliptica. Phytochem.* 66: 1804-1811.

Mead, P.S., Slutsker, L., Dietz, V., McCaig, L.F., Bresee, J.S., Shapiro, C., Griffen, P.M. and Tauxe, R.V. (1999). Food-Related Illness and Death in the United States. *Emerg. Infect. Dis.* 5: 607-625.

Mitchell, B.A., Brown, M.H. and Skurray, R. A. (1998). QacA Multidrug Efflux Pump from *Staphylococcus aureus*: Comparative Analysis of Resistance to Diamidines, Biguanidines, and Guanylhydrazones, *Antimicrob. Agents Chemother.* 42(2): 475-477.

Morosini, M.I., Canton, R., Martinez-Beltran, J., Negri, M.C.,Perez-Diaz, J.C., Baquero, F. and Blazquez, J. (1995). New extended-spectrum TEM-type beta-lactamase from *Salmonella enterica subsp. enterica* isolated in anosocomial outbreak. *Antimicrob. Agents Chemother.* 39:458-61.

Mulvey, M.R., Soule, G., Boyd, D., Demczuk, W., Ahmed,R. and the Multi-provincial *Salmonella Typhimurium* casecontrol study group. (2003). Characterization of the first Extended-spectrum beta-lactamase-producingSalmonella isolate identified in Canada. *J. Clin. Microb.*41: 460-462.

Nakaido, H. (1994). Prevention of drug access to bacterial targets: Permeability barriers and active efflux. *Science.* 264: 382-388.

Neidhardt, F.C. (1996). *Escherichia coli* and *Salmonella: Cellular and Molecular Biology.* ASM press, Washington, D.C.

Neu HC, Wilson AP, Gruneberg RN (1993). Amoxycillin/clavulanic acid: a review of its efficacy in over 38,500 patients from 1979 to 1992. *J. Chemother.* 5(2): 67-93.

Nkuo-Akanji, T., Ndip, R., Mcthomas, A. and Fru, E. C. (2001). Anti-salmonella activity of medicinal plants from Cameroon. *Central Afr. J. Med.,* 47(6): 155-158.

Norrby, R.S., Nord, C.E. and Finch, R. (2005). Lack of development of new antimicrobial drugs: a potential serious threat to public health. *The Lancet Infect. Dis.* 5(2): 115-119.

Nostro, A., Germarno, M.P., D'Angelo, V., Marino, A. and Canatelli, M.A. (2000). Extraction methods and bioautography for evaluation of medicinal plant antimicrobial activity. *Lett.Appl. Microbiol.* 30: 379-384.

Odeh, R., Kelkar, S., Hujer, A.M., Bonomo, R.A.,Schreckenberger, P.C. and Quinn, J.P. (2002). Broadresistance due to plasmid-mediated AmpC betalactamasesin clinical isolates of *Escherichia coli. Clin.Infect. Dis.* 35: 140-145.

Ohanu, M.E., Mbah, A.U., Okonkwo, P.O. and Nwagbo, F.S. (2003). Interference by malaria in the diagnosis of typhoid using Widal's test alone. *West Afr J Med.;* 22: 250-252.

Olowe, O.A., Olayemi, A.,B., Eniola, K.I.T. and Adeyeba, A.O. (2003). Aetiological agents of iarrhoea in children under 5 years of age in Osogbo. *African Journal of Clinical and Experimental Microbiology.* 4(3):62 – 66.)

Olubuyide I.O. (1992). Factors that may contribute to death from typhoid infection among Nigerians. *West Afri J Med;* 11:112-115.

Oluduro, A. and Omoboye, O. (2010). *In Vitro* Antibacterial Potentials and Synergistic Effect of South-Western Nigerian Plant Parts Used in Folklore Remedy for *Salmonella typhi* infection. *Nature and Science;* 8 (9): 52-59.

Oluwatuyi, M, Kaatz, G.W., and Gibbons, S. (2004). Antibacterial and resistance modifying activity of *Rosmarinus officinalis. Phytochemistry.* 65 (24): 3249-3254.

Onayade, O.A., Onayade, A.A., Soforawa, (1996). Wound healing with plants. The African perspective. In: Hostettmann, K., Chinyanganya, F., Maillard, M., Wolfender, J.-L. (Eds.), Chemistry, Biological and Pharmacological properties of African Medicinal plants. Proceedings of the First International IOCD-Symposium. University of Zimbabwe Publications, Harare, Zimbabwe. 77-110.

Otegbayo, J. A. (2005). Typhoid Fever: The Challenges of Medical Management. *Annals of Ibadan Postgraduate Medicine.* 3(1): 60-62.

Otegbayo, J.A., Daramola, O.O., Onyegbutulem, H.C., Balogun, W.F. and Oguntoye, O.O. (2003). Retrospective analysis of typhoid fever in a tropical tertiary health facility. *Trop Gastroenterol.* 23: 9-12.

Over, U., Gur, D., Unal, S., Miller, G.H. and Aminoglycoside Resistance Study Group (2001). The changing nature of aminoglycoside resistance mechanisms and prevalence of newly recognized resistance mechanisms in Turkey. *Clin. Microbiol. Infection.* 7(9): 470-478.

Pagani, L., Perilli, M., Migliavacca, R., Luzzaro, F. and Amicosante, G. (2000). Extended-spectrum TEM- and SHV-type beta-lactamase-producing Klebsiellapneumoniae strains causing outbreaks in intensive careunits in Italy. *Eur. J. Clin. Microbiol. Infect. Dis.* 19: 765-772.

Panhotra, B.P. and Agarwal, K. C. (1982). Urinary tract infection caused by *Salmonella typhimurium* and *Salmonella barielly. Ind. J. Med. Res.,* 76: 62-64,

Pankey, G., Ascraft, D. and Patel, N. (2005). *In vitro* synergy of daptomycin plus rifampin against *Enterococcus faecium* resistant to both linezolid and vancomycin. *Antimicrob. Agents Chemother.* 49(12): 5166-5168.

Parry, C.M., Hien, T.T., Dougan, G., White, N.J. and Farrar, J.J. (2002). Typhoid fever. *N. Engl. J. Med.* 347 (22): 1770-1782.

Paulsen, I.T., Sliwinski, M.K. and Saier, Jr. M.H. (1998). Microbial genome analyses: global comparisons of transport capabilities based on phylogenies, bioenergetics and substrate specificities. *J. Mol. Biol.* 277:573-592.

Pei, S.J. (2001). Ethno botanical approaches of traditional medicine studies. Some experiences from Asia. *Pharmaceutical Biology.* 39: 74 – 79.

Petit, A., Gerbaud, G., Sirot, D., Courvalin, P. and Sirot, J. (1990). Molecular epidemiology of TEM-3 (CTX-1) β-lactamase. *Antimicrob. Agents Chemother.* 34: 219-224.

Pohl, P., Glupczynski, Y., Marin, M., VanRobaeys, G., Lintermans, P. and Couturier, M. (1993). Replicon typingcharacterization of plasmids encoding resistance togentamicin and apramycin in *Escherichia coli* and *Salmonella typhimurium* isolated from human and animal sources in Belgium. *Epidemiol. Infect.* 111: 229-238.

Pradel, E. and Pages, J.M. (2002). The AcrAB-TolC efflux pump contributes to multidrug resistance in the nosocomial pathogen *Enterobacter aerogenes. Antimicrob. Agents Chemother.* 46(8): 2640 - 22643.

Pretto, J.B., Cechinel-Filho, V., Noldin, V.F., Sartori, M.R.K., Isaias, D.E.B. and Cruz, A.B. (2004). Antimicrobial activity of fractions and compounds from *Calophyllum brasiliense* (Clusiaceae/Guttiferae). *J. Biosci.* 59(9-10): 657-662.

Preston, K.E., Kacica, M. A., Limberger, R. J., Archinal, W.A. and Venezia, R.A. (1997). The resistance and integrasegenes of pACM1, a conjugative multiple-resistance plasmid, from *Klebsiella oxytoca. Plasmid* 37: 105-118.

Preston, K.E., Radomski, C.C.A. and Venezia, R.A. (1999).The cassettes and 3' conserved segment of an integronfrom *Klebsiella oxytoca* plasmid pACM1. *Plasmid* 42: 104- 114.

Prodinger, W.M., Fille, M., Bauernfeind, A., Stemplinger,I., Amann, S., Pfausler, B., Lass-Florl, C. and Dierich, M.P. (1996). Molecular epidemiology of *Klebsiella pneumoniae* producing SHV-5 β-lactamase: parallel outbreaks due to multiple plasmid transfer. *J. Clin.Microbiol.* 34:564-568.

Reading, C. and Cole, M. (1977). Clavulanic Acid: a Beta-Lactamase- Inhibiting Beta-Lactam from *Streptomyces clavuligerus. Antimicrob. Agents Chemother.* 11(5): 852-857.

Rossato, S.C., Leitao – Filho, H. and Gegossi, A. (1999). Ethno botany of caicaras of the Atlantic forest coast (Brazil). *Economic Botany.* 53: 387 – 395.

Sanchez, L., Pan, W., Vinas, M. and Nikaido, H. (1997). The acrAB homolog of *Haemophilus influenzae* codes for a functional multidrug efflux pump. *J. Bacteriol.* 179(21): 6855-6857.

Seppala, H., Skurnik, M., Soini, H., Roberts, M.C. and Huovinen, P. (1998). A Novel Erythromycin Resistance Methylase Gene (*ermTR*) in *Streptococcus pyogenes. Antimicrob. Agents Chemother.* 42(2): 257- 262.

Shibata, H., Kondo, K., Katsuyama, R., Kawazoe, K., Sato, Y., Murakami, K., Takaishi, Y., Arakaki, N. and Higuti, T. (2005). Alkyl Gallates, Intensifiers of ß-Lactam Susceptibility in Methicillin-Resistant *Staphylococcus aureus Antimicrob. Agents Chemother.* 49(2): 549-555.

Si, W., Gong, J., Tsao, R., Kalab, M., Yang, R. and Yin, Y. (2006). Bioassay-guided purification and identification of antimicrobial components in Chinese green tea extract *J. Chromatogr. A*: 1125(2): 204-210.

Sibanda, T. and Okoh, A. I. (2007). The challenges of overcoming antibiotic resistance: Plant extracts as potential sources of antimicrobial and resistance modifying agents, *African J. Biotechnology,* 6(25):2886-2896.

Smith, E.C.J., Williamson, E.M., Wareham. N., Kaatz, G.W. and Gibbons, S. (2007). Antibacterials and modulators of bacterial resistance from the immature cones of *Chamaecyparis lawsoniana. Phytochem.* 68(2): 210-217.

Smith, A. (2004). Bacterial resistance to antibiotics, In: Denyer, S. P., Hodges, N. A., and Gorman, S.P. (Eds). Hugo and Russell's Pharmaceutical microbiology, 7th ed. Blackwell Science, Massachusetts, USA.

Spratt, B.G. (1994). Resistance to antibiotics mediated by target alterations. *Science.* 264: 388-393.

Stapleton, P.D., Shah, S., Anderson, J.C., Hara, Y., Hamilton-Miller, J.M.T. and Taylor, P.W. (2004). Modulation of beta-lactam resistance in *Staphylococcus aureus* by catechins and gallates. *Int. J. Antimicrob. Agents.* 23(5): 462-467.

Su, L.H., Chiu, C.H., Chu, C., Wang, M.H., Chia, J.H. and Wu, T.L. (2003). In vivo acquisition of ceftriaxone resistance in *Salmonella enterica* serotype Anatum. *Antimicrob. Agents Chemother.* 47: 563-567.

Stermitz, F.R., Lorenz, P., Tawara, J.N., Zenewicz, L.A. and Lewis, K. (2000a). Synergy in a medicinal plant: Antimicrobial action of berberine potentiated by 5'-methoxyhydnocarpin, a multidrug pump inhibitor. *Appl. Biol. Sci.* 97(4): 1433-1437.

Stermitz, F.R., Tawara-Matsuda, J., Lorenz, P., Mueller, P., Zenewicz, L. and Lewis, K (2000b). 5'-Methoxyhydnocarpin-D and Pheophorbide A: *Berberis* Species Components that Potentiate Berberine Growth Inhibition of Resistant *Staphylococcus aureus. J. Nat. Prod.* 63(8):1146 -1149.

Takashima, J., Ikeda, Y., Komiyama, K., Hayashim, M., Kishida, A. and Ohsaki, A. (2007). New constituents from the leaves of *Morinda citrifolia. Chempharm Bull.* 55: 345-5.

Takahashi, T., Kokubo, R. and Sakaino, M. (2004). Antimicrobial activities of eucalyptus leaf extracts and flavonoids from *Eucalyptus maculata*. *Lett. Appl. Microbiol.* 39(1): 60-64.

Tassios, P.T., Markogiannakis, A., Vatopoulos, A.C., Velonakis, E.N., Katsanikou, K., Papadakis, J.A., Kourea- Kremastinou, J. and Legakis, N.J. (1997). Epidemiology of antibiotic resistance of *Salmonella enteriditis* during the period 1987-1993 in Greece. *J. Clin. Microbiol.* 35: 1316-1321.

Tassios, P.T., Gazouli, M., Tzelepi, E., Milch, H., Kozlova, N.S., Sidorenko, S.V., Legakis, N.J. and Tzouvelekis, L.S. (1999). Spread of a *Salmonella typhimurium* clone resistant to third generation cephalosporins in three European countries. *J. Clin. Microb.* 37: 3774-3777.

Taylor, P.W., Stapleton, P.D. and Luzio, J.P. (2002). New ways to treat bacterial infections. *DDT.* 7(21): 1086-1091.

Tegos, G., Stermitz, F.R., Lomovskaya, O. and Lewis, K. (2002). Multidrug Pump Inhibitors Uncover Remarkable Activity of Plant Antimicrobials, *Antimicrob. Agents Chemother* 46(10): 3133- 3141.

Teran, W., Antonia, F., Segura, A., Rojas, A., Ramos, J.L. and Gallegos, M.T. (2003). Antibiotic-dependent induction of *Pseudomonas putida* DOTT1E TtgABC efflux pump is mediated by the Drug Binding Repressor TtgR. *Antimicrob. Agents Chemother.* 47(10): 3067-3072.

Tshikalange, T.E., Meyer, J.J.M. and Hussein, A.A. (2005). Antimicrobial activity, toxicity and the isolation of a bioactive compound from plants used to treat sexually diseases. *J. Ethnopharmacol.* 96(3): 515- 519.

Threlfall, E.J., Ward, L.R., Skinner, J.A. and Rowe, B. (1997). Increase in multiple antibiotic resistance in nontyphoidal salmonellas from humans in England and Wales: a comparison of data for 1994 and 1996. *Microb. Drug. Resist.* 3: 263-266.

Thuluvath, P.J. and McKendrick, M.W. (1988). *Salmonella* and complications related to age-Sheffield experience. *Q J Med*; 67:497503.

Ulubelen, A., Oksuz, S., Kolak, U., Bozok-Johansson, C., Celik, C. and Voelter, W. (2000). Antibacterial diterpenes from the roots of *Salvia viridis*. *Planta Med.* 66(5): 458-462.

Varma, J.K., Molbak, K., Barrett, T.J., Beebe, J.L., Jones, T.F., Rabatsky-Her, T., Smith, K.E., Vugia, D.J., Chang, H.G. and Angulo, F.J. (2005). Antimicrobial-resistant nontyphoidal Salmonella is associated with excess bloodstream infections and hospitalizations. *J Infect Dis.* 191(4):554-61.

Verdet, C., Arlet, G., Barnaud, G., Lagrange, P.H. and Philippon, A. (2000). A novel integron in *Salmonella enteric serovar Enteritidis*, carrying the bla(DHA-1) gene and its regulator gene ampR, originated from *Morganella morganii*. *Antimicrob. Agents Chemother.* 44: 222-225.

Villa, L., Mammina, C., Miriagou, V., Tzouvelekis, L.S., Tassios, P.T., Nastasi, A. and Carattoli, A. (2002a). Multidrug and broad-spectrum cephalosporin resistance among *Salmonella enterica serotype Enteritidis* clinical isolates in Southern Italy. *J. Clin. Microb.* 40: 2662-2665.

Villa, L., Pezzella, C., Tosini, F., Visca, P., Petrucca, A. and Carattoli, A. (2000). Multiple-Antibiotic Resistance Mediated by Structurally-Related IncL/M Plasmids Carrying an Extended-Spectrum β-Lactamase Gene and a Class 1 Integron. *Antimicrob Agents Chemother* 44: 2911-2914.

Wilke, M.S., Lovering, A.L. and Strynadka, N.C.J. (2005). Beta-Lactam antibiotic resistance: a current structural perspective. *Curr. Opin. Microbiol.* 8(5): 525-533.

Winokur, P.L., Brueggemann, A., DeSalvo, D.L., Hoffmann, L., Apley, M.D., Uhlenhopp, E.K., Pfaller, M.A. and Doern, G.V. (2000). Animal and human multidrug-resistant, cephalosporin-resistant Salmonella isolates expressing a plasmid-mediated CMY-2 AmpC beta-lactamase. *Antimicrob. Agents Chemother.* 44: 2777-2783.

Winokur, P.L., Vonstein, D.L., Hoffman, L.J., Uhlenhopp, E.K. and Doern, G.V. (2001). Evidence for transfer of CMY- 2 AmpC β-lactamase plasmids between *Escherichia coli* and *Salmonella* isolates from food animals and humans. *Antimicrob. Agents Chemother.* 45: 2716-2722.

World Health Organization (WHO) (2002). Antimicrobial resistance. Fact sheet No. 194.

Wray, C., Hedges, R.W., Shannon, K.P. and Bradley, D.E. (1986). Apramycin and gentamicin resistance in *Escherichia coli* and *salmonellas* isolated from farm animals. *J. Hyg.* 97: 445-456.

Wright, G.D. (2005). Bacterial resistance to antibiotics: Enzymatic degradation and modification. *Adv. Drug Deliv. Rev.* 57(10): 1451-1470.

Yam, T.S., Hamilton-Miller, J.M. and Shah, S. (1998). The effect of a component of tea (*Camellia sinensis*) on methicillin resistance, PBP2' synthesis, and beta-lactamase production in *Staphylococcus aureus*. *J. Antimicrob Chemother.* 42(2): 211-216.

Yang, Z.C., Wang, B.C., Yang, X.S., Wang, Q. and Ran, L. (2005). The synergistic activity of antibiotics combined with eight traditional Chinese medicines against two different strains of *Staphylococcus aureus*. Colloids and surfaces B: *Biointerfaces*,41(2-3): 79-81.

Zhao, W.H., Hu, Z.Q., Okubo, S., Hara, Y. and Shimamura, T. (2001). Mechanism of synergy between Epigallochatechin gallate and _-Lactams against methicillin resistant *Staphylococcus aureus*. *Antimicrob. Agents Chemother.* 45(6): 1737-1742.

5

Salmonellas Antibiotic Sensibility and Resistance; The Sensitivity to Herb Extracts and New Synthesize Compounds

Birol Özkalp

Department of Medicinal Laboratory
Vocational School of Health Services of Selçuk University, Konya
Turkey

1. Introduction

Although the use of antibiotics in the developed countries is done under certain principles, there is an opposite situation at the use of antibiotics in the underdeveloped countries which is irregular. The prevalence of antibiotic use in the fight against infectious diseases additionally raises the problem of increasing resistance of the micro-organisms against antibiotics. There is also an increase in the resistance of antibiotics which is used against *Salmonella*. The newly synthesized chemical compounds and extracts derived from plants which are alternative to existing antibiotics determining the sensitivity to *Salmonella*, plays an important role for increasing the options of alternative antibiotics.

1.1 Appearance and staining characteristics

Salmonella bacteria are asporogenic, capsule-free, motile via peritrichous cilium (*Salmonella gallinarium* or *Salmonella pulorum* are immotile), rod-shaped bacteria with an approximate length of 2,0-5,0 μm, width of 0,7-1,5 μm. They are stained well with bacteriologic stains and they are gram-negative (Picture 1). Most of them have type 1 (mannose sensitive (ms), hemagglutinating); S. Gallinarium and some origins have type 2 fimbriae. S. paratyphi As do not have fimbriae.

1.2 Reproduction and biochemical characteristics

Salmonella bacteria reproduce in many ordinary mediums. They are aerobe and facultative anaerobe. Their reproduction temperature limit is very wide even they reproduce at 37° C best. (20°C- 42°C). This is extremely important for reproduction of *Salmonellas* which cause food intoxication at room temperature. They like to produce at average pH of 7,2. They make homogenous turbidity in bouillon and similar liquid medium. They make round, slab sided, mostly tumescent colonies with a diameter of 2-3 mm, regular surface. In colonies of various *Salmonellas*, some differences may exist in terms of size, protuberance, surface and side. *Salmonella typhi* may also make gnome colonies which may reach to 0,2-0,3 mm diameter within the first 24 hours. Biochemical characteristics of bacteria which are obtained from these colonies are same as normal colonies; and they are agglutinated with O serums only antigenically and they differ from bacteria in S colonies in terms of not reacting with anti H, anti Vi serums. If they are reproduced in mediums including sulfurous compounds, sulfates and tiosulfates which may be assimilated, normal colonies occur from bacteria that make gnome colonies.

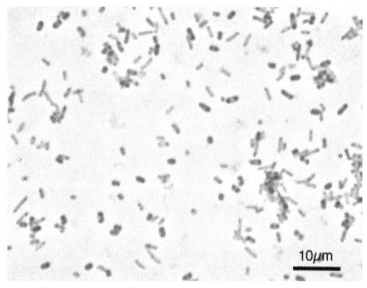

Fig. 1. Microscopic View of *Salmonella*

Some of *Salmonellas*, S. Schottemuelleri (s. paratyphi) in particular and some others form M colonies in appropriate mediums. It is detected that these bacteria have M antigens and agglutination is prevented by anti O and anti H serums. Furthermore, R colonies are formed by *Salmonella* which reproduce in inappropriate mediums (Picture 2).

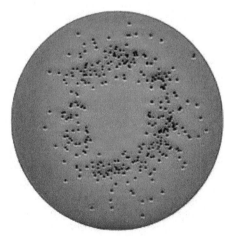

Fig. 2. *Salmonella* colonies

Salmonellas are not effective on lactose. This characteristics is important in first differentiation from Escherichias. As these bacteria which are planted in a separator plaque medium (endo, EMB) including lactose and an appropriate reagent are not effective on lactose, they make colorless colonies; however those effective on lactose make dark red, black, greenish bright colonies (Picture 3).

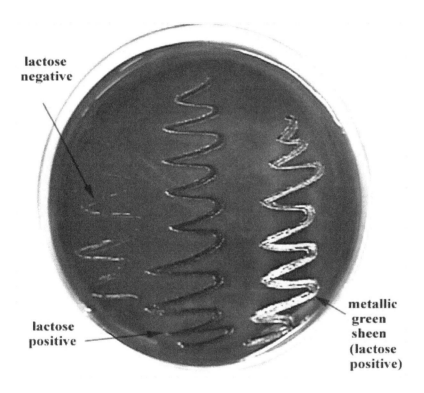

Fig. 3. View of *Salmonella* and Lactose Positive Colonies

Salmonellas do not effect on sucrose, adonitole and salicin in usual other than lactose. They digest glucose, mannite and maltose by producing acid and gas except *Salmonella typhi* and S. gallinarum; and *Salmonella typhi* and gallinarum digest them by producind acid only. They produce H2S in general (except S. paratyphi A); they are indole negative, methyl red positive, Vogesproskauter negative and they reproduce in citrated mediums (Simmon), they do not digest urea. They could not be produced in KCN (potassium cyanide 0,5%) mediums. ONPG (orthonitro phenyl galactopyranoside) assay is negative. (They do not have beta galactosidase enzymes that may digest lactose). This assay is positive in Arizona. Biochemical characteristics of *Salmonellas* were shown in Table 1.

Motility	+
Indole	-
H2S	+
Oxidase	-
Urease	-
Nitrate reduction	+
Citrate Utilization	+
MR	+
VP	-
Lysine decarboxilation	+
Ornithine decarboxilation	+
Phenylalanine deamination	-
Malonate Utilization	-
Lactose	-
Sucrose	-
Salicine	-
Inositol	-
Amygdalin	-
Gas Production from glucose	+
β-galaktosidase (ONPG Test)	-
Reproduction in KCN	-
	(-) Negative (+) Positive

Table 1. Biochemical characteristics of *Salmonellas*

2. Resistence in *Salmonellas*

Salmonella bacteria are resistless to heat. They die at 55°C in 20 minutes. They do not have resistance to dryness. But they may stay alive in humid environments away from daylight, sewages, well water and soil for a long time. They are very resistant to cold. Staying alive in cold food and drinks has an epidemiological importance. They may stay alive in liyophilized situations for years.

Antiseptics effect rapidly providing the direct contact. Chlorine within normal concentrations kill *Salmonellas* in the water. However, effect of these agents to *Salmonellas* in stool particles and other organic substances lower.

To differ them from other bacteria in terms of accompanying coliform and intestinal settlement habits, their status against various chemical agents and stains were examined and consequently special cases of *Salmonellas* appeared against some of them.

Malachite green, a stain kills E.coli or slower their reproduction within suitable densities, however it does not effect S. typhi. Similarly, while paratyphoid bacillus are very resistant and typhoid bacillus is quite resistant to Brillant green stain, dysentery bacillus and coliforms are very sensitive. Although lithium chloride inhibits E.coli similarly, it s inefficient to *Salmonella typhi*. Differently, sodium tetrathionate increases Slamonella reproduction although it is noneffective to coliforms.

To preserve *Salmonellas*, after making immersing culture to a vertical agar including 1% agar and 2% Na H PO$_4$ 12H$_2$O and reproduction is provided, they may stay alive in tightly closed agar in the dark for months. Furthermore, *Salmonellas* may be lyophilized.

Although sodium deoxycholate reduces reproduction of coliforms, they do not effect reproduction of *Salmonellas* and Shigellas.

Resistance against chemotherapeutics in *Salmonellas* has appeared quite lately. Although an important resistance has not been observed in *Salmonellas* against antimicrobic agents such as chloramphenicol, tetracyclines, kanamycin, ampicilin, streptomycin, sulfonamids until 1960, increasingly resistant strains against tetracyclins, ampicilin, streptomycin, sulphonamids have been detected from that day to this. Resistance to chloramphenicol develops slower than others. Resistance depends on plasmids.

3. Antibiotic resistance in microorganisms

Microorganisms may become resistant to antibiotics that they are in effect spectrum in time due to any reason. Antibiotic resistance in bacteria is called as inability to treat the infection disease against treatment doses of antibiotics that are in effect spectrum of microorganisms naturally and used routinely.

There are various predictions recently about where antibiotic resistance genes which are thought to be appeared after wide usage of antibiotics in medicine come from. According to an opinion, resistance genes has developed as a protection mechanism in bacteria species that produce antibiotic first. Resistance genes in these species are in the same group with antibiotic production genes. Another possibility is that antibiotic resistance genes has developed from normal genes present in the microorganism. However, possibility of development of antibiotic resistance due to spontaneous mutation is very low. (10^{-5}-10^{-10}).

3.1 Resistance problem against antibiotics

Antibiotics which has been started to be used within last 50-60 years provided the most important contribution in human life and enabled treatment of many infectious diseases successfully. Antibiotics which are one of the most important inventions of humanity lost their effects significantly because of resistance developed after inappropriate and unnecessary usage. Microorganisms gain a sustaining power, namely resistance eventually against antimicrobic agents which are used to destroy themselves. The resistance developed against antimicrobic agents is a very important problem which will threat all humanity today. Hospital infections which develop by resistant origins against many drugs mainly in hospitals increase hospitalization and death rates and cause more additional cost. Today, resistance not only in hospital sources, but also in sources acquired from the society increases significantly and this case augments the problem and carries it to serious levels.

Resistance may also develop to other antimicrobis which are close in terms of structure or effect form to a chemotherapeutic agent in a microorganism species which has become resistant to such antimicrobic agent; and this condition is called as cross resistance. The condition that a microorganism becomes resistant to many antimicrobic agents with different structure and effect is called as multiple-drug resistance.

3.2 Natural (intrinsic) resistance

It is a resistance type without hereditary characteristic. This means becoming resistant of a microorganism due to its structure. Absence of the target molecule that the antimicrobic agent is effective by binding is responsible from the natural resistance in general. Any origin of the resistant species to an antimicrobic agent is not effected from such antibiotic.

Many gram negative bacteria show resistance to vancomycine and methicillin and enterococcus show resistance to cephalosporin because of their cell wall structure. As passage of aminoglycosides into cell membrane is an oxygen dependent, energy requiring case, aminoglycosides are not effective on anaerobes since sufficient drug can not enter into the cell in obligate anaerobe bacteria that oxidative phosphorylation is not present.

3.3 Acquired (hereditary) resistance

It is the resistance type acquired. Here, the drug is effective when bacteria population contacts the antimicrobic agents first; however, resistance develops against the antimicrobic agent in the microorganism population during the contact period or repeated treatments. The resistance developed against antimicrobics occurs by this manner and resistant origins appear and diffuse by selection after genetic change. Genetic resistance is under control of chromosome, plasmide, transposon. Microorganisms become resistant against antimicrobics by using one or more resistance mechanisms.

4. Mutation

Changes which has been occurred in alignment or structure of nucleotide pairs forming gene structure in DNA and modified protein structure coded and function as well are called as mutation. As a result of mutations, mutants which show resistance to various drugs, disinfecting agents, chemotherapeutics, inhibitors, ultraviolet beams, phages and such agents may appear.

5. Genetic material transfer between bacteria

In bacteria, genetic material may be transferred into another bacteria partially and three basic mechanisms play role in genetic material transfer after these transfers.

5.1 Transformation

It is recombination of free DNA fragment which was left into the environment by the donor bacteria without any mediator (another bacteria or bacteriophage) in the environment with the receiver's own genetic elements. In another words; if a microorganism is produced in an environment including genetic material (DNa fragments) of another microorganism which is very close in terms of DNA composition and planting from this liquid medium into a solid medium after a certain period, it is seen that some colonies have different morphologies and genetic material of them are similar to original colonies of dead microorganisms which give the genetic material.

Some specific characters has been able to be transferred to receiver bacteria by transformation. These include lactose and galactose positive genes, resistance against antibacterial substances, virulence etc. These factors are transferred into mRNA by transcription after combination with donor DNA and it is transloced from here and causes appearance of new characters in the cell. Although transformation is not successful among enterobacters, it is reported recently that transformation was detected in high calcium ion concentrations in Escherichia coli.

5.2 Transduction

Transfer of genetic material from a donor bacteria into a receiver bacteria via bacteriophages is called as transduction. Gen transfer via transduction is detected in Gram negative (*Salmonella*, E.coli, Shigella, Proteus, Vibrio, P. Aeruginosa etc.) and Gram positive (staphylococcus and bacillus) microorganisms.

5.3 Conjugation

Conjugation is a method of gene transfer that genetic material transfer realized as a result of physical relation of donor cell with the receiver cell. For realization of conjugation between two cells, cells should contact with each other. Generally, contact occurs via sex pilus which are synthesized by special genes in sex factors in the cell. These are longer and thicker tan other normal pilus (fimbria). They serve as a pipe or passage bridge as they are hollow. Genetic material passes from here and is transmitted into the receiver. Normal pilus do not have roles on gene transmission.

Another way in conjugation is direct contact. In this way, membranes of two bacteria combine and form a pore on the combination point. And DNA transmission occurs via this pore. Receiver population is very wide in conjugation and it may occur between different species and genus.

There are two types of conjugative structures consisting of chromosomal elements known as conjugative transposones and plasmides.

5.3.1 Conjugative transposons

Conjugative tronsposons are genetic elements of a bacteria which may replace from chromosome of a bacteria to its plasmid or to a chromosome or plasmids. They show similar characteristics with plasmids and bacteriophages. The difference from plasmids is inability to replicate by themselves. These structures are in Gram positive bacteria and Bacteroides species and cause diffusion of antibiotic resistance genes among various bacteria groups. Conjugative transposons form a circular intermediate form by binding covalently after leaving from bacterial DNA. This intermediate form may be transmitted to another region in the same cell or to another cell and be bounded to genomes or plasmids of the receiver cell.

5.3.2 Plasmids

Genes that make bacteria resistant against antimicrobial agents may be present on chromosomes of bacteria as well as they are carried on small DNA fragments called as plasmid. The term plasmid was first used by Lederberg to define all extrachromosomal hereditary elements. Today, this phrase was limited by extrachromosomal DNA which replicates independent from chromosome. Plasmids are circular extrachromosomal DNA molecules with fibril pairs and they code various activities which are not required for aliveness of the bacteria in natural environments and conditions.

The most common antibiotic resistance in bacteria caused by plasmid. Genes that control antibiotic resistance of bacteria exist on R (resistance) plasmids and these plasmids cause spreading of the resistance by being transmitted to other bacteria. Existence of resistance genes on plasmids and their ability to be transmitted showed that plasmids are basic vectors for spresding of resistance genes among bacteria populations. Plasmids on bacteria of enteobacteriacea family are related with transmission of various genetic characteristics such as drug resistance, hemolysin, enterotoxin and co-lysin production, tolerance to heavy metals, resistance to ultraviolet beams, carbohydrate fermentation and H2S production.

Plasmids acquire their multiple drug genes via 2 paths.

1. To form plasmids that show multiple drug resistance via subsequent transposon insertion (This path is not used by many plasmids).
2. To acquire multiple drug resistance genes by receiving linear DNA fragments that may be inserted into DNA like transposons called as integrons.

Integrons are integrated only to a single point differently from transposons and they do not code transposase. Integrase enzymes are coded by plasmids that they are integrated. Many integrons carry promoter-free antibiotic resistance genes and are integrated to plasmids specific to direction. Many resistance gene may be present consecutively on a plasmid.

6. Resistance mechanisms of bacteria against antibiotics

Various resistance mechanisms has developed in bacteria which are coded by any of abovementioned paths against antibiotics which are compounds with low molecular weight and suppress bacteria reproduction or kill them.

6.1 Resistance dependent on external membrane in gram negative bacteria

Target regions of beta lactam antibiotics are on the outer surface of the cytoplasmic membrane and targets of many other antibiotics are in the cell. Therefore, all antibiotics should pass the external membrane barrier to reach to target regions in gram negative bacteria. Passage from the external membrane is via pores. Requirement of antibiotics diffusing from pores gives a minor resistance to all Gram negative bacteria (5 to 10 times) and mutations that may occur in pores may increase this resistance ratio. Some of external membrane pores are specific and some of them are non-specific pores. Mutations on non-specific pores may provide resistance to more than one antibiotic type.

6.2 Enzymatic inactivation of antibiotics

An important resistance mechanism especially in gram negative bacteria against beta-lactam antibiotics is beta-lactamase. The beta-lactamase enzyme which is present on some microorganisms (Staphylococcus aureus, E. coli, *Salmonella* spp., Shigella spp., etc.) and is coded by R-plasmids hydrolyzes C-N bound in beta-lactam bond in the structure of antibiotics such as penicillin, cephalosporin, ampicillin, cloxacilin etc. and inactivates antibiotics (Arda, 2000). Beta lactamases are released into periplasmic space in Gran negative bacteria and into extracellular environment in Gram positive bacteria. Since external membrane pores limits antibiotic passage and beta lactamases are released into periplasmic space, resistance of gram negative bacteria can be obtained with a lower enzyme level than gram positive bacteria.

The most important mechanism of the resistance against aminoglycosides is enzymatic inactivation of the antibiotic. Aminoglycoside modifying enzymes inactivates antibiotic by addition of a group such as phosphoryl, adenyl or acetyl. These enzymes are on the outer side of the cytoplasmic membrane in gram negative bacteria. Therefore, a decrease occurs both in passage of antibiotics from cytoplasmic membrane and in their power to inhibit protein synthesis. Although an enzyme which inactivates tetracycline in aerobe conditions recently, it was understood that this enzyme is not effective clinically and in terms of total resistance of tetracycline.

6.3 Active ejaculation of antibiotics

Ejaculation type resistance is first observed in tetracycline. A cytoplasmic membrane protein catalyzes energy dependent ejaculation of bacteria of tetracycline. Genes that code this protein are both in Gram negative (tet A- tet G) and in Gram positive bacteria (tet K - Tet L). This type of resistance was detected in staphylococcus species for macrolide antibiotics. Another ejaculation system with a low efficiency was also found for fluoroquinolones.

6.4 Modification of target areas of antibiotics

The resistance mechanism which plays role in the resistance against beta lactam antibiotics is changing specific binding regions of penicillin binding proteins (PBP) which are the target region of the antibiotic. This type of resistance is common among Gram positive bacteria. The "mec" gene which gives methicillin resistance is the best defined gene among these resistance genes. Furthermore, this type of resistance was also detected against glycopeptides, tetracyclines, macrolides, lincosamide, quinolon, rifampicin, trimethoprim and sulphonamides.

7. Antibiotic resistance conditions of *Salmonella* origins

Resistance against chemotherapeutics in *Salmonellas* has appeared quite lately. While no resistance has been observed in *Salmonellas* against chloramphenicol, kanamycine, ampicillin, streptomycin and sulphonamides until 1960, an increasing resistance has been developed against tetracyclins in particular, ampicillin, streptomycin and sulphınamides since 1960s. However, the most effective chemotherapeutic agents on *Salmonella* species are chloramphenicol, tetracyclines, ampicillin and gentamicin.

S. typhi origins were sensitive to all antibiotics including chloramphenicol in particular until 1970s. After a wide epidemia created by chloramphenicol resistant S. typhi origins in 1972, these resistant sources were found in many countries mostly in India, Mexico, Thailand and Vietnam.

First trimethoprim resistance for S. typhimurium has been detected in 1973 and then it is reported in 7% of human sources. No trimethoprim resistance was detected in S. typhi until 1980. During past 20 years, antibiotic resistance and multiple drug resistance has increased in *Salmonella* species. Resistance cases were reported in non-typhoidal *Salmonellas* in many countries in South America, Middle-east and South Asia. *Salmonella* sources that show aggressive multiple drug resistance is a big problem in many countries. Antibiotic resistance is commonly under control of plasmids in *Salmonella* sources. Plasmids are gained as a result of antibiotic pressure which is used common in feeds of livestock and in medicine and in veterinary. Resistance plasmids gains resistance genes from other plasmids in the same source or plasmids that is carried by other bacteria origins involved in chromosome or host organism. Resistance may also develop spontaneous mutation of a chromosomal gene as a response to selective antibiotic pressure.

8. Determination of antimicrobial characteristics of essential oils and methods used

Herbal extracts and essential oils have been used for long years for different purposes. However, their use for wider purposes by utilizing from their different features and studies related with them are continued rapidly. The most emphasized subject is antimicrobial characteristics. By using these characteristics, essential oils are started to be used in protection of raw and processed food, as additives in modern drugs and in natural treatments. There are many articles published associated with research of antimicrobial characteristics of essential oils and herbal extracts. In these studies, a kind of essential oil is studies by targeting against a pathogen microorganism.

This information is usually useful, however every study has procedural differences. Antimicrobial test methods used differ from each other. Furthermore, there are differences between selected oils or plants that they were extracted in terms of place where they are picked up and extraction methods. It is more likely that study results may differ due to these factors.

Until 1960s, many methods for drug, especially antibiotic sensitivity tests of microorganisms and many different modifications of these methods were reported. Superiority and area of use of every methods are limited. To interpret results with a highest level, all characteristics of the method should be understood well. There are two basic methods that are used to detect antibiotic sensitivity of bacteria. These methods are "Titration (Dilution) Methods" that antibiotics interact with microorganisms after serial dilution and "Diffusion Methods" which are performed by placing a test substance impregnated paper disc to medium surface after planting the culture to be tested.

Essential oils have some characteristics such as volatility, hydrophobicity and having special odors activating in respiratory system. Essential oils are heterogenous mixtures that organic substances are present as a mixture. These final characteristics reveal that especially odorous oils may be biologically active. In fact, essential oils have various pharmacological activities. The most reported characteristic is antimicrobial effect. Tests that reveal such characteristics does not depend on a certain standardization and they may be performed in any laboratory randomly. Techniques used are generally agar diffusion and dilution methods.

Dilution techniques were developed to detect sensitivity of a microorganism to antibiotics. However, they are also used to determine antimicrobial characteristics in plant extracts or essential oils. It depends on serial dilution of antimicrobial agent and inoculating bacteria culture. After incubation, the effect of antimicrobial agent tested with which concentration against the microorganisms is determined according to presence and absence of reproduction. Presence and absence of reproduction is performed via turbidity detection and low final concentration value that no reproduction occurs is defined as Minimum Inhibiting Concentration(MIC) value. This technique is macro-both dilution technique which has been performed in standard assay tubes for a long time. Recently, a method that acts with the principle of this method but requires very less medium and test material was started to be used in testing synthetic and natural antimicrobial agents excepts antibiotics. The method which is more advantageous mostly than other diffusion techniques and reveal MIC value accurately is micro tube dilution or microbroth dilution methods. In this methods, plates including 80, 96 or more pits which were developed commercially are used. Material dilutions are prepared in these pit series and agent and the microorganism are activated by adding less culture. After the incubation, absence and presence of reproduction by turbidity determination. Turbidity detection procedure may be performed by simply observation or by using special turbidity readers. Although this method is mostly used for antibiotics, they are also used for plant extracts and essential oils. The most important advantage is to perform the assay with 10 to 25 µl. Because it is very hard to obtain essential oils plenty. Another advantage is to allow testing many agents at the same time.

Another method used in antimicrobial tests is agar diffusion method. This technique is commonly preferred due to easy testing of essential oils. Agar diffusion technique has been used to determine antimicrobial characteristics of various agents since early 1940s. Qualitative and quantitative information may be revealed by this method. In agar diffusion technique, an appropriate medium including test organism is used by a hole system including the substance to be tested. The essential oil which was dissolved homogenously is put on medium with certain volumes. Holes contact with the medium. By this method, sometimes essential oil impregnated paper discs are used instead of making a hole on the medium. Consequently, essential oil is diffused to previously microorganism vaccinated medium from holes or paper discs. Structural characteristics of the agent used may be effective on diffusing percentage or period and may effect test results. If the agent used is effective at the end of incubation period, significant, inhibition zones occur around holes indicating no reproduction. Oil quantity applied and diameter of the disc or hole used are important parameters in this method. Because, diameters of inhibition zones occurred at the end of incubation are under the control of these parameters. Thickness of the medium that the hole is opened effects the diameter of inhibition zone. A certain period should pass for formation of inhibition zone. This period is called as "critical period" (T erit). Before this period, inhibition zones may not become significant or when incubation is performed over this period, occurred zones start to disappear. Furthermore, density of inoculum used should be certain and fixed. Because, an agent which be effective normally will be seemed as ineffective due to microorganism concentration and will not form an inhibition zone or will not be within real sizes. Therefore, inoculum concentration should be hold in critical level. If density of microorganism is at the required level, duration of the incubation period is not very important. Diameters of inhibition zones formed is measured via a scale and recorded. Increasing or decreasing concentrations of the agent is put into the pits and increase or decrease in diameters of zones proportionally are expected. However, it is reported that there is a definite parallelism between zone diameter values obtained via agar diffusion method and corresponding agent concentrations and real MIC values, but zone diameters obtained are not compatible with MIC values.

Another method which has been started to be used frequently to determine microbiological activity of essential oils recently is bioautography. Bioautography method is very easy and correct results giving method to try plant extracts or pure substances against botanic and human pathogens. In this method, it is revealed that which of organic compounds forming the essential oil is responsible from the activity as well as antibacterial characteristics of the essential oil. This method bases on principles of agar diffusion technique. However, it may differ in terms of application of the substance to be tested and evaluation of results. The biggest difference is that thin layer chromatography (TLC) is used in the method and the essential oil is activated by test microorganisms after application to TLC plates. By the help of TLC technique, compounds in the essential oil are removed roughly and the compound which is responsible from the activity is revealed. Test substance is applied to both TLC plates in the method and one of plates are accepted as reference plate. The other is the plate that microorganisms are applied. Fractions are marked by making the reference plate colored with reagents or examining under 254 or 336 nm UV light. After incubation of the plate used in the assay, the inhibition zone presence is determined by such substance and Rf value of that substance is calculated. Rf value (Retention Factor) is found by calculation the ratio of the distance that such substance has moved on the plate to the distance that dissolving agent has moved. Referans olarak saklanan İTK plağındaki maddeler ile inhibisyon zonlarının oluşturduğu maddelerin R_f değerleri karşılaştırılarak zonu oluşturan madde işaretlenmekte ve bu aşamadan sonra zonu oluşturan madde çeşitli yöntemlerle referans plaktan izole edilerek tayin yoluna gidilmektedir. In fact, bioautography method is suitable to reveal compounds with high antimicrobial activity such as antibiotics. It reveals the most active components among plant extracts or similar organic compounds. Three bioautography was reported since today. These are;

a. Direct bioautography method that microorganism is directly placed on TLC plate,
b. Contact biouatography method that the substance moved on TLC plate is isolated and transferred into inoculated medium by the microorganism.
c. Immersion bioautography or "Agar-overlay biouautography" that is performed by pouring the medium which was inoculated by a certain microorganism on TLC plate.

The last method is formed by combining direct bioautography and contact bioautography. Direct biouatography is especially used for bacteria and spore producing fungus. This method is quite sensitive and very net inhibition zones may be observed at the end of the assay. However, disadvantage of this method is difficulty of development of microorganisms on TLC plate. This problem is not present in contact biouatography method, but isolation and transfer of the substance from TLC plate cause some problems. Bigger than required inhibition zones appear and this causes difficulty in discrimination between active components. It is reported that immersion bioautography which is the combination of both methods are generally used for yeasts and bacteria. By pouring certain medium on TLC plate, active substances are tested in place and no reproduction problem occurs because sufficient medium is used. The problem of this technique is different diffusing coefficients of active substances. As a solution, a softer medium is obtained by decreasing the agar amount on the medium which is poured on TLC plate and diffusions of components on the plate into inoculated agar is facilitated. Whether which of these three methods is used, tetrazolium salts are used in general to provide determination of inhibition zones that are expected to form after incubation or to visualize. These reactive substances provide microorganisms to be colored purple and formation of colorless inhibition zones are provided at backstage.

There are various methods for determination of antimicrobial effects of volatile substances, especially essential oils in steam phase. The most applicable method is micro-atmosphere method. It is reported that this method is more appropriate for filament fungus in general. Petri plates with a size of 120 mm including medium is used in the method. Fungal spores which are prepared in sterile distilled water are inoculated on the centre of the medium as final concentration to be 10^4 spore/spot. The oil to be tested is impregnated purely with different amounts to the filter paper which has the same size with the petri plate. Sole microorganism is inoculated to one of the plates to use as control, no test substance is applied. Separate petri plates are used to try different quantities of the essential oil. Prepared filter paper is placed into the cap of the petri and petri plate is closed and incubated in a reverse position for 2 to 12 days. Plates are controlled during this 12 days and development status of inoculated microorganisms are controlled. Usability of essential oils in elimination or disarming of fungal or bacterial load in the air has been searched. So, it has been stated that there would be chance to protect the atmosphere of library, museum, hospital, cinema etc places without damaging to people against to microbial flora through essential oil. Some of researchers observed the inhibition instead of emission the pure essential oil into filter paper to put directly into cover section of petri cup. To prevent lost of essential oil, around of petri cup is covered with parafilm and is let to incubation in converse type. At the end of 3 days incubation period, petri cup is opened and steam of essential oil is released and it is left 3 days more to incubation. This second time applied incubation provides to determine whether antimicrobial affect of essential oil is at bactericidal feature or not. If essential oil steam only inhibited development of microorganism and it has been informed that the microorganisms at the plate will be developed at the end of this period. Some of the researchers searched whether steam pressure has any affect or not through emission the essential oil in various diameter filter papers and applying different quantities. For this purpose, volatile pure materials have been tested and consequently it has been set forth that steam pressure doesn't support steam activity of essential oil.

9. The methods that are used to determine antimicrobial features

9.1 Resuscitation of microorganisms

The bacteria occurring in lyophilize cultures, are extracted from their tubes under aseptic terms and they are transmitted into Nutrient Broth tubes to resuscitate. At the end of incubation period at 37ºC for 24 hours, the cultures are planted in sole colony into Mueller Hinton Agar (MHA) plates and then they are left into incubation again. At the end of this period the purity of microorganisms are checked and they are transmitted to 2 ml micro-reaction tubes (eppendorf) including 1,5 ml 15% sterile glycerol solution which was prepared previously. Those tubes are kept in -85ºC as to be used later.

9.2 Agar diffusion method

For the Agar diffusion method, first of all 25 ml Mueller Hinton Agar (MHA) has been poured into sterile glass petris in 120 mm diameter and the petris have been left for frozen on a smooth surface. The bacteria that will be used are developed in 5 pursuant to McFarland cloudy gauge at Müller Hinton Broth (MHB). Then 1 ml is taken from bacteria solution and poured into petri cups in 120 mm diameter that were prepared previously and the bacteria is provided to spread over medium through sterile drigalski spatula. Then those petris are left to dry in sterile cabinet with semi opened cover. Approximately six reservoirs at per petri is opened on drying surface medium through sterile corc-borer with 6 mm diameters. Essential oil are weighted in 1 mg and solved in 1 ml DMSO and double layer serial dilutions are prepared. Essential oil concentration from 1 mg/ml to 15.6 µl is obtained. Those dilutions are pipette in 50 µl into opened holes on mediums. After completing pipette process, the materials are stored in fridge for 30 minutes for diffusion of materials into medium and then they are left into incubation at 37°C for 24-48 hours.

9.3 Microbroth dilution method

The cultures are cultivated in petries which include MHA to regenerate them and incubated for a period of 24 hours after they are taken out from -85°C. After incubation, some colonies are developed in medium and these colonies are transferred to 10ml of tubes that contain Mueller Hinton Broth (MHB) and again incubated at 37°C for a period of 24 hours. After a period of 18-24 hours of incubation, the cultures developed in liquid medium are again transferred to double power of MHB tubes in definite amounts after they are tuned up in haze as per Mc Farland No: 0.5 (approximately 10^8 cfu/ml).

The essential oils to be tested are transferred to 4mg of sterilized flakons and these are dissolved by adding 2ml of sterilized Dimethyl-Sulfoxide(DMSO) in 25% ratio. The essential oils should be fully dissolved in DMSO and a homogenous mixture should be obtained. Beginning from stocked solutions obtained, essential oils in micro-reaction tubes (eppendorf) are diluted with sterile distilled water and a series of combinations (1, 1/2, 1/4, 1/8, ...) are prepared from 2mg/ml to 1.95 µg/ml.

Micro titration petries (Brand) having 96"U" type bores are used for experiment. 100 µl of diluted mixtures are transferred to each bore column series via micropipettes. In addition to the essential oils to be tested, to control the solvents the DMSO and Standard antibiotic chloramphenicol (Sigma) are tested as positive control. After all concentrations are transferred to the bores, the microorganisms are added to them. To do this, the microorganism cultures whose haze is tuned up as per McFarland No:05 before are transferred to the reservoirs that are produced proper to multichannel automatic pipetors and 100 µl of mixture are pipeted to each line of bores as one microorganism is in the same eline. After these procedures, the cover of micro titration petries are closed and they are incubated at 37°C for period of 24 hours. At the end of incubation period, to be able detect the regeneration some TTC salt is sprayed over petries. Later it is left to incubation and coloring process at 37°C for a period of 3 hours. At the end of incubation, the areas that are not colored are the ones that no regeneration is obtained.

9.4 Detection of antimicrobial compounds in essential oils by bio-autography method

9.4.1 Thin layer chromatography (TLC) system

Thin layer chromatography plaques (Aldrich) that are coated with silica gel 60 GF$_{254}$ adsorbent and in 0.2mm of height are used in appropriate sizes over aluminum supports. These plaques are kept in their special protected boxes in room temperature, droughty, and dark places. Pure essential oils are applied over plaques as 1 µl by using capillary tubes. 9:1 (v/v) proportion Hexane-Ethyl Acetate dissolvent as 20 ml is prepared and filled in a covered rectangular glass container. The TLC plaque where essential oil applied parts are marked is immersed vertically into dissolvent, but one should be careful as the dissolvent not to reach the oily parts. The TLC plaques are developed in this system as twins. While one of the plaques is taken out and kept aside for experiment, the other one is cleaned out from dissolvent by evaporating it and analyzed under 254/364nm wave lengths and UV active spots are marked. To be able to detect the compounds that do not absorb UV, anisaldehyde / H_2SO_4 color reactive is sprayed over the plaque and it is heated at 110°C for a period of 1-3 minutes.

9.4.2 Preparation of microorganisms

A cell suspension of approximately 1.5×10^9 cfu/ml in MHB is prepared for bio-autography method which contains bacterial culture that is prepared the day before in MHB medium and tuned up in haze as per McFarland No:5.

9.4.3 Assessment of activity

Pre-prepared molten agar that is delivered in 20ml of Erlenmeyer is kept in water-bath and then 2ml of bacterial suspension culture is added into agar and a final concentration of 1.5×10^8 cfu/ml is obtained. TLC plaques which are pre-prepared and not processed with reactive are placed in a petri that is filled with Nutrient Agar. Then molten agar that is kept in hot water-bath and inoculated with microorganisms in a well stirred situation is added into the petri as a thin layer not much than 1mm. These plaques are incubated for a period of 24-48 hours at 37°C. At the end of incubation period, formed inhibition zones and their R_f values of corresponding reference plaque fractions are measured and registered.

9.4.4 Gas chromatography (GC)

The compounds in essential oil are isolated and evaluated under following conditions by taking into account their attachment periods in gas chromatography (R_t) and their relative rates.

GC Analysis Conditions	
System	Shimadzu GC-17A
Column	CPSil 5CB
Carrying Gas	Nitrogen (1ml/min.)
Splitting Rate	50:1
Detector	FID
Temperatures	
Injection	250°C
Column	60°C//5°C/min//260°C-20 min
Detector	250°C,

Table 2.

9.4.5 Gas chromatography / mass spectrometer(GC/MS)

After isolating compounds in essential oil within a gas chromatography column, spectrums of each compound is determined by using a mass spectrometer. All evaluations are made primarily by using "TBAM Essential Oil Compounds Library". Furthermore, Wiley and Adams-LIBR library scanning software" and "The Wiley/NBS Registry of Mass Spectral Data" systems are used.

GC/MS Analysis Conditions	
System	Shimadzu GCMS-QP5050A
Column	CPSil 5CB (25mx0.25mm i.d.)
Carrying Gas	Helium (1ml/min)
Temperatures	
Injection	250°C
Column	60°C//5°C/min.//260°C-20 min.
Splitting Rate	50:1
Power of Electrons	70 eV
Mass Spectrum	35-400 m/z

Table 3.

9.4.6 Dispense of preparative TLC plaque

Readymade TLC plaque coated in size of 20x20 is used for preparative purposes to isolate active compounds in essential oil. Neutral essential oil is applied over plaque as a thin layer and eliminated in hexane: ethylacetat (9:1 v/v) dissolvent system within a TLC vessel. At the end of this process, a narrow fraction of TLC plaque is cut out vertically and this sample is analyzed under 254-364nm UV light and R_f value of this to be isolated fraction is determined by the application of anisaldehyde/ H_2SO_4 reactant. Subsequent to this process, sample plaque fraction is laid alongside of TLC plaque that is not processed with reactant yet and by taking into account the R_f value, the area on the plaque where the target substance is present is determined and silica gel over the aluminum support is rubbed out. Silica gel and substance compound is transferred to a glass funnel where glass wool is present and acetone is poured over them as a solvent. The substance is dissolved with acetone and resolving silica gel in the funnel gathers in the Erlenmeyer downside. Later on by using a vacuumed rotavapor, the solvent is removed in 40°C.

10. References

[1] Anderson, E.S., Threlfall, E.J. (1974). The Characterization of Plasmids in the Enterobacteria. J. Hyg. Camb., 72: 471-487

[2] Arda, M., Minbay, A., Leloğlu, N., Aydin, N., Akay, Ö. (1992). Özel Mikrobiyoloji. Erzurum: Atatürk Üniversitesi Yayınları No: 741

[3] Arda, M. (2000). Temel Mikrobiyoloji, Ankara: Medisan Yayınevi.

[4] Ben Hassen, A., Meddeb, M., Ben Chaabane, T., Zribi, M., Ben Redjeb, S.(1994). Characteristics of the antibiotic resistance plasmid in Salmonella typhi isolated in Tunis in 1990. Ann. Biol. Clin., 52(2):133-136

[5] Bilgehan, H. (1986). Klinik mikrobiyoloji. İzmir: Bilgehan Basımevi.

[6] Bilgehan, H. (1992). Salmonella, Klinik Mikrobiyoloji, Özel Bakteriyoloji ve Bakteriyel İnfeksiyonlar, İzmir: Barış Yayınları, Fakülte Kitabevi, s.: 25-45

[7] Cisse, M.F., Sow, A.I., Dieye-Sarr, E., Boye, C.S., Gaye-Diallo, A., Diop, D., Mboup, S., SAMB, A. (1993). Antibiotic sensitivity of Salmonella strains isolated in a pediatric population in Dakar. Research of Beta-Lactamase and Plasmids. Bull. Soc. Pathol. Exot., 86(1):43-47

[8] Crosa, J.H., Tolmasky, M.E., Actis, L.A., Falkow, S. (1994). Plasmids. In: Metods for General and Molecular Bacteriology , Ed.: P. Gerhardt, R.G.E. Murray, W.A. Wood, Washington, Massachusetts: American Society for Microbiology, p.: 365-386

[9] Dalhoff, A. (1994). Quinolone Resistance in Pseudomonas aeruginosa and Staphylococcus aureus. Development during Therapy and Clinical Significans. Infection, 22(suppl 2): 111-121

[10] Dib,C., Trias, J., Jarlier, V. (1995). Lack of additive effect between mechanism of resistance to carbapenems and other Beta-Lactam agents in Pseudomonas aeruginosa. Eur. J. Clin. Microbiol. Infect Dis., 14(11): 979-986

[11] Elwell, L.P., Fling, M.E. (1989). Resistance to Trimethoprim. In: Handbook of Experimental Pharmacology Vol 91, Ed.: L.E. Bryan, Berlin, Heildberg: Springr-Verlag, p.: 249-290.

[12] Filetici, E., Martini, A., Magni, L., Fantasia, M. (1988). R-plasmid in Salmonella isolates from sporadic cases of gastroenteritis. Eur. J. Epidemiol., 4(3): 366-370

[13] Garau, J. (1994). Beta-Lactamases, Current Situation and Clinical Importance. Intensive Care Med., 3(suppl 20): 5-9

[14] Gericke, B., Rische, H., Schmidt, E., Tschape, H. (1988). Salmonella montevideo from a hospital outbreak with different antibiograms determined by aplasmid zentrabl. Bacteriol. Microbiol. Clin., 186(2): 180-188

[15] Gyles, C.L., Thoen, C. (1986). Pathogenesis of bacterial infections in animals. Iowa: Iowa State University Pres

[16] Kamili, M.A.(1993). Multiple drug resistant typhoid fever out break in Kashmir Valley. Indian J. Med. Sci., 47(6): 147-151

[17] Kapil, A., Ayyagari, A., Grag, R.K., Agarwal, K.C. (1994). S. Typhi with transferable chloramphenicol resistance isolated in chandigarh during 1983-1987. Indian J Pathol. Microbiol., 37(2): 179-183

[18] Karmaker, S., Biswas, O., Shaikh, N.M., Chatterjee, S.K., Kataria, V.K., Kumar, R. (1991). Role of a large plasmid of Salmonella typhi encoding multiple drug resistance. J. Med. Microbiol., 34(3): 149-151

[19] Le Minor, L. (1992). The Genus *Salmonella*. In: A Handbook on the Biology of Bacteria: Ecophysiology, Isolation, Identification, Application 2.th Ed. Vol III , Ed.: A. Balows, H.G. Truper, M. Dworkin, W. Harder, K.H. Schleifer, New York, Berlin, Heidelberg: Springer-Werlag, p.: 2760-2774

[20] Lee, L.A., Threatt, V.L., Puhr, N.D., Levine, P., Ferris, K., Tauxe, R.V. (1993). Antimicrobial-resistant *Salmonella* spp. isolated from 49 healty broiler chickens after slaughter. J. Am. Vet. Med. Assoc., 202(5): 752-755

[21] Ling, J.M., Zhou, G.M., Woo, T.H., French, G.L. (1991). Antimicrobial Susceptibilities and Beta-Lactamase production of Hong Kong isolates of gastroenteric *Salmonella* and *Salmonella typhi*. J. Antimicrob. Chemother., 28(6): 877-885

[22] Maiorini, E., Lopez, E.L., Marrow, A.L., Ramirez, F., Procopio, A., Furmanski, S., Woloj, G.M., Miller, G., Cleary, T.G. (1993). Multiple resistant nontyphoidal *Salmonella gastroenteritis* in children. Pediatr. Infect Dis. J., 12(2): 204-205

[23] Nordmann, P. (1996). Novel Beta Lactamases in Gram Negative. 27. Türk Mikrobiyoloji Kongresi; 4. Konferans

[24] O'brien, T.F., Digiorgio, J., Parsonnet, K.L., Kass, E.H., Hopkins, J.D. (1993). Plasmid diversity in E.coli isolated from processed poultry and poultry processors. Vet. Microbiol, 35:243-255.

[25] Oppezzo, O.J., Avanzati, B., Anton, D.A. (1991). Increased Susceptibility to Beta-Lactam Antibiotics and Decreased Porin Content Caused by env B Mutations of *Salmonella typhimurium*. Antimicrob. Agents. Chemother., 35: 1203-1207

[26] Philippon, A., Arlet, G., Lagrange, P.H. (1994). Origin and impact of Plasmid-Mediated extended-spectrum Beta-Lactamases. Eur. J. Clin. Microbiol. Infect. Dis., 13(suppl 1): 17-29

[27] Provence, D.L., Curtiss III, R. (1994). Gene Transfer in Gram-Negative Bacteria. In: Methods for General and Molecular Bacteriology, Ed.: P. Gerhardt, R.G.E. Murray, W.A. Wodd, Washington, Massachusestts: American Society for Microbiology, p.: 317-347

[28] Rao, R.S., Amarnath, S.K., Sujatha, S. (1992). An outbreak of tyhoid due to multidrug resistant *Salmonella typhi* in pondicherry. Trans R. Soc. Trop. Med. Hyg., 86(2): 204-205

[29] Rathish, K.C., Chandrashekar, M.R., Nagesha, C.N. (1994). Multidrug Resistant *Salmonella typhi* in Bangalore, South India. Indian J. Med. Sci., 48(4): 85-88

[30] Rathore, M.H., Bux, D., Hasan, M. (1996). Multidrug-Resistant *Salmonella typhi* in Pakistani children: Clinical features and treatment. South Med. J., 89(2): 235-237

[31] Rodrigue, D.C., Tauxe, R.V., Rowe, B. (1990). International Increase in *Salmonella enteritidis*. A New Pandemic. Epidemiol. Infect., 105: 21-27

[32] Rydberg, J., Larsson,C., Miorner, H. (1994). Resistance to Fluoroquinolones in Pseudomonas aeruginosa and Klebsiella pneumoniae. Scand. J. Infectd. Dis., 26: 317-320

[33] Salyers, A.A., Whitt, D.D. (1994). Antibiotics of Action and Mechanism of Bacterial Resistance. In: Bacterial Pathogenesis, A Molecular Approach, Ed.: A.A. Salyers, D.D. Whitt, Washington DC: ASM Press , p.: 97-110

[34] Salyers, A.A., Shoemaker, N.B., Stevens, A.M., Yew LI, L. (1995). Conjugative Transposons an Unusual and Diverse Set of Integrated Gene Transfer Elements. Microbiol. Rev., 59(4): 579-590

[35] Sanders, C.C. (1989). The Chromosomal Beta – Lactamases. In: Handbook of Experimental Pharmacology Vol 91, Ed.: L.E. Bryan, Berlin, Heilderberg: Springr-Verlag, p.: 129-150

[36] Secmeer, G., Kanra, G., Cemeroğlu, A.P., Ozen, H., Ceyhan, M., Ecevit, Z. (1995). *Salmonella typhi* Infections. A 10-year retrospective study. Turk J. Pediatr., 37(4): 339-341

[37] Threlfall, E.J., Frost, J.A. (1990). The Identification, Typing and Fingerprinting of *Salmonella*: Laboratory Aspects and Epidemiological Applications. J. Appl. Bacteriol., 68: 5-16

[38] Trias, J., Nikado, H. (1990). Quter Membrane Protein O2 Catalyzes Facilitated Diffusion of Carbapenems and Penems Through The Quter Membrance of Pseudomonas aeruginosa. Antimicrob. Agents Chemother., 34(1): 52-57

[39] Tünger, A., Çavusoğlu, C., Korkmaz, M. (2002). Asya Mikrobiyoloji, İzmir: Asya Tıp Yayıncılık Ltd.Şti.

[40] Wattal, C., Kaul, V., Chugh, T.T., Kler, N., Bhandari, S.K. (1994). An outbreak of multidrug resistant *Salmonella typhimurium* in Delhi (India). Indian J. Med. Res., 100: 266-341

[41] Wu, S.X., Tang, Y. (1993). Molecular epidemiologic study of an outbreak of *Salmonella typhimurium* infection at a Newborn Nursery. China Med. J., 106(6): 423-427

[42] Baytop, T., Türkiye'de Bitkiler ile Tedavi Geçmişten Bugüne,2. Baskı, Nobel Tıp Basımevi, İstanbul, Türkiye (1999)

[43] Hammer, K.A., Carson, C.F., ve Riley, T.V., Antimicrobal Activity of Essential Oils and Other Plant Extracts, J. Appl. Mikrobiol., 85, 985-990 (1999).

[44] Duke, A.J., Handbook of Medicinal Herbs, CRC Press, Florida, USA (1985).

[45] Baytop, A., Farmasotik Botanik Ders Kitabı, İ.Ü. Basımevi, İstanbul, Türkiye (1991).

[46] Vanden Berge, D.A ve Vlietinck, A.J., Screening Methods for Antimicrobal and Antiviral Agents from Higher Plants, Methods in Plant Biochemistry, (Eds: Harborne, J.B., Dey, P.M.) Academic Press, London, England, 37-53 (1991).

[47] Cowan, M.M., Plan Products As Antimicrobial Agents, Clin. Microbiol. Rev., 12, 564-582 (1999).

[48] Dorman, H.J.D. ve Deans, S.G., Antimicrobial Agents from Plants: Antibacterial Activity of Plant Volatile Oils, J. Appl. Microbiol., 88, 308-316 (2000).

[49] BEŞE, M., Mikrobiyolojide kulklanılanAntibiyotik Duyarlılık ve Deneme

[50] Yöntemleri, Kardeşler Basımevi, İstanbul (1998).

[51] Koneman, E.W., Allen, S.D., Janda, W.M., Schreckenbeger VE P.C. ve Winn W.C., Color Atlas and Textbook of Diagnostic Microbiology, Lippincot-Raven Pub, Philadelphia, USA, 785-856 (1997)

[52] Elof, J.N., A Sensitive Quick Microplate Method to Determine the Minimal İnhibitory Concentration of Plant Extracts for Bacteria, Planta Med., 64, 711-713 (1998).

[53] Lambert, R.J.W., Suspectibilty Testing: İnoculum Size Dependency of İnhibition Using the Colworth MIC Technique, J. Appl. Microbiol., 89, 275-279 (2000).

[54] Sökmen, A., Jones, B.M. ve Ertürk, M., The in Vitro Antibacterial Activity of Turkish Medicinal Plants, J. Ethnopharm., 67, 79-86 (1999).

[55] Mehrabian, S., Majd, A. Ve Majd, I., Antimicrobial Effects of Three plants on Some Airborne Microorganisms, Aerobiologia, 16, 455-458 (2000).

[56] Hostettmann, K., Strategy fort he Biological Evaluation of Plant Extracts, Pure App. Chem., 70, 1109-1113 (1998).

[57] Cannel, J.P.R., Natural Products Isolation, Humana Press, New Jersey, USA, 240-241 (1998).

6

Antibiotic Susceptibility Patterns of *Salmonella* Typhi in Jakarta and Surrounding Areas

Lucky H. Moehario[1], Enty Tjoa[2],
Veronica N. K. D. Kalay[3] and Angela Abidin[4]
[1]Department of Microbiology Faculty of Medicine University of Indonesia, Jakarta
[2]Department of Microbiology Faculty of Medicine Catholic
University of Atmajaya Indonesia, Jakarta
[3]Department of Microbiology Faculty of Medicine Christian University of Indonesia
Jakarta and Division of Microbiology, Siloam Hospital Kebon Jeruk, Jakarta
[4]Division of Microbiology, St. Carolus Hospital, Jakarta
Indonesia

1. Introduction

Typhoid fever, also known as enteric fever, is a potentially fatal multi systemic illness caused primarily by *Salmonella* enterica serotype Typhi (*S.* Typhi). The classic presentation of the disease includes fever, malaise, diffuse abdominal pain, and constipation. Untreated, typhoid fever may progress to severe condition like delirium, intestinal hemorrhage, bowel perforation, and death. The disease remains a critical public health problem in developing countries. In 2000, it was estimated that over 2,16 million of typhoid occurrences worldwide, resulting in 216,000 deaths, and that more than 90% of this morbidity and mortality occurred in Asia [1]. A report from World Health Organization in 2008 on typhoid fever in five Asian countries showed the annual typhoid incidence (per 100,000 person years) among 5-15 years age group varied from 24.2 and 29.3 in Vietnam and China, to 180.3 in Indonesia; and to 412.9 and 493.5 in Pakistan and India, respectively; multidrug resistant *S.* Typhi were 23% (96/413) [2]. Further, unlike *S.* Typhi originated from Pakistan, Vietnam and India, those from Indonesia collected in North Jakarta, were all susceptible to antibiotic tested, i.e. Chloramphenicol, Ampicillin, Trimethoprim-Sulfamethoxazole; none of multidrug resistance were found. Nalidixic acid resistance was rather high in Pakistan, India and Vietnam, but none was found in Indonesia [2].

In Indonesia the prevalence of typhoid fever was 358-810 per 100,000 populations in 2007, with 64% of the disease was found in people aged 3-19 years. Mortality rate varies from 3.1-10.4% among hospitalized patients. Hatta and Ratnawati, 2008 reported a rise of resistance of *S.* Typhi to 6.8% to all three of first line drugs (Chloramphenicol, Ampicillin, Co-trimoxazole) in South Sulawesi (East of Indonesia) [3]. Antibiotics Fluoroquinolone and 3rd generation of Cephalosporin are frequently used for therapy of patients suspected typhoid fever in the past decade in many places especially in endemic countries including Indonesia due to resistance issues against conventional antibiotics [4, 5, 6, 7]. This study aimed to overview antibiotic susceptibility of *S.* Typhi originated from Jakarta and surrounding areas in particular to those recommended by Performance Standard for Antimicrobial Susceptibility Testing for Clinical and Laboratory Standard Institute within 9 years period up to 2010.

2. Materials and methods

2.1 Specimens

Specimens used in the study were blood received in our laboratory i.e. Laboratory of Clinical Microbiology Faculty of Medicine University of Indonesia (CML-FMUI) Jakarta between 2002-2007. Our laboratory accepted specimens from hospitals, mainly the National Hospital Cipto Mangunkusumo (a tertiary general public hospital), primary health cares, private practices, and individuals. As in CML-FMUI, blood specimens from in and outpatients as well as other sources examined in Siloam Hospital Kebon Jeruk and St. Carolus Private Hospital in 2008-2010 were also included.

2.2 Culture and antibiotic susceptibility tests

Culture and antibiotic susceptibility tests were established in each of the above-mentioned institutions. Microbiology tests were performed according to microbiology standard practices and Performance Standards for Antimicrobial Susceptibility Testing for the Clinical and Laboratory Standards Institute (CLSI) [8]. Cultures were performed using Bac-T Alert™ (Enseval)/Bactec™ 9050 (Becton Dickinson), and sub cultured were on Salmonella-Shigella and MacConkey agar. Microorganism identification was determined using conventional biochemical reactions i.e. acid production from glucose, lactose, maltose, mannitol and saccharose, IMViC tests (Indole, Methyl Red, Voges Proskauer and Citrate) and H2S production in TSI agar. In recent years, API20E biochemical identification system (BioMerieux, Paris, France) was used instead. Susceptibility of microorganisms to antibiotics was assessed using the disc diffusion method. Antimicrobial susceptibility results were categorized in to three groups: Sensitive (S), Intermediate (I) and Resistant (R) according to CLSI guidelines. The antibiotics susceptibility data was then entered into the WHO-Net 5.4 program.

2.3 Antibiotics

Standard disc diffusion method was employed. The following antibiotic discs Chloramphenicol (CHL) 30µg, Amoxycillin (AMX) 25µg, Trimethoprim-Sulfamethoxazole (SXT) 1.25/23.75µg, Ceftriaxone (CRO) 30µg, Ciprofloxacin (CIP) 5µg, and Levofloxacin (LVX) 5µg were included in the study. These antibiotics are frequently used to treat typhoid fever in Jakarta, Indonesia. Susceptibility of S. Typhi to antibiotics was tabulated, and good activity in-vitro was defined by antimicrobial susceptibility of 80% or greater. Minimal inhibitory concentration was not examined.

3. Results

During nine years period from 2002-2010, 247 isolates of S. Typhi were collected, in which 35 isolates were from CML-FMUI, 73 and 139 isolates came from Siloam Kebon Jeruk and St. Carolus Hopitals respectively. In 2002-2007, all S. Typhi isolated in CML-FMUI was susceptible to antibiotics Levofloxacin, Ciprofloxacin, Trimethoprim-Sulfamethoxazole and

Amoxycillin (Figure 1). In 2008-2010, all of *S.* Typhi isolated in Siloam Hospital Kebon Jeruk was susceptible to Levofloxacin and Ciprofloxacin as also found in St. Carolus Hospital except antibiotic Levofloxacin was not tested on isolates from St. Carolus Hospital (see Figure 1). Susceptibility of the microorganism to Trimethoprim-Sulfamethoxazole showed almost similar pattern to those of CML-FMUI in the earlier years ranging from 98.6% to 100%, and so Amoxycillin that was 98.5% to 100%. Susceptibility of these microorganisms to Ceftriaxone seemed to increase from 92.6% in 2002-2007 to 98.6% or greater in 2008-2010. Lastly, although antibiotic Chloramphenicol was scarcely used in the treatment of typhoid fever compared to Fluoroquinolones, this antibiotic was still effective. The susceptibility of *S.* Typhi isolates to Chloramphenicol was 94.1% in 2002-2007, and was apparently increase to 98.6% or greater in 2008-2010 (Figure 1). Overall, during 9 years period up to 2010, antibiotic Chloramphenicol, Amoxycillin, Trimethoprim-Sulfamethoxazole, Ceftriaxone, Ciprofloxacin and Levofloxacin showed good activity in-vitro against *S.* Typhi originated from Jakarta and adjacent areas.

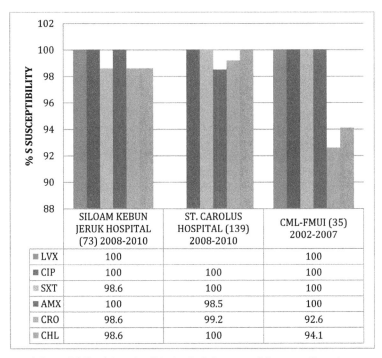

	SILOAM KEBUN JERUK HOSPITAL (73) 2008-2010	ST. CAROLUS HOSPITAL (139) 2008-2010	CML-FMUI (35) 2002-2007
LVX	100		100
CIP	100	100	100
SXT	98.6	100	100
AMX	100	98.5	100
CRO	98.6	99.2	92.6
CHL	98.6	100	94.1

Fig. 1. Susceptibility of *S.*Typhi to Antibiotics in Jakarta and Surrounding Areas

4. Discussions

Jakarta, as a capital city, is the biggest urban area in Indonesia. The city connects to five other satellites cities i.e. Bogor, Depok, Tangerang, Bekasi, Karawang, and it is densely populated where some districts are still impoverished. Health care system in the country does not support extensive program that covers laboratory examinations. This reflected in the limited samples received in the laboratories. Clinical Microbiology Laboratory of Faculty of Medicine University of Indonesia is located in the center of Jakarta and known as the referral Microbiology Laboratory as it is part of the Department of Microbiology FMUI. The laboratory receives specimens from other laboratories, hospitals, primary health cares and also individuals. Furthermore, Siloam Kebon Jeruk in West Jakarta and St. Carolus in Central Jakarta are private hospitals; their laboratories serve for the hospitalized, outpatients and also other sources. In this study, 247 isolates of *S.* Typhi examined during 2002-2010 were susceptible to antibiotics Chloramphenicol, Amoxycillin, Trimethoprim-Sulfamethoxazole, Ceftriaxone, Ciprofloxacin, and Levofloxacin. In addition, *S.* Typhi was in fact still susceptible to antibiotic Tetracycline (Data not shown).

Some reports on antibiogram of *S.* Typhi from other institutions within the country were similar i.e. Central Laboratory of Cipto Mangunkusumo Hospital reported all *S.* Typhi isolated from hospitalized typhoid fever cases in 2009 were all susceptible to antibiotics (Chloramphenicol, Cotrimoxazole, Ceftriaxone, Cefuroxime, Ampicillin-Sulbactam, Amoxycillin-Clavulanic acid, Ciprofloxacin, Levofloxacin) [9]. In 2003-2005, ninety isolates of *S.* Typhi were collected from some districts in East Jakarta. All of these isolates were susceptible to first line drugs (Chloramphenicol, Trimethoprim-Sulfamethoxazole, Ampicillin) and Tetracycline except for only 1 isolate which was resistant to Chloramphenicol [10].

Data obtained from outside Jakarta such from Pakanbaru in Sumatra Island in 2009-2010 also showed very similar susceptibility patterns. Those *S.* Typhi isolates were all susceptible to Chloramphenicol, Tetracycline, Trimetoprim-Sulfamethoxazole, Amoxycillin-Clavulanic Acid, Cefotaxim, Cefepime, Ceftazidime, Cefazolin, Ceftriaxone and Ciprofloxacin [11]. *S.* Typhi isolated from South Sulawesi, however, showed an increase resistant against Chloramphenicol and Ciprofloxacin between 2001-2007, which was 1.04% to 7.84% and 0.11% to 6.83% respectively [3]. In 2003, a collaborated study on enteric bacteria in patients with diarrhea had been carried out in United States Naval Medical Research Unit, Jakarta that involved many health institutions from many cities in Indonesia including Medan, Padang, Batam, Jakarta, Pontianak, Denpasar and Makassar. A total of 111 *S.* Typhi had been isolated from feces, and all were susceptible to antibiotic tested i.e. Ampicillin, Trimethoprim-Sulfamethoxazole, Chloramphenicol, Tetracyclin, Cephalotin, Ceftriaxone, Norfloxacin and Ciprofloxacin. Nalidixic acid resistance was not found [12].

Resistance to Chloramphenicol was reported to emerge in only two years after its introduction in 1948, and was not until 1972 that typhoid fever caused by Chloramphenicol-resistant *S.* Typhi became a major problem; outbreaks occurred in Mexico, India, Vietnam, Thailand, Korea, and Peru (cited from Parry et al, 2002 [13]). Toward the end of the 1980s and the 1990s, *S.* Typhi developed resistance simultaneously to all the drugs that were then used as first line treatment (Chloramphenicol, Trimethoprim, Sulfamethoxazole, and Ampicillin). Despite multidrug-resistant *S.* Typhi are still common in many areas of Asia, strains that are fully susceptible to all first line antibiotics have reemerged in some areas [13]. Chau et al, 2007 reported that of eight endemic countries from Indian continent to China, multidrug resistance (MDR) *S.* Typhi varied from 16 to 37% and Nalidixic acid resistance were 5 to 51% [7]. In some places Nalidixic Resistant *S.* Typhi (NRST) was reported to cause more complication and poorer outcome of the disease; the presence of NRST is critical and influenced the successful rate of therapy with Fluoroquinolone [14].

Despite an increase resistance elsewhere, certain areas especially in Northern India, reported that Chloramphenicol resistance has reduced from a high of 18% to only 2% [15].

Differences in the antibiotic susceptibility profiles as well as clinical appearance of typhoid fever cause by S. Typhi rely on many factors. Conditions such as disease control programs, inadequate policy of using antibiotics, local conditions that include personal hygiene, availability of clean drinking water, food handling and sanitation contribute to the complexity and outcome of the disease. Nevertheless, one important element of the diverse clinical manifestation of typhoid fever is the presence of genome plasticity of S. Typhi. Many studies had been conducted in the late nineties on genome profile of S. Typhi and showed the diversity of the genome. Some of strains originated from a certain region or country share some degree of similarity to other strains from different places. Our earlier study on genetic relationship using pulsed-field gel electrophoresis (PFGE) found that S. Typhi originated from five cities i.e. Medan, Jakarta, Pontianak, Makassar and Jayapura in Indonesia expanding from west to east part of the country, had clusters of endemic strains in certain geographic areas [16]. The presence of specific strains in localized area might have been the reason of varied symptoms of the disease and, possibly their susceptibility to antibiotics. Despite heterogeneity and different clonality of the 33 isolates of S. Typhi used in the study, these endemic strains were in fact all susceptible to Chloramphenicol, Ampicillin and Cotrimoxazole [16].

Some investigators reported a correlation of certain genome profiles of S. Typhi strains and their ability to cause a fatal typhoid fever [17]. Others reported specific flagellar types were associated with severe outcome of the disease [18]. A study carried out in our laboratory by Tjita in 2000 showed that 3 S. Typhi strains had identical genome profiles deduced from PFGE [19]. Each of the strains showed different susceptibility against several antibiotics i.e. one strain was resistant to Tetracycline; another two were multi resistant to Chloramphenicol/Tetracycline, and Ampicillin/Chloramphenicol/ Tetracycline respectively (see Figure 2). In addition to the findings, two other S. Typhi strains which resistant to Ampicillin and Tetracycline were found to be an identical strains [19]. The mentioned conditions could have been the reasons that conventional drugs such as Chloramphenicol, Amoxycillin, Trimethoprim- Sulfamethoxazole and Tetracyclin still have good activity in-vitro against S. Typhi strains in Jakarta and surrounding areas. In recent years, antibiotic Fluoroquinolone has been widely used in the treatment of typhoid fever in Indonesia, and it showed superiority in term of efficacy and safety [20]. Previous reports by Ochiai et al, 2008 [2] and Tjaniadi et al, 2003 [12] showed Nalidixic acid resistant S. Typhi was not found thus far in Indonesia. Cautious is adviced, however, since quinolone resistant S. Typhi strains have been an important issue in regional and global [6, 21, 22, 23].

In conclusion, despite resistance issue of S. Typhi from other countries, this study showed that most of all S. Typhi isolated in certain places in Jakarta and neighboring areas were susceptible to antibiotic tested (Chloramphenicol, Amoxycillin, Trimethoprim-Sulfamethoxazole, Ceftriaxone, Ciprofloxacin, Levofloxacin). This information is important since antimicrobial therapy plays a key role in management of typhoid fever disease. The susceptibility profiles, however, were only derived from certain strains, which may not represent all strains, which present in Indonesia. Therefore it is necessary to perform cultures and antibiotic sensitivity tests on patients with suspected typhoid fever, and so the patients can be treated with definitive antibiotic therapy. Needless to say, adequate antibiotic therapy will prevent the spread of antibiotic-resistant *S.* Typhi strains. Lastly, promotion of public health such as personal hygiene, sanitations, clean drinking water, food handlings and also vaccination are equally important as prevention of the disease.

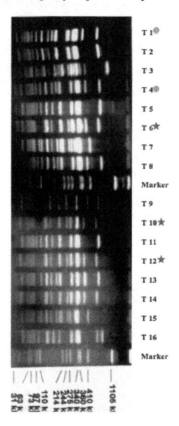

Sixteen *S.* Typhi isolates were originated from patients with typhoid fever in Jakarta. The genome was digested *Xba*I restriction enzyme. T6, T10 and T12 [★] were identical strains, which showed different susceptibility patterns i.e. T6 resistant to Tetracycline, T10 resistant to Chloramphenicol/Tetracycline, T12 resistant to Ampicillin/Chloramphenicol/Tetracycline. T1 and T4 [●] had identical genome profiles but each showed different susceptibility patterns i.e. T1 resistant to Ampicilin, and T4 resistant to Tetracycline. Modified from Tjita, 2000 [19].

Fig. 2. Genome profile of *S.* Typhi isolates from Jakarta using Pulsed-Field Gel Electrophoresis

5. Acknowledgments

We thank the staffs and technicians at the Clinical Microbiology Laboratory Department of Microbiology Faculty of Medicine University of Indonesia for their works and contributions to the study.

6. References

[1] Crump JA, Luby SP, Mintz ED. The global burden of typhoid fever. Bull World Health Organ 2004; 82: 346-53.

[2] Ochiai RL, Acosta CJ, Danovaro-Holliday MC, Baiqing D, Bhattacharya SK, Agtini MD, Bhutta ZA, Canh DG, Ali M, Shin S et al. A study of typhoid fever in five Asian countries: disease burden and implications for controls. Bull World Health Organization 2008; 86: 260–268.

[3] Hatta M and Ratnawati. Enteric Fever in endemic areas of Indonesia: an increasing problem of resistance. J Infect Developing Countries 2008; 2(4): 279-282.

[4] Hasan R, Zafar A, Abbas Z, Mahraj V, Malik F and Zaidi A. Antibiotic resistance among *Salmonella enterica* serovars Typhi and Paratyphi A in Pakistan (2001-2006). J Infect Developing Countries 2008; 2(4): 289-294.

[5] Kumar S, Rizvi M and Berry N. Rising prevalence of enteric fever due to multidrug-resistant Salmonella: an epidemiological study. Journal of Medical Microbiology 2008; 57: 1247-1250.

[6] Jin Yang. Enteric Fever in South China: Guangxi Province. J Infect Developing Countries 2008; 2(4): 283-288.

[7] Chau TT, Campbell JI, Galindo CM, Hoang NVM, Diep TS, Thi Nga TT, Chau NVV, Tuan PQ, Page AL, Ochiai RL et al. Antimicrobial Drug Resistance of *Salmonella enterica* Serovar Typhi in Asia and Molecular Mechanism of Reduced Susceptibility to the Fluoroquinolones. Antimicrobial Agents and Chemotherapy 2007; 51 (12): 4315-4323.

[8] Performance Standards for Antimicrobial Susceptibility Testing. Clinical and Laboratory Standard Institute (CLSI); year up to date.

[9] Loho T. Microorganism mapping and their Susceptibility to antibiotics in National Hospital Cipto Mangunkusumo Jakarta in 2009. (Unpublished).

[10] Ali S. Antibiotic Susceptibility of *Salmonella* Typhi in Jakarta in 2001-2003. (Unpublished).

[11] Anggraini D. Susceptibility of *Salmonella* Typhi in Private Hospital in Pakanbaru Sumatra in 2010. (Unpublished).

[12] Tjaniadi P, Lesmana M, Subekti D, Machpud N, Komalarini S, Santoso W, Simanjuntak CH, Punjabi N, Campbell JR, Alexander WK, Beecham III HJ, Corwin AL and Oyofo BA. Antimicrobial Resistance of Bacterial Pathogens Associated with Diarrheal Patients in Indonesia. Am J Trop Med Hyg 2003; 68 (6): 666-670.

[13] Parry CM, Hien TT, Dougan G, White NJ and Farrar JJ. Typhoid Fever. Medical progress. N Engl J Med 2002; 347: 1770-1782.

[14] Kadhiravan T, Wig N, Kapil A, Kabra SK, Renuka K and Misra A. Clinical outcomes in typhoid fever: adverse impact of infection with nalidixic acid-resistant *Salmonella* Typhi. BMC Infectious Diseases 2005; 5:37. http://www.biomedcentral.com/1471-2334/5/37

[15] Shrikala Baliga. Drug Resistance in *Salmonella typhi:* Tip of the Iceberg. *Online J Health Allied Scs.* 2004; 4: 1.

[16] Moehario LH. The molecular epidemiology of *Salmonella typhi* across Indonesia reveals bacterial migration. J Infect Dev Ctries 2009; 3(8): 579-584.

[17] Thong KL, Passey M, Clegg A, Comb BG, Yassin RM, and Pang T. Molecular analysis of isolates of *Salmonella* Typhi obtained from patients with fatal and non fatal typhoid fever. J Clin Microbiol 1996; 34: 1029-1033.

[18] Franco A, Gonzales C, Levine OS, Lagos R, Hall RH, Hoffman SL, Moechtar MA, Gotuzzo E, Levine M and Hone DM. Further consideration of the clonal nature of *Salmonella typhi*: Evaluation of molecular and clinical characteristics of strains from Indonesia and Peru. J Clin Microbiol 1992; 30(8): 2187-2190.

[19] Tjita PJ. Analysis of genetic diversity of *Salmonella typhi* resistant against first line drugs using Pulsed-Field Gel Electrophoresis. (Magister Thesis in Biomedical Science, 2000. Unpublished).

[20] Nelwan RH, Chen K, Nafrialdi and Paramita D. Open study on efficacy of Levofloxacin in treatment of uncomplicated typhoid fever. Southeast Asian J Trop Med Public Health 2006; 37(1): 126-130.

[21] Accou-Demartin M, Gaborieau V, Song Y, Roumagnac P, Marchou B, Achtman M, and Weill FX. *Salmonella enterica* Serotype Typhi with Non-classical Quinolone Resistance Phenotype Emerging Infectious Diseases 2011; 17(6): 1091-1094.

[22] Keddy KH, Smith AM, Sooka A, Ismail H, and Oliver S. Fluoroquinolone-Resistant Typhoid, South Africa. Emerging Infectious Diseases 2010; 16(5): p. 880.

[23] Joshi S and Amarnath SK. Fluoroquinolone resistance in *Salmonella typhi* and *S. paratyphi* A in Bangalore, India. Transactions of the Royal Society of Tropical Medicine and Hygiene 2007; 101(3): 308-310.

7

Nanotechnology Tools for Efficient Antibacterial Delivery to *Salmonella*

Ali Nokhodchi[1,2], Taravat Ghafourian[1] and Ghobad Mohammadi[3]
[1]Medway School of Pharmacy, Universities of Kent and Greenwich, Chatham
[2]Drug Applied Research Center and Faculty of Pharmacy
Tabriz University of Medical Sciences, Tabriz
[3]Faculty of Pharmacy, Kermanshah University of Medical Sciences, Kermanshah
[1]UK
[2]Iran

1. Introduction

In recent years, an increasing number of salmonellosis outbreaks have been recorded around the world, and probably there should be more cases that were not detected or reported (1). Many different types of *Salmonella* exist, some of which cause illness in both animals and people, and some types cause illness in animals but not in people. The various forms of *Salmonella* that can infect people are referred to as serotypes, which are very closely related microorganisms that share certain structural features. Some serotypes are only present in certain parts of the world (1). *Salmonella* spp are gram negative anaerobic and intracellular bacteria. Salmonellosis, mainly due to *Salmonella typhimurium*, occurs more frequently in HIV-infected patients than in healthy individuals and the frequency of bacteraemia is much higher in such patients (2).

Despite the discovery of new antibiotics, treatment of intracellular infections often fails to eradicate the pathogens completely. One major reason is that many antimicrobials are difficult to transport through cell membranes and have low activity inside the cells, thereby imposing negligible inhibitory or bactericidal effects on the intracellular bacteria (3). In addition, antimicrobial toxicity to healthy tissues poses a significant limitation to their use (3). Therefore, the delivery of the drug to the bacterial cells is currently a big challenge to the clinicians. This is on top of the problems posed by the emerging Multi-Drug Resistant species. Moreover, the reduced membrane permeability of microorganisms has been cited as a key mechanism of resistance to antibiotics (4).

Indeed, the challenge is to design the means of carrying an antibiotic into bacterial cells. The pioneer concept of targeted drugs was developed by Ehrlich in 1906 and defined as the 'magic bullet'. Since then targeted drug delivery has involved design and development of small molecule drugs that can specifically interact with the intended receptors in intended tissues. For example prodrugs can be designed for brain delivery of the active drug (5). Another common example is colon delivery of prodrugs designed to release the drug by taking advantage of the bacterial reductase enzymes in colon (6).

However, the drug development process is inevitably lengthy and breakthroughs are quite scarce which has led to the ever increasing cost of discovery and development of new drugs (7). On the other hand, nanotechnology offers a more convenient method for targeted therapy.

Logistic targeting strategies can be employed to enable the drug to be endocytosed by phagocytic cells and then released into the bacteria. To reach the above goal, a drug carrier is generally needed for a drug to arrive at the target site (8). The first study employing a drug carrier for targeted drug delivery was published approximately 40 years ago, using antibodies as carriers of radioactivity for the specific recognition of tumor cells (9). The ideal drug carrier ensures the timely release of the drug within the therapeutic window at the appropriate site, is neither toxic nor immunogenic, is biodegradable or easily excreted after action, and is preferably cheap and stable upon storage (10). Out of different types of drug carriers that have been investigated, many are soluble macromolecular carriers or liposomes (11-15).

By searching all published work on drug carriers it can be concluded that "the ideal drug carrier" does not exist. The suitability of a drug carrier is determined by the disease that will be targeted, its access to the pathological site, and the carriers' ability to achieve appropriate drug retention and timely drug release (16). When these types of formulations are administered by the intravenous route, phospholipidic, polymeric or metal particles are localized preferentially in organs with high phagocytic activity and in circulating monocytes, ensuring their clearance (8). The ability of circulating carriers to target these cells is highly dependent on tissue characteristics and on the carrier's properties. The liver rather than the spleen or bone marrow captures the submicronic particles (8). Immediately after injection, the foreign particles are subjected to opsonization by plasma proteins. This is the process by which bacteria are altered by opsonins so as to become more readily and more efficiently engulfed by phagocytes. In this way, 'classical' or 'conventional' carriers are recognized by the mononuclear phagocytic system (8).

The approaches for drug carrier to improve the drug's antibacterial efficacy are shown in Figure 1. In most cases, i.v. administration of the formulation is needed particularly for passive and active targeting.

The local administration of drug/carriers will increase the residence time of antibiotics at the site of infection (17-19). These carriers are generally investigated with the intention to treat local infections in body parts with limited blood flow as in bone, joint, skin, and cornea.

In passive targeting after i.v. administration of carriers which tend to be taken by phagocytic cells, drug-carrier complex will target intracellular infections. These infections are often difficult to treat as a result of limited ability of the antimicrobial agent to penetrate into cells. This approach makes use of the recognition of drug carriers (nanoparticles) as foreign material in the bloodstream by the phagocytic cells of the mononuclear phagocyte system, the cell type often infected with microorganisms (20, 21).

Regarding the other two approaches (passive targeting with long-circulation time, and active targeting) the targeting of infectious foci is not restricted to mononuclear phagocyte system tissues. In passive targeting a drug carrier with long duration of circulation is used and this is an area which has extensively been investigated, whereas in active targeting carriers specifically bind to the infectious organism or host cells involved in the inflammatory response.

Fig. 1. Drug carrier approaches targeting bacterial infections to improve antibacterial efficacy of drugs.

This chapter focuses mainly on the current research for increasing anti-*salmonella* performance of antibiotics by means of liposomes and nanoparticle systems. Structure, properties, advantages and disadvantages of these drug delivery systems have been discussed. It is clear that such systems may improve the antibiotic efficacy by increasing the drug concentration at the surrounding of the bacteria.

2. Liposomes for antisalmonellosis drug delivery

2.1 Introduction

Liposomes are composed of small vesicles of a bilayer of phospholipid, encapsulating an aqueous space ranging from about 30 to 10000 nm in diameter (Figure 2). They are composed of one or several lipid membranes enclosing discrete aqueous compartments. The enclosed vesicles can encapsulate water-soluble drugs in the aqueous spaces, and lipid soluble drugs can be incorporated into the membranes. They are used as drug carriers in the cosmetic and pharmaceutical industry. The main routes of liposome administration are parenteral, topical and inhalation, and, in a few occasions, possibly other routes of administration can be used. Majority of current products are administered parenterally (22).

Liposome structure was first described in 1965, and they were proposed as a drug delivery nanoparticle platform in 1970s. In 1995, Doxil (doxorubicin liposomes) became the first liposomal delivery system approved by the Food and Drug Administration (FDA) to treat AIDS associated Kaposi's sarcoma (23). Liposomal drug delivery systems can be made of either natural or synthetic lipids. The main building blocks of some liposomal formulations are phospholipids (22). These are natural biomacromolecules that play a central role in human physiology as they are structural components of biological membranes and support organisms with the energy (24). They are amphiphilic molecules, poorly soluble in water, consisting of a hydrophilic part containing hydroxyl groups (the polar head), a glycerol backbone and two fatty acid chains, which form the hydrophobic part. One of the most commonly used lipids in liposome preparation is phosphotidylcholine, which is an electrically neutral phospholipid that contains fatty acyl chains of varying degrees of saturation and length. Cholesterol is normally incorporated into the formulation to adjust membrane rigidity and stability (8). Liposomes can be characterized in terms of size and lamellarity as small unilamellar vesicles (SUV), large unilamellar vesicles (LUV) and multi lamellar vesicles (MLV). MLVs are usually considered large vesicles and aqueous regions exist in the core and in the spaces between their bilayers. The structure of these liposomes is shown in Figure 2.

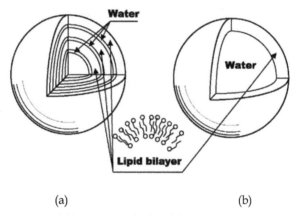

(a)   (b)

Fig. 2. Schematic structures of (a) multilamellar and (b) unilamellar liposomes (the picture was taken from http://what-when-how.com/nanoscience-and-nanotechnology/nanoencapsulation-of-bioactive-substances-part-1-nanotechnology).

The main advantages of liposomes as drug delivery systems can be in their versatile structure that can be easily modified according to experimental needs; they can also encapsulate hydrophilic drugs in their aqueous compartments and hydrophobic drugs in their bilayers, while amphiphilic drugs will be partitioned between the two. Moreover, being mainly made of phospholipid, they are non-toxic, non-immunogenic and fully biodegradable. Methods for preparing liposomes can take into consideration parameters such as the physicochemical characteristics of the liposomal ingredients, materials to be contained within the liposomes, particle size, polydispersity, surface zeta potential, shelf time, batch-to-batch reproducibility, and the possibility for large-scale production of safe and efficient products (23).

2.2 Preparation of liposoms

Liposome formation happens spontaneously when phospholipids are dispersed in water. However, in order to obtain the desired formulation with particular size and structure, various methods such as thin film method (24), sonication (25), extrusion (26), injection methods (27), dehydrated-rehydrated vesicles (28), reverse phase evaporation (29) and one step method (30) have to be used.

Each technique is briefly described below, but for more details, it is recommended to refer to the cited references. In brief, in thin film method liquids are dissolved in organic solvents and the solvent is removed under vacuum or nitrogen stream to form a thin film on the wall of a flask or test tube. In order to complete the formation of liposomes aqueous phase is added to the lipid film at a temperature above the phase transition of the lipid (24).

The sonication method is usually used to reduce the particle size and lamellarity of MLVs. In case of using the probe sonicator, the reduction in size of the liposomes can be guaranteed (25).

In order to get very homogeneous vesicles with a predetermined size, the extrusion technique is used. MLVs are extruded under pressure through particular filter with well-defined pore sizes from 30 nm to several micrometers. If the extrusion is repeated several times unilamellar liposomes can be formed (26).

Very small unilamellar vesicles with a particle size of 30 nm can be prepared using the ethanol injection method. Generally, lipids are dissolved in ethanol and injected rapidly into the aqueous solution, under stirring. At the end, the injected ethanol has to be removed from the system (27).

As dehydrated-rehydrated vesicles are able to hold high amounts of hydrophilic drugs under mild conditions, therefore this method is suitable for the drugs that are losing their activity under harsh conditions (28). Empty liposomes, usually unilamellar vesicles, are disrupted during a freeze drying step in the presence of the drug meant to be encapsulated. A controlled rehydration is obtained in the presence of concentrated solution of the drug. This technique can produce large oligolamellar liposomes of a size around 400 nm to several micrometers. It has been shown that in case of producing smaller liposomes (100-200 nm) sucrose can be added (31).

In the reverse phase evaporation technique which is similar to thin film technique, lipids are dissolved in organic solvent and the solvent is removed by evaporation (29). The thin film is resuspended in diethyl ether followed by the addition of third of water and the suspension is sonicated in a bath sonicator. The emulsion is evaporated until a gel is formed and finally the gel is broken by the addition of water under agitation. The traces of organic solvent should be removed by evaporation (29).

Finally, in the one-step method, lipid dispersion should be hydrated at high temperatures under nitrogen gas stream. This method has the capability to produce liposomes in the range of 200-500 nm (30).

2.3 Targeted delivery by liposomes

The main methods of delivery from liposome to cytoplasm include the exchange of membrane and lipids, contact release, adsorption, fusion and endocytosis. Through these processes, drugs can be released into the bacterial or eukaryotic cells. Liposomal formulations have been used for the delivery of antitumor anthracyclines such as doxorubicin (23) and antifungal agent amphotericin B. Targeted delivery of liposomes to tumor cells has been explored through arsenoliposomes (32). Liposomes for antibacterial chemotherapy are under intensive research to enhance the antibacterial activity and improve pharmacokinetic properties. Advantages of liposomal antibiotics include improved pharmacokinetics, decreased toxicity, enhanced activity against intracellular pathogens, target selectivity and as a tool to overcome bacterial drug resistance (3).

Some liposomes are unique because they can be selectively absorbed by tissues rich in reticuloendothelial cells, such as the liver, spleen and bone marrow. This can serve as a targeting mechanism, but it also removes liposomes from the circulation rather rapidly. Although the poor stability of liposomes, particularly the rapid uptake from the body is not desirable, it could be useful for eradicating the infection by 'passive targeting' through macrophage activation and killing or elimination of parasitic infections.

On the other hand, surface charge and phospholipid composition can affect the interactions of liposomes with bacterial cell surface. For example it has been shown that cationic liposome formulations are more efficient in binding to skin bacterial cells (33).

Moreover, by attaching targeting ligands such as immunoglobulines (34), antibody segments, aptamer (35), peptides and small molecule ligands, and oligosaccharide chains (36), to the surface of the liposomes, they can selectively bind to microorganisms or infected cells and then release the drug payloads to kill or inhibit the growth of the microorganisms (23). The highly specific liposomes are those containing antibodies or immunoglobulin fragments which have affinity to specific receptors on the surface of the infected tissue cells or pathogens (3).

Biofilm surface characteristics have also been used for targeted delivery. Biofilms are microbial aggregations that are covered in an extracellular matrix of polymeric substances. The matrix is usually composed of complex mixture of oligomeric and polymeric molecules such as proteins, lipids and polysaccharides which, as Microbial Associated Molecular Patterns (MAMPs), elicit host defenses (37). Pathogens are much more difficult to control when living in biofilms. This is partly due to the matrix preventing drug transport to the microbial cells. Moreover, bacteria in biofilms grow slower and have reduced metabolic activity, and therefore they are expected to be less susceptible to the antibiotics (38). Currently a great deal of research is focused on exploring new chemotherapeutic targets in biofilms (37). On the other hand liposomes have proven efficient in targeting and eradication of various types of biofilms. Examples are immunoliposomes with high affinity to various oral bacteria including Streptococcus oralis (34) and polysaccharide-coated liposomes for the efficient delivery of metronidazol to periodontal pocket biofilm (39).

pH-sensitive liposomes offer another method for targeting and efficiently delivering the liposomal content into cytoplasm. Such liposomes are stable at physiological pH but undergo destabilization under acidic conditions. Therefore, they are able to promote fusion of target plasma or endosomal membranes, the so called 'fusogenic' properties, at acidic pH (40). Several mechanisms can trigger pH-sensitivity in liposomes. One of the most widely used methods is the use of a combination of phosphatidylethanolamine (PE) or its derivatives with compounds containing an acidic group that act as a stabilizer at neutral pH (41). Other more recent methods include the use of novel pH-sensitive lipids, synthetic fusogenic peptides/proteins (42) and association of pH-sensitive polymers with liposomes (43). pH-sensitive liposomes have found applications in many therapeutic area including the antibiotic delivery to intracellular infections (44).

2.4 Pharmacokinetics consideration of liposomal drug delivery

Liposomal carriers can lead to sustained release of antibiotics during drug circulation in the body. Thus, appropriate levels of drug will be available for a longer duration in comparison with the conventional antibiotic formulations where the outcome is a quick and short effect (45). However, conventional liposomes are quickly opsonized after intravenous administration and therefore they are taken up by the mononuclear phagocyte as foreign antigens. As a consequence blood circulation time is lowered. By controlling the physicochemical properties of the vesicles (size and charge distribution, membrane permeability, tendency for aggregation or fusion, drug encapsulation efficiency, membrane rigidity) and therefore their interaction with the biological environment, many different types of liposomes with the aim of obtaining longer circulation half-lives can be developed (8).

The plasma circulation time of antibiotics can be improved by encapsulation in polyethylene glycol-coated (pegylated) (STEALTH) liposomes. The PEG coating forms a hydration layer that retards the reticuloendothelial system recognitions of liposomes through sterically inhibiting hydrophobic and electrostatic interactions with plasma proteins (46). Other methods that can confer hydrophilicity or steric repulsion are by the use of compounds having sialic residues, or through MLVs containing phospholipids with long saturated chains and negative surface charge (47). The increased half lives of stealth liposomes increase their ability to leave the vascular system into some extravascular regions.

2.5 Antibiotic loaded liposomes against *Salmonella* spp

One of the distinguishing features of liposomes is their lipid bilayer structure, which mimics cell membranes and can readily fuse with the cell membrane and deliver the antibiotic contents into the cellular cytoplasm. As a result, drug delivery may be improved to bacterial and eukaryotic cells alike. By directly fusing with bacterial membranes, the drug payloads of liposomes can be released into the cell membranes or to the interior of the bacteria. In terms of extracellular pathogens, improved antibiotic delivery into the bacterial cells is of particular importance especially since it can interfere with some of the bacterial drug-resistance mechanisms which involve low permeability of the outer membrane or efflux systems (48).

Liposomes are particularly successful in eradicating intracellular pathogens. Examples of these include liposomal formulations of antituberculosis agents isoniazid and rifampin (49), and ampicillin loaded liposomes for eradication of Listeria monocytogenes (50). This is partly due to improved drug retention in the infected tissue and the decreased toxicity as a result of sustained release of drug from liposomes. Moreover, liposomal formulations often have improved antibiotic pharmacokinetics with extended circulation time and prolonged tissue retention.

Liposomal chemotherapeutics for the treatment of salmonellosis may employ some of the conventional antibiotics with proven inhibitory or cidal activity *in vitro*. Bacterial gastro-intestinal infections with *Salmonella typhi* may be treated with chloramphenicol. Alternatives to chloramphenicol include amoxicillin, co-trimoxazole and trimethoprim (51). Recently treatment with cephalosporins and fluoroquinolones has become popular, as several members of these antibiotic families have been shown to be effective. The treatment of paratyphoid fever is the same as that for typhoid (51). *Salmonella* food-poisoning is self-limiting and does not require antibiotic therapy, unless the patient is severly ill or blood cultures indicate systemic infection. In this case, third generation cephalosporins or fluoroquinolones are the most reliable agents (51). Ceftriaxone or a first generation fluoroquinolone such as ciprofloxacin, ofloxacin or pefloxacin but not norfloxacin have been recommended as the first choice in typhoid and paratyphoid by The Sanford Guide to Antimicrobial Therapy (52). The improved efficiency of liposome formulations of antibiotics has been shown *in vitro* and *in vivo*. The *in vitro* infection models utilize macrophages infected with *salmonella*.

2.5.1 Penicillin loaded liposomes

The tissue distribution of ampicillin loaded liposomes was studied in normal noninfected mice and showed that ampicillin concentrated mostly in the liver and spleen (53). The Liposome formulation of ampicillin was significantly more effective than free ampicillin in reducing mortality in acutely infected mice with *Salmonella typhimurium* C5. These liposomes were quite efficient in targeting ampicillin to the spleen but were less effective in targeting ampicillin to the liver and reducing mortality in acute salmonellosis (53).

2.5.2 Cephalosporine loaded liposomes

Third generation cephalosporines have been indicated as suitable candidates for the treatment of *Salmonella* infections (52). Liposome formulations of these antibiotics may improve pharmacokinetics and also the targeted delivery to the intracellular infections. In a study with cephalotin, treatment of infected macrophages with multilamellar liposome-encapsulated cephalothin enhanced the intraphagocytic killing of *Salmonella typhimurium* over that by macrophages treated with free cephalothin (54). Resident murine peritoneal macrophages were shown to be capable of interiorizing the liposome-antibiotic complex leading to a relatively high intracellular concentration of cephalothin. The intracellular killing of the bacteria was maximal at 60 min of incubation; at this time, 60% of the interiorized organisms had been killed (54).

Desiderio & Campbell infected mice with *Salmonella typhimurium* to investigate the effectiveness of liposome-encapsulated cephalothin treatment (55). In the study they also compared the results with formulations containing free cephalothin. They showed that following intravenous administration, liposome-encapsulated cephalothin was cleared from the circulation more rapidly and concentrated in the liver and spleen. Treatment of infected mice with the liposome antibiotic complex was more efficacious in terms of reducing the number of *Salmonella typhimurium* in these organs compared to the injection of free antibiotic, although treatment did not completely eliminate the bacteria from this site (55).

Another study showed that egg phosphatidylcholine liposomes containing cephapirin were relatively stable in serum, and provided acceptable serum levels of cephapirin for 24 hr after i.v. administration while free drug at a similar dosage was undetectable in 3-5 hr. Moreover, the liposome formulation, as opposed to the free drug, could be used successfully for prophylaxis. Cephapirin activity in the spleen and liver was greatly increased and persisted for at least 24 hr when iv injections of the liposome formulation was used. This formulation of liposome, in contrast with the other liposome formulation containing tris salt of cholesterol hemisuccinate, could prolong survival in mice infected with *Salmonella typhimurium* (56).

Ceftiofur sodium is a third generation broad spectrum cephalosporin widely used clinically to treat respiratory diseases and mastitis. Its spectrum also covers *Salmonella* spp. The liposome formulations of ceftiofur were prepared in order to increase drug half life *in vivo* for veterinary purposes (57). The pharmacokinetic study in healthy cows showed that liposome preparations provided therapeutically effective plasma concentrations for a longer duration (elimination half life of more than double) than with the drug alone. These liposomes were stable and the minimum inhibitory concentrations against *Salmonella enteritidis* were 1/4th that of free ceftiofur sodium (57).

2.5.3 Aminoglycoside loaded liposomes

Despite the susceptibility of *Salmonella* spp to aminoglycosides, their use against many important intracellular bacterial infections has been limited due to the cell membrane permeability problems. Lutwyche et al. prepared several liposomal encapsulation formulations including pH-sensitive DOPE-based carrier systems containing gentamicin in order to achieve intracellular antibiotic delivery and therefore increase the drug's therapeutic activity against intracellular pathogens (58). They reported the superiority of some of the pH-sensitive liposomes over conventional liposome formulations, which was associated with the intracellular delivery of the antibiotic and was dependent on endosomal acidification. This liposomal carrier demonstrated pH-sensitive fusion that was dependent on the presence of unsaturated phosphatidylethanolamine (PE) and the pH-sensitive lipid N-succinyldioleoyl-PE. These formulations also efficiently eliminated intracellular infections caused by a recombinant hemolysin-expressing *Salmonella typhimurium* strain which escape the vacuole and reside in the cytoplasm. Moreover, *in vivo* pharmacokinetics and biodistribution tests confirmed that encapsulation of gentamicin in pH-sensitive liposomes significantly increased the concentrations of the drug in plasma compared to those of free gentamicin. Furthermore, liposomal encapsulation increased the levels of accumulation of drug in the infected liver and spleen by 153- and 437-folds, respectively (59).

Other investigations have indicated that even with conventional liposomes, liposome encapsulated gentamicin is less toxic in mice than is free gentamicin and is extremely effective-therapy for disseminated *Salmonella* infections in mice. For example when gentamicin sulfate was encapsulated in liposomes composed solely of egg phosphatidylcholine, the mean half-lives of the encapsulated drug in serum were around four times that of free (nonencapsulated) gentamicin in mice and rats following i.v. administration. Moreover, liposome encapsulation led to higher and more prolonged activity in organs rich in reticuloendothelial cells especially in spleen and liver. In acute septicemia infections in mice, the liposomal formulation showed enhanced prophylactic activity when compared with the free drug. In a model of murine salmonellosis, liposomal gentamicin greatly enhanced the survival rate (60). Similarly, a single iv injection of low dose gentamicin loaded multilamellar liposomes (composed of egg phosphatidylcholine, egg phosphatidylglycerol, cholesterol and alpha-tocopherol) resulted in 80% survival of mice infected with *Salmonella Dublin,* while zero survival was observed when treated with the same amount of free gentamicin. Higher concentrations of free gentamicin led to neuromuscular paralysis, while the slow release of this dose from liposomes increased the survival rate to 100%. After the single dose treatments with liposomes, high concentrations of the drug were detectable for 10 days (61). The liposome-encapsulated gentamicin has also been proven successful in the treatment of Mycobacterium Avium-M intracellular complex (MAC) bacteremia in AIDS patients. In this case, MAC colony counts in blood fell by 75% or more when given intravenously twice weekly for 4 weeks (62).

Another effective antibiotic for liposomal formulation which attracted the interest of researchers is streptomycin. Conventional liposomal formulation of streptomycin made with egg yolk phosphatidylcholine was investigated using *in vivo* model of *Salmonella* infection in mice. Liposome-entrapped streptomycin prolonged the survival to more than 15 days for all mice infected with the virulent strain of *Salmonella enteritidis*, while treatment with the same dose of free streptomycin resulted in all of the mice dying between days 5 and 7. The prolongation of survival was due to suppression of the multiplication of *S. enteritidis*. Furthermore, the liposome-entrapped drug was less toxic than the free drug when applied at high doses. A tissue distribution study in various organs demonstrated that liposomal streptomycin was selectively accumulated in the spleen and liver with concentrations in these organs about 100 times higher than those in mice receiving the free drug (63).

In contrast to this, another investigation on *S. enteritidis* indicated a less concentration of streptomycin administered using some of the liposome formulations in the liver and spleen in comparison with the free drug (9). In this study, several formulations of streptomycin sulfate liposomes, prepared from a mixture of L-a-dipalmitoy phosphatidyl choline (DPPC) and cholesterol with or without a charge inducing agent, were used in drug targeting experiments using Swiss mice. The biodistribution results indicated that although, in comparison with the free drug, some of the liposome formulations exhibited 2-3 times higher concentration of streptomycin in the liver and spleen, this effect decreased over time from one to seven days. Despite this, the survival rate experiments indicated a definite protection against *Salmonella enteritidis* exhibited by the liposome-encapsulated streptomycin compared to the free drug (64). Therefore, it seems that the liposome formulation plays the major role in the targeting effect and the delivery efficiency of the liposomes for intracellular infections.

2.5.4 Fluoroquinolone loaded liposomes

Ciprofloxacin is a synthetic bactericidal fluoroquinolone which inhibits the activity of bacterial DNA gyrase, resulting in the degradation of bacterial DNA by exonuclease activity. Consequently, ciprofloxacin has broad-spectrum efficacy against a wide variety of bacteria, including the family *Enterobacteriaceae* of which *Salmonella* spp is a member of (65). It has been used in the treatment of individuals with *Salmonella* infections, including those with typhoid fever and chronic typhoid carriers (52). Despite the enormous success with ciprofloxacin, there are some factors which limit the drug's clinical utility, such as its poor solubility at physiological pH and rapid renal clearance. Several investigations have focused on the formulation of this drug as liposomes, in order to improve the drug delivery.

Ciprofloxacin loaded liposomes, consisting of dipalmitoyl-phosphatidylcholine, dipalmitoyl-phosphatidylglycerol and cholesterol, were used to treat *Salmonella Dublin* infected mice (66). It has been reported that a single injection of liposome formulation was 10 times more effective than a single injection of free drug at preventing mortality. Treatment with liposomal ciprofloxacin produced dose-dependent decreases in bacterial counts in spleen, stool, and Peyer's patches, indicating that the drug had distributed to all areas of inflammation, not just to the major reticuloendothelial system organs. Although liposome formulation was cleared rapidly from the blood, drug persisted in the liver and spleen for at least 48 h after administration of a dose (66).

In a similar study, Webb et al. encapsulated ciprofloxacin into large unilamellar liposomes. The LUVs composed of dipalmitoylphosphatidylcholine-cholesterol, distearoylpho-sphatidylcholine-cholesterol, or sphingomyelin-cholesterol. In comparison with the free drug, the liposomal formulations increased the circulation lifetime of the drug by >15 fold and resulted in 10^3 to 10^4 fold fewer viable *Salmonella typhimurium* in the livers and spleens after intravenous administration (67). These results show the utility of liposomal encapsulation in improving the pharmacokinetics, biodistribution, and antibacterial efficacy of ciprofloxacin.

3. Polymeric nanoparticles for antisalmonellosis drug delivery

3.1 Introduction

Nanoparticles (NP) are solid colloidal particles with particle sizes smaller than 1000 nm. However, most nanoparticles utilized in drug delivery are in the size range of 100–200 nm. Nanoparticles can be classified into two main subgroups: nanospheres and nanocapsules. Nanospheres have a matrix-type structure, and drug molecules can be adsorbed on their surface or entrapped inside their matrix. Nanocapsules have a capsule-like structure and possess the capability of encapsulating the drug molecules inside the capsule or adsorbed to them externally. Because these systems have unique characteristics, such as very small particle size, high surface area, and possibility of surface modification, they have been attracting much interest for drug-delivery purposes during recent years. Nanoparticles are able to adsorb and/or encapsulate a drug, thus protecting it against chemical and enzymatic degradation. Generally, the drug is dissolved, entrapped, encapsulated or attached to a NP matrix and depending upon the method of preparation, nanoparticles, nanospheres or nanocapsules can be obtained. Owing to their polymeric nature, nanoparticles (Figure 3) may be more stable than liposomes in biological fluids and during storage.

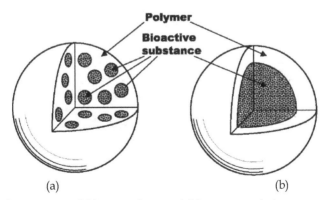

(a)   (b)

Fig. 3. Schematic structures of (a) nanosphere and (b) nanocapsule type nanoparticles (the picture was taken from http://what-when-how.com/nanoscience-and-nanotechnology/nanoencapsulation-of-bioactive-substances-part-1-nanotechnology).

Nanocapsules are vesicular systems in which the drug is confined to a cavity surrounded by a unique polymer membrane, while nanospheres are matrix systems in which the drug is physically and uniformly dispersed. In order for nanoparticles to minimize the side effects, the polymers associated with nanoparticles must be degraded *in vivo* due to intracellular polymeric overloading. Thus in recent years, biodegradable polymeric nanoparticles have attracted considerable attention as potential drug delivery devices in view of their applications in the controlled release of drugs, their ability to target particular organs, as carriers for DNA in gene therapy, and their ability to deliver proteins, peptides and genes through a peroral route of administration (68). Most polymers PLGA, chitosan, gelatin, alginate, and poly cyanoacrylate can be used in the formulation of nanoparticles.

It is believed that nanoparticles could be effective in increasing drug accumulation at the site of infection with reduced toxicity and side effects after parenteral or oral administration (69, 70). Polymeric nanoparticles have been explored to deliver a variety of antimicrobial agents to treat various infectious diseases and have shown great therapeutic efficacy (71).

3.2 Antibiotic loaded cyanoacrylate nanoparticles

The polymers involved in nanoparticle structure should be degraded in order to release the drug, therefore, there should be a direct correlation between the rate of degradation and the drug release rate. If degradation happens in the presence of esterase, it was shown that the degradation of the polymer in esterase-free medium is low, therefore, the drug release rate is low accordingly. The drug release was increased when the medium contained carboxyesterase (72).

The *in vitro* interaction between [3H]ampicillin-loaded polyisohexylcyanoacrylate nanoparticles and murine macrophages infected with *Salmonella typhimurium* was investigated and the results showed that the uptake of nanoparticle-bound [3H]ampicillin by non-infected macrophages was six- and 24-fold greater respectively compared to free [3H]ampicillin. However, there was no difference between nanoparticle-bound ampicillin and free ampicillin in terms of bactericidal activity against intracellular *Salmonella typhimurium*. This unexpected observation might be accounted for by bacterium-induced inhibition of phagosome-lyosome fusion within the macrophages, thereby preventing contact between the bacteria in the phagosomes and the nanoparticles in the secondary lysosomes (73).

In another study the intracellular distribution of (3H)ampicillin-loaded polyisohexylcyanoacrylate nanoparticles in the same cells using ultrastructural autoradiography was investigated by the same authors (74). Ampicillin penetration and retention into the cells obviously increased by means of nanoparticles. After 2-4 h treatment with the nanoparticle formulation, numerous intracellular bacteria were seen to be in the process of destruction. After 12 h treatment, numerous spherical bodies and larger forms were seen in the vacuoles and it was an indication of marked damaging action of the ampicillin on the bacterial walls. The targeting of ampicillin therefore allowed its penetration into the macrophages and vacuoles infected with *Salmonella typhimurium* (74).

Pinto-Alphandary et al. used transmission electron microscopy to prove that ampicillin which usually penetrates into cells at a low level is directly carried in when loaded on nanoparticles, and brought into contact with intracellular bacteria (75). They concluded that ampicillin loaded polyisohexylcyanoacrylate nanoparticles is an ideal formulation when an intracellular targeting for ampicillin is needed.

Page-Clisson et al. (76) investigated the antibacterial efficiency of polyalkylcyanoacrylate nanoparticles loaded with ciprofloxacin and ampicillin against *Salmonella typhimurium*. It was shown that *in vivo* treatment with ciprofloxacin led to a significant decrease of bacterial counts in the liver whatever the stage of infection and the form used. However, none of the treatments were able to sterilize the spleen or the liver (76).

Ampicillin was also attached to nanoparticles of polyisohexylcyanoacrylate (PIHCA) for the treatment of C57BL/6 mice experimentally infected with *Salmonella typhimurium* C5. The injection of the nanoparticles containing ampicillin treated all mice, whereas by the injection of non-loaded nanoparticles all mice died within 10 days (77).

3.3 Antibiotic loaded PLGA nanoparticles

Some polymeric nanoparticles may be more effective than liposomes in acute salmonellosis model due to better stability of nanoparticles in serum compared to liposomes. Therefore it is believed that antibiotic loaded nanoparticles can improve the targeting, particularly in the case of intracellular bacteria. For example, gentammicin (78), azithromycin and clarithromycin loaded nanoparticles using poly(lactide-co-glycolide) [PLGA] (79, 80) were more effective than corresponding intact drug against *Salmonella typhimurium*.

As mentioned before, nanoparticles should be degraded *in vivo* to avoid side effects and it has been shown that PLGA nanoparticles fulfill such requirements. Therefore, in most cases for antibiotics such as rifampcin (81), amphotericin (82), azithromycin (79) and clarithromycin (80) PLGA nanoparticle preparations have been recomended.

Mohammadi *et al.*, showed that azithromycin and clarithromycin-loaded (PLGA) nanoparticles (NPs) prepared with three different ratios of drug to polymer have better antibacterial activity against *Salmonella typhi* (79). In other words, the nanoparticles were more effective than pure azithromycin and clarithromycin against *Salmonella typhi* and *S. aureus*, respectively, with the nanoparticles showing equal antibacterial effect at 1/8 concentration of the intact drug. Both studies on azithromycin and clarithromycin proved that the antibacterial activity of nanoparticles were about 8-fold more than the free azithromycine and clarithromycin (Figure 4). The higher antibacterial effect of clarithromycin and azithromycin may have resulted from higher bacterial adhesion of the nanoparticles. For example, an adhesion of Eudragit nanoparticles containing PLGA to the *S. aureus* bacteria was reported (83). Although, Figure 4 shows that the ratio of drug:PLGA has no significant effect on antibacterial activity of azithromycin and clarithromycin, Table 1 shows that the particle size of nanoparticles, their zeta potential and the encapsulation efficiency are remarkably dependent on the ratio of drug:polymer used in the formulations. This indicates that by controlling the ratio of drug:carrier the desirable particle size and zeta potential could be achieved. As it is shown in Figure 5 all nanoparticles were spherical in appearance.

Several investigations have shown that nanoparticles could not be very effective on all different types of bacteria and that the antibacterial effect depends on bacterial type (84). For example, recently Martins et al. evaluated the antibacterial activity of PLGA nanoparticles containing violacein against different bacteria (84). Although, they showed that the MIC with nanoparticles is 2-5 times lower than free violacein against *Staphylococcus aureus*, the results failed to show any significant activity against *Escherichia coli* and *Salmonella enterica*.

Formulations	Encapsulation efficiency (%)	Mean particle size (nm)	Zeta potential (mV)
AZI:PLGA (1:1)	50.5 ± 3.4	252 ± 5	-5.6 ± 2.15
AZI:PLGA (1:2)	66.8 ± 2.8	230 ± 7	-11.10 ± 1.87
AZI:PLGA (1:3)	78.5 ± 4.2	212 ± 4	-15.56 ± 2.53
CLR:PLGA (1:1)	57.4 ± 4.3	280 ± 15	-6.3 ± 1.70
CLR:PLGA (1:2)	72.9 ± 3.2	223 ± 12	-10.08 ± 1.63
CLR:PLGA (1:3)	80.2 ± 4.0	189 ± 10	-14.26 ± 1.92

Table 1. Encapsulation efficiency, mean particle size and zeta potential of various formulations containing Azithromycine and clarithromycin (data taken from references 79, 80)

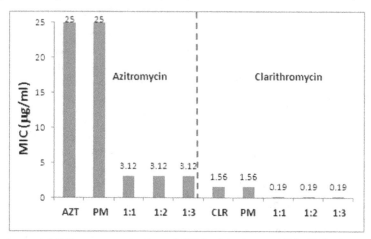

Fig. 4. Minimum inhibitory concentrations (MICs) of the intact AZT, CLR, physical mixtures (PM) and drug-loaded nanoparticles suspensions with different drug:PLGA ratios (data are reproduced from references 79, 80).

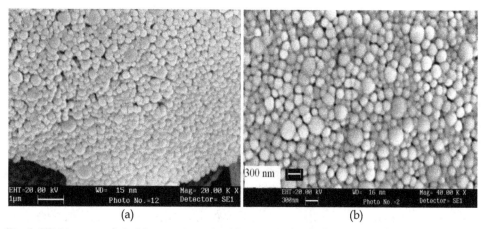

Fig. 5. SEM images of clarithromycin and azithromycin-loaded nanoparticles with the ratio of drug:PLGA 1:2 (SEM taken from ref. 80).

3.4 High loading antibiotic nanoparticles

One of the problems with antibiotic loaded nanoparticles is that in some cases the capacity of a polymeric drug carrier should be engineered to incorporate high concentrations of antibiotics to achieve the required dosage, yet avoid side effects that may be associated with higher amounts of carriers. This seems a difficult task, however, Ranjan *et al* introduced two novel technologies by which high concentrations of gentamicin could be incorporated into the formulations (85).

In the first technology, Ranjan *et al.*, made an attempt to enhance antibacterial efficacy of gentamicin using a new technology called core-shell nanostructures (78). In this research pluronic based core-shell nanostructures encapsulating gentamicin were prepared. The maximum antibiotic loading was 20% in their formulation with a zeta potential of -0.7. It was shown that when using core-shell nanostructures containing gentamicin, not only that significant reduction in toxicity and side effects was evident, but also the percentage of viable bacteria in the liver and spleen was significantly reduced (78).

In the second technology, Ranjan et al (85) incorporated gentamicin into macromolecular complexes with anionic homo- and block-copolymers via cooperative electrostatic interactions between cationic drugs and anionic polymers (Figure 6). They showed the possibility of incorporating 26% by weight of gentamicin in the nanoplexes with average diameter of 120 nm and zeta potential of -17 (85). This was 6% more drug loading compared to their previous study. Their study showed that in addition to the high loading of drug carried by these polymeric nanoplexes, the nanoplexes can potentially improve targeting of interacellular pathogens such as *salmonella*.

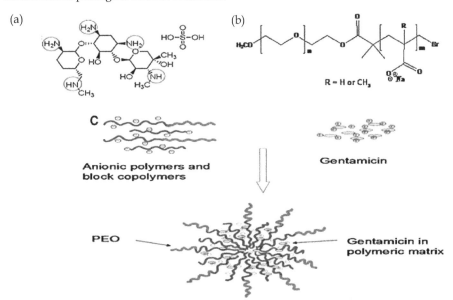

Fig. 6. (a) Gentamicin is cationic aminoglycoside antibiotic with five amino groups, (b) anionic block copolymers for electrostatic complexation to gentamicin, (c) strategy to incorporate gentamicin within polymeric nanoplex (Figure was taken from ref 85).

3.5 Xerogel systems containing antibiotic

During the last fifteen years, a special attention has been dedicated on silica xerogel system to treat diseases due to intracellular pathogens (86-92). The properties of silica xerogel systems such as size, zeta potential, pore structure, and the surface characteristics make them suitable carriers for therapeutics to target the replicative niche of intracellular pathogen. These are ideal systems for the delivery of gentamicin as this antibiotic does not kill intracellular *Salmonella* due to the polar nature of the drug which is associated with low level of intracellular penetration. A study showed that when gentamicin was incorporated into silica xerogel formulations, 31% of the drug entrapped in the matrix system remained biologically active and the bactericidal effect was retained after drug release. The results showed that by incorporation of PEG the drug release can be modulated. Administration of two doses of the xerogel formulations showed a remarkable reduction in the load of *Salmonella entrica* in the spleen and liver of the infected mice (86). A similar study was performed by another group on gentamicin silica xerogel systems showing that the silica xerogel was more effective in clearing the infection in the liver compared to the same dose of the free drugs (87).

3.6 Vaccine delivery by polymeric nanoparticles

Ochoa et al (93) made an attempt to use nanoparticle for the delivery of vaccines. An immunogenic subcellular extract obtained from whole *Salmonella Enteritidis* cells (HE) was encapsulated in nanoparticles made with the polymer Gantrez (HE-NP). When they studied the immunogenicity and protection of HE-loaded nanoparticles against lethal *Salmonella Enteritidis* in mice, an increase in survival was observed compared to a control group (80% of the mice immunized with the HE-loaded nanoparticle formulation survived even when administered 49 days before the lethal challenge). They noticed that the cytokines released from *in vitro*-stimulated spleens showed a strong gamma interferon response in all immunized groups at day 10 post-immunization. However, the immunity induced by HE-loaded nanoparticles at day 49 post-immunization suggests the involvement of a TH2 subclass in the protective effect. It can be concluded from their study that, HE-nanoparticles may represent an important alternative to the conventional attenuated vaccines against *Salmonella Enteritidis* (93).

4. Metal nanoparticles as antisalmonellosis agents

In the fast-developing field of nanotechnology, metal nanoparticles are of great interest due to their multiple applications as chemical catalysts, adsorbents, biological stains, and building blocks of novel nanometer scale optical, electronic, and magnetic devices. Metal nanoparticles are pure metal nano sized material (Figure 7) with the size of usually up to 200 nm. They have been suggested to be suitable for biological applications. It was shown that if the size of these nanoparticles is less than 50 nm they are the most suitable particles as therapeutic agents as the biosystem fails to detect them (94).

Different types of nanometals including copper, magnesium, zinc, titanium, gold, and silver have been investigated but silver nanoparticles have been employed and investigated most extensively compared to the other metals since ancient times to fight infections and control spoilage (95-97). A large number of successful *in vitro* studies were performed for the evaluation of the antisalmonella effect of metal nanoparticles. These nanoparticles are usually nonspecific and are broad spectrum antibacterial. It is also reported that silver can cause argyrosis and argyria and is toxic to mammalian cells (98). As silver attacks a broad range of targets in the microbes, therefore it is difficult for microbes to develop resistance against silver (99). This property of silver makes it an excellent candidate for antimicrobial effect.

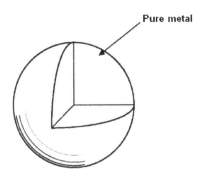

Fig. 7. Schematic structure of a metal nanoparticle

In terms of production, it is suggested that monodispersed particles (very narrow particle size distribution) rather than polydispersed nanoparticles (broad particle size distribution) are preferred. This is because the former distribution is believed to be more effective against microbes due to the high surface/volume fraction so that a large proportion of silver atoms can be in direct contact with their environment (100).

Recently, the potential use of silver nanoparticles on pathogenic bacteria was reviewed (101). There are various physical, chemical or biological methods which can be used to produce metallic nanoparticles. Among these, it seems, the biological method is popular due to the reliability and being eco-friendly. This method has attracted the attention of researchers in the field (102-108). In fact, a number of different species of bacteria and fungi are able to reduce metal ions producing metallic nanoparticles with antimicrobial properties. Recently, it has been shown that silver nanoparticles produced by the fungus *F. acuminatum* have efficient antibacterial activity against multidrug resistant and highly pathogenic, *Salmonella typhi* (109). Additionally, plant extracts can also be used to obtain metallic nanoparticles (110). Metal nanoparticles were also modified to be used in the prevention of biofilm formation on the implanted devices (111-114), however, care must be taken when this type of metal nanoparticles are used due to potential risk on patient's health (115-117).

Researchers suggested that to achieve a better utilization of the antimicrobial activity, metal nanoparticles may be combined with nontoxic and biocompatible polymers. For example, in an attempt NaPGA- (poly (g-glutamic acid)) and CaPGA-coated magnetite nanoparticles were synthesized (118) and their antibacterial activity against *Salmonella enteritidis, Staphylococcus aureous* and *Eschercia coli* were tested. The results showed that both produced nanoparticles were more effective against *Salmonella enteridis* compared to commercial antibiotics, linezolid and cefaclor. In addition, these nanoparticles showed no toxicity toward human skin fibroblast cells.

In few cases polymers such as PVP have been used as steric stabilizers to obtain monodispersed silver nanoparticles (119, 120). Although silver nanoparticles have the capability to remain dispersed in liquids without major signs of agglomeration, in case of the appearance of aggregation hydrophilic surfactants, proteins, amino acids and PVA (poly vinyl alcohol) can be used (121-125). Metal nanoparticles have also found application in various other fields, i.e. catalysis and sensors as mentioned before (126-128). However, their undesirable and unforeseen effects on the environment and in the ecosystem should not be ignored (129, 130). The antibacterial effect of silver and copper nanoparticles was also investigated on *Escherichia coli, Bacillus subtilis* and *Staphylococcus aureus* (131). The results showed that the efficiency of silver and copper nanoparticles were different on different bacteria. Among the bacteria used, *B. subtilis* showed the highest sensitivity to copper nanoparticles compared to silver, whereas silver nanoparticles were more effective on the other two bacteria compared to copper nanoparticels (133).

Interesting results were reported by Patil *et al* when they synthesized and tested chloramphenicol loaded nano-silver particles against *Salmonella typhi* (97). For the first time they used PVP in their formulations containing silver as a carrier for chloramphenicol. In the formulation, PVP played a dual role. It acts as a stabilizer and linker for binding chloramphenicol to the silver nanoparticles (Figure 8). The nanoparticles showed considerably enhanced activity against clinically isolated *Salmonella typhi*.

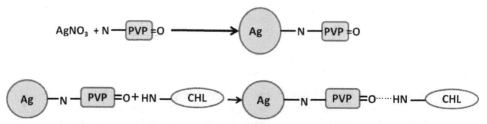

Fig. 8. Top: schematic representation of the synthesis of silver nanoparticles (PVP as a stabilizer); bottom: schematic representation of the synthesis of chloramphenicol loaded silver nanoparticles (PVP as a linker) (figure was reproduced from ref 97).

The summary of some of metal nanoformulations are listed in Table 2.

Gold and platinum nanoparticles have also attracted the attention of researchers due to their antibacterial activity (132, 133). Several research groups studied the cytotoxicity of gold nanoparticles in different cell types (134, 135). It was shown that citrate-capped gold nanoparticles were not cytotoxic to baby hamster kidney cells and human hepatocellular liver carcinoma cells, but cytotoxic to human carcinoma cells at certain concentrations (135).

Despite all research data about the toxicity of gold nanoparticles, still more research for better understanding of gold nanoparticles toxicity is reqired. Recently, Wang *et al* prepared 16 nm gold nanospheres stabilized with citrate ions and their antimicrobial activity was tested against *Salmonella typhi* bacteria strain TA 102 (133). The results showed that gold nanoparticles are not mutagenic or toxic in *Salmonella*, but is photomutagenic to the bacteria. The photomutogenicity was due to the presence of citrate and Au^{3+} ions used during the preparation of gold nanoparticles. Their final results showed that although there was a good surface interaction between gold nanoparticles and the bacteria, the gold nanoparticles were not able to penetrate into the bacteria.

Type of nanoparticle	Type of *Salmonella*	Reference
ASAP Nano-silver Solution	*Salmonella typhi*	(136)
silver colloid nanoparticles	*Salmonella enteric*	(137)
silver–silicon dioxide hybrid	*Salmonella enteric*	(137)
ZnO nanoparticles	*Salmonella typhimurium*	(138)
Spherical silver nanoparticles	*Salmonella typhimurium*	(139)
Zinc oxide QuantumDots	*SalmonellaEnteritidis*	(140)
Silver nanoparticles	*Salmonella typhi*	(141)
Silver nanoparticles	*Salmonella typhimurium*	(142)
Silver bionanoparticles	*Salmonella typhi*	(143)
Silver bionanoparticles	*Salmonella paratyphi*	(144)
TiO2 nanoparticles	*Salmonella typhimurium*	(145)
ZnO nanoparticles	*Salmonella typhimurium*	(145)
Silver nanoparticles	*Salmonella typhus*	(146)
Iron nanoparticles	*Salmonella paratyphi*	(147)
silver nanoparticles	Not specified	(148)
Silver Nanoparticles	*Salmonella typhimurium*	(149)
Silver bionanoparticles	*Salmonella typhi*	(150)
Ag–SiO2 anoparticles	*Salmonella typhimurium*	(151)
$Zn_{1-x}Ti_xO$ (x = 0, 0.01, 0.03 and 0.05) nanoparticles	*Salmonella typhi*	(152)
platinum nanoparticles	*Salmonella Enteritidis*	(153)
CuO nanoparticles	*Salmonella paratyphi*	(154)

Table 2. Various metal naoparticles used against different microbes

Similar study was carried out on gold and platinum nanoparticles (132) and the results showed that gold nanoparticles can interact with *Salmonella Enteritidis* but did not penetrate the bacterial cell, whereas platinum nanoparticles were observed inside bacterial cells due to binding to DNA. They concluded that gold nanoparticles can be used alongside with bacteria to deliver the nanoparticles to specific points in the body for targeted delivery.

A major controversy with metal nanoparticles is that whether they are toxic to bacteria or bacteria develops resistance mechanism against these nanoparticles. If the former is true, there might be a devastating effect to the ecosystem which will lead to a global destabilization. Nanoparticles have a greater potential to travel through an organism and could be more toxic due to their larger surface area and specific structural/chemical properties.

Although the evolution of nanotechnology is about to bring various advantages to our lives over conventional formulations but the lung toxicity of metal nanoparticles (155)

should be carefully considered as these nanoparticles are very small and light, and they have larger surface area with a greater potential to travel through an organism or tissues (156). These small particles can travel via nasal nerves to the brain (156, 157). It has been shown that most of metallic nanoparticles such as TiO_2, Ag, Al, Zn, Ni exhibit cellular toxicity on human alveolar epithelial cells (158). The results reported by Park et al (158) showed that these metal nanoparticles could damage the cell directly or indirectly. The cell damage is probably dependent on the size, structure, and composition of the nanoparticles, yet more studies are needed for better understanding of the toxicity mechanism of the metal nanoparticles.

5. References

[1] Brands D. *Salmonella*, Deadly disease and epidemics. Philadelphia: Chelsea House publication. 2005, 16.

[2] Gordon MA. *Salmonella* infections in immunocompromised adults. J. Infection, 2008, 56: 413-422.

[3] Z. Drulis-Kawa, A. Dorotkiewicz-Jach, Liposomes as delivery systems for antibiotics, Int. J. Pharm., 2010, 387: 187-198.

[4] Davin-Regli A, Bolla JM, James CE, Lavigne JP, Chevalier J, Garnotel E, Molitor A, Pagès JM. Membrane Permeability and Regulation of Drug "Influx and Efflux" in Enterobacterial Pathogens, Current Drug Targets, 2008, 9:750-759.

[5] Pavan B, Dalpiaz A, Ciliberti N, Biondi C, Manfredini S, Vertuani S. Progress in drug delivery to the central nervous system by the prodrug approach. Molecules, 2008, 13: 1035-1065.

[6] Patel M, Shah T, Amin A. Therapeutic opportunities in colon-specific drug-delivery systems. Critical Reviews In Therapeutic Drug Carrier Systems, 2007, 24: 147-202.

[7] Peck RW. Driving earlier clinical attrition: if you want to find the needle, burn down the haystack. Considerations for biomarker development. Drug Discovery Today, 2007, 12: 289 – 294.

[8] Couvreur P, Fattal E, Andremont A. Liposomes and nanoparticles in the treatment of intracellular bacterial infections. Pharm. Res. 1991;8:1079–86.

[9] Ghose T, Cerini M. Radiosensitisation of Ehrlich ascites tumour ceUs by a specific antibody.Nature, 1969, 263: 993-995.

[10] Tomlinson E, In: Tomlinson E, Davis S5 (eds), site-spedfic drug delivery, cell biology, medical and pharmaceutical aspects, John Wiley, London, UK, (Patho) physiology and the temporal and spatial aspects of drug delivery. 1986, pp 1-26.

[11] Pouton ON, Drug targeting--current aspects and future prospects. J. Clin. Hosp. Pharm. 1985, 10:45-58.

[12] Langer R. Drug delivery and targeting. Nature 1998, 392:5-10.

[13] Duncan R, Drug targeting: where are we now and where are we going? J Drug Target 1997, 5:1-4.

[14] Allen TM. Liposomal drug formulations. Rationale for development and what we can expect for the future. Drugs 1998, 56:747-756.

[15] Jones M, Leroux J. Polymeric micelles - a new generation of colloidal drug carriers. Eur. J. Pharm. Biopharm. 1999, 48:101-111.

[16] Tomlinson E. Theory and practice of Site-specific drug delivery. Adv. Drug Del. Rev. 1987, 1:187-198.

[17] Fujimoto K, Yamamura K Osada T, Hayashi T, Nabeshima T, Matsushita M, Nishikimi N, Sakurai T, Nimura Y. Subcutaneous tissue distribution of vancomycin from a fibrin glue/Dacron graft carrier. J. Biomed. Mater. Res. 1997, 36:564-567.

[18] Hamouda T, Hayes MM, Cao Z, Tonda R, Johnson K, Wright DC, Brisker J, Baker JR Jr, A novel surfactant nanoemulsion with broad-spectrum sporicidal activity against BaCIllus species. J. Infect. Dis. 1999, 180:1939-1949.

[19] Lichtenstein A, Margallt R. Uposome-encapsulated silver sulfadiazine (SSD) for the topical treatment of infected burns: thermodynamics of drug encapsulation and kinetics of drug release. J. Inorg. Biochem. 1995, 60:187-198.

[20] Peters K, Leitzke S, Diederichs JE, Borner K, Hahn H, Mulier RH, Ehlers. Preparation of a clofazimine nanosuspension for intravenous use and evaluation of its therapeutic efficacy in murine Mycobacterium avium infection. Antimicrob. Chemother. 2000, 45:77-83.

[21] Couvreur P, Fattal E, Andremont A. Uposomes and nanoparticles in the treatment of intracellular bacterial infections. Pharm. Res. 1991, 8:1079-1086.

[22] Allen, Jr LV, Popovich NG, Ansel HC. Ansel's pharmaceutical dosage forms and drug delivery systems 8th edition. Lippincott Williams & Wilkins, 2005: 665.

[23] Zhang L, Pornpattananangkul D, Hu CMJ, Huang CM. Development of nanoparticles for antimicrobial drug delivery. Current Medicinal Chemistry. 2010, 17: 585-594.

[24] Bangham AD, Standish MM, Watkins JC. Diffusion of univalent ions across the lamellae of swollen phospholipids. J. Mol. Biol. 1965, 13: 238-252.

[25] Sharma A, Sharma US, Liposomes in drug delivery: progress and limitations. Int. J. Pharm. 1997, 154: 123-140.

[26] Olson F, Hunt CA, Szoka FC, Vail WJ, Papahadjopoulos D. Preparation of liposomes of defined size distribution by extrusion through polycarbonate membrane. Biochim. Biophys. Acta, 1979, 557:9-23.

[27] Batzri S, Korn ED. Single bilayer lipsomes prepared without sonication. Biochim. Biophys. Acta, 1973, 298:1015-1019.

[28] Seltzer SE, Gregoriadis G, Dick R. Evaluation of the dehydration-rehydration method for production of contrast-carruing liposomes. Invest. Radiol. 1988, 23:131-138.

[29] Ugwu S, Zhang A, Parmar M, Miller B, Sardone T, Peikov V et al., Preparation, charcaterization, and stability of liposome-based formulations of mitoxantrone. Drug Dev. Ind. Pharm. 2005, 31:223-229.

[30] Szoka FJ, Papahadjopoulos D, Procedure of preparation of liposomes with large internal aqueous space and high capture by reverse-phase evaporation. Proc. Nat. Acad. Sci. USA, 1978, 75:44194-4198.

[31] Talsma H, Van Steenbergen MJ, Borchert JC, Crommelin DJ, A novel technique for one-step preparation of liposomes and non-ionic surfactant vesicles without the use of organic solvents. Liposome formation in a continuous gas stream: the bubble method. J. Pharm. Sci. 1994, 83:276-280.

[32] Zagana P, Haikou M, Klepetsanis P, Giannopoulou E, Ioannou PV, Antimisiaris SG. In vivo distribution of arsonoliposomes: Effect of vesicle lipid composition Original Research Article Int. J. Pharm. 2008, 347:86-92.

[33] Robinson AM, Bannister M, Creeth JE. The interaction of phospholipid liposomes with mixed bacterial biofilms and their use in the delivery of bactericide, Colloids Surf. A Physicochem. Eng. Asp., 2001, 186, 43-53.

[34] Robinson AM, Creeth JE, Jones MN. The specificity and affinity of immunoliposome targeting to oral bacteria. Biochim. Biophys. Acta, 1998, 1369, 278-286.

[35] Kang HZ, O'Donoghue MB, Liu HB, Tan WH. A liposome-based nanostructure for aptamer directed delivery. Chem. Communic. 2010, 46, 249-251.

[36] Zhu JM, Yan F, Guo ZW, Marchant RE. Surface modification of liposomes by saccharides: Vesicle size and stability of lactosyl liposomes studied by photon correlation spectroscopy. J. Colloid Interface Sci. 2005, 289, 542-550.

[37] Estrela AB, Heck MG, Abraham WR. Novel Approaches to Control Biofilm Infections. Curr. Med. Chem. 2009, 16, 1512-1530.

[38] Payne DJ. Microbiology. Desperately seeking new antibiotics, Science, 2008, 321, 1644-1645.

[39] Vyas SP, Sihorkar V, Dubey PK. Preparation, characterization and *in vitro* anti-microbial activity of metronidazole bearing lectinized liposomes for intraperiodontal pocket delivery. Pharmazie, 2001, 56, 554-560.

[40] Simoes S, Moreira JN, Fonseca C, Duzgunes N, de Lima MCP. On the formulation of pH-sensitive liposomes with long circulation times. Adv. Drug Del. Rev., 2004, 56, 947- 965.

[41] Düzgüneş N, Straubinger RM, Baldwin PA, Papahadjopoulos D. pH-sensitive liposomes: introduction of foreign substances into cells, in: J. Wilschut, D. Hoekstra (Eds.), Membrane Fusion, Marcel Dekker, New York, 1991, pp. 713-730.

[42] Mastrobattista E, Koning GA, Van Bloois L, Filipe AC, Jiskoot W, Storm G. Functional characterization of an endosome-disruptive peptide and its application in cytosolic delivery of immunoliposome-entrapped proteins, J. Biol. Chem., 2002, 277, 27135- 27143.

[43] Roux E, Francis M, Winnik FM, Leroux JC. Polymer based pH-sensitive carriers as a means to improve the cytoplasmic delivery of drugs. Int. J. Pharm. 2002, 242: 25- 36.

[44] Cordeiro C, Wiseman DJ, Lutwyche P, Uh M, Finlay JC, Finlay BB, Webb MS. Antibacterial efficacy of gentamicin encapsulated in pH-sensitive liposomes against an in vivo *Salmonella enterica serovar typhimurium* intracellular infection model. Antimicrob. Agents Chemother., 2000, 44, 533-539.

[45] Hamidi M, Azadi A, Rafiei P. Pharmacokinetic consequences of pegylation. Drug Deliv. 2006, 13: 399-409.

[46] Ceh B, Winterhalter M, Frederik PM, Vallner JJ, Lasic DD. Stealth ® liposomes from theory to product. Adv. Drug Deliv. Rev. 1997, 24: 165-177.

[47] Immordino ML, Dosio F, Cattel L. Stealth liposomes: review of the basic science, rationale, and clinical applications, existing and potential. Int. J. Nanomedicine, 2006, 1: 297-315.

[48] Mugabe C, Halwani M, Azghani AO, Lafrenie RM, Omri A. Mechanism of enhanced activity of liposome-entrapped aminoglycosides against resistant strains of Pseudomonas aeruginosa. Antimicrob. Agents Chemother. 2006, 50: 2016-2022.

[49] Labana S, Pandey R, Sharma S, Khuller GK. Chemotherapeutic activity against murine tuberculosis of once weekly administered drugs (isoniazid and rifampicin) encapsulated in liposomes. Int. J. Antimicrob. Agents, 2002, 20: 301-304.

[50] Bakker-Woudenberg IAJM, Lokerse AF, Vink-van den Berg JC, Roerdink FH, Michel MF. Effect of liposome-entrapped ampicillin on survival of Listeria monocytogenes in murine peritoneal macrophages, Antimicrob. Agents Chemother., 1986, 30: 295-300.

[51] Kelly P, Farthing MJG. Infections of the gastrointestinal tract, in: F. O'Grady, H.P. Lambert, R.G. Finch and D. Greenwood (eds), Antibiotic and Chemotherapy, Churchill Livingstone Inc., New York, 1997, pp. 708-720.

[52] Gilbert DN, Moellering RC, Eliopoulos GM, Sande MA. The Sanford Guide to Antimicrobial Therapy 2005, 35th Edn., Antimicrobial Therapy Inc., USA.

[53] Fattal E, Rojas J, Youssef M, Couvreur P, Andremont A. Liposome-entrapped ampicillin in the treatment of experimental murine listeriosis and salmonellosis. Antimicrob Agents. Chemother 1991, 35:770–772.

[54] Desiderio JV, Campbell SG. Intraphagocytic killing of *Salmonella typhimurium* by liposome-encapsulated cephalothin. J. Inf. Dis. 1983, 148:563–5770.

[55] Desiderio JV, Campbell SG. Liposome-encapsulated cephalothin in the treatment of experimental murine salmonellosis. J. Reticuloendothel. Soc. 1983, 34:279–287.

[56] Christine ES, Kathy AS, Sharma M. Formulation and *in Vivo* Activity of Liposome-Encapsulated Cephapirin. Journal of Liposome Research1989; 1(3): 379-392.

[57] Liu S, Guo D, Guo Y., Zhou W. Preparation and pharmacokinetics of ceftiofur sodium liposomes in cows. J. Vet. Pharmacol. Therap. 2010, 34: 35–41.

[58] Lutwyche P, Cordeiro C, Wiseman DJ, ST-Louis M, UH M, J. Hope M, Webb MS, Finlay BB. Intracellular Delivery and Antibacterial Activity of Gentamicin Encapsulated in pH-Sensitive Liposomes. Antimicrob. Agents Chemothe. 1998, 2511-2520.

[59] Cordeiro C, Wiseman DJ, Lutwyche P, UH M, Evans JC, Finlay BB, Webb MS. Antibacterial efficacy of gentamicin encapsulated in pH-sensitive liposomes against an *in vivo Salmonella enterica* serovar Typhimurium intracellular infection model. Antimicrob. Agents Chemother. 2000, 44: 533-539.

[60] Swenson CE, Stewart KA, Hammett JL, Fitzsimmons WE, Ginsberg RS. Pharmacokinetics and in vivo activity of liposome- encapsulated gentamicin. Antimicrob. Agents Chemother. 1990, 34: 235–240.

[61] Fierer J, Hatlen L, Lin JP, Estrella D, Mihalko P, Yau-Young A. Successful treatment using gentamicin liposomes of *Salmonella dublin* infections in mice. Antimicrob. Agents Chemother. 1990, 34:343–348.

[62] Nightingale SD, Saletan SL, Swenson CE, Lawrence AJ, Watson DA, Pilkiewicz FG, Silverman EG, Cal SX., Liposome-encapsulated gentamicin treatment of mycobacterium-avium-mycobacterium-intracellulare complex bacteremia in aids patients, Antimicrob. Agents Chemothe. 1993, 37: 1869-1872.

[63] Tadakuma T, Ikewaki N, Yasuda T, Tsutsumi M, Saito S, Saito K. Treatment of experimental salmonellosis in mice with streptomycin entrapped in liposomes. Antimicrob. Agents Chemother. 1985, 28:28–32.

[64] Khalil RM, Murad FE, Yehia SA, El-Ridy MS, Salama HA. Free versus liposome-entrapped streptomycin sulfate in treatment of infections caused by *Salmonella enteritidis*. Pharmazie 1996, 51: 182-184.

[65] Paganoni R, Herzog C, Braunsteiner A, Hohl P. Fleroxacin: in-vitro activity worldwide against 20,807 clinical isolates and comparison to ciprofloxacin and norfloxacin, J. Antimicrob. Chemother. 1988, 22: 3–17.

[66] Magallanes M, Dijkstra J, Fierer J. Liposome-incorporated ciprofloxacin in treatment of murine salmonellosis. Antimicrob. Agents Chemother. 1993, 37: 2293–2297.

[67] Webb MS, Boman NL, Wiseman DJ, Saxon D, Sutton K, Wong KF, Logan P, Hope MJ. Antibacterial efficacy against an in vivo *Salmonella typhimurium* infection model and pharmacokinetics of a liposomal ciprofloxacin formulation. Antimicrob. Agents Chemother. 1998, 42: 45–52.

[68] Soppimath KS, Aminabhavi TM, Kulkarni AR, Rudzinski WE. Biodegradable polymeric nanoparticles as drug delivery devices. J. Controlled Release 2001, 70: 1–20.

[69] Pandy R, Khuller GK, nanoparticles-based oral drug delivery system for an injectable antibiotic-streptomycin. Evaluation in a murine tuberculosis model. Chemotherapy 2007, 53:437-441.

[70] Fattal E, Rojas J, Robolt-Treupel L, Andremont A, Courvreur P, Ampicilin loaded liposomes and nanoparticles: comparison of drug loading, drug release and *in vitro* antimicrobial activity. J. Microencapsul. 1991, 8:29-36.

[71] Pinto-Alphandary H, Andremont A, Couvreur P. Targeted delivery of antibiotics using liposomes and nanoparticles: research and applications. Int. J. Antimicrob. Agents 2000, 13: 155-168.

[72] Lenaerts V, Couvreur P, Christiaens-Leyh D, Joiris E, Rolland M, Rollman B, Speiser P. Degradation of polyisobutylcyanoacrylate nanoparticles. Biomaterials 1984, 5:65–68.

[73] Balland O, Pinto-Alphandary H, Pecquet S, Andremont A, Couvreur P. The uptake of ampicillin-loaded nanoparticles by murine macrophages infected with *Salmonella typhimurium*. J. Antimicrob. Chemother. 1994, 33:509–522.

[74] Balland O, Pinto-Alphandary H, Viron A, Puvion E, Andremont A, Couvreur P. Intracellular distribution of ampicillin in murine macrophages infected with *Salmonella typhimurium* and treated with (3H) ampicillin-loaded nanoparticles. J. Antimicrob. Chemother. 1996, 37:105–15.

[75] Pinto-Alphandary H, Balland O, Laurent M, Andremont A, Puisieux F, Couvreur P. Intracellular vizualization of ampicillin loaded nanoparticles in peritoneal macrophages infected *in vitro* with *Salmonella typhimurium*. Pharm. Res. 1994, 11: 38–46.

[76] Page-Clisson ME, Pinto-Alphandary H, Chachaty E, Couvreur P, Andremont A. Drug targeting by polyalkylcyanoacrylate is not efficient against persistent *Salmonella*. Pharm. Res. 1998, 15:542–547.

[77] Fattal E, Youssef M, Couvreur P, Andremont A. Treatment of experimental salmonellosis in mice with ampicillin-bound nanoparticles. Antimicrob. Agents Chemother. 1989, 33:1540–1543.

[78] Ranjan A, Pothayee N, Seleem MN, Taler RD, Brenseke B, Sriranganathan N, Riffle JS, Kasimanickam R, Antibacterial efficacy of core-shell nanostructures encapsulating gentamicin against an in vivo intracellular *Salmonella* model. Int. J. Nanomed. 2009, 4:289-297.

[79] Mohammadi G, Valizadeh H, Barzegar-Jalali M, Lotfipour F,Adibkia K, Milani M, Azhdarzadeh A, Kiafar F, Nokhodchi A. Development of azithromycin–PLGA

nanoparticles: Physicochemicalcharacterization andantibacterial effect against *Salmonella typhi*. Colloids and Surfaces B: Biointerfaces 2010; 80: 34–39.

[80] Mohammadi G,, Nokhodchi A, Barzegar-Jalali M, Lotfipour F, Adibkia K, Ehyaei N, Valizadeh H, Physicochemical and anti-bacterial performance characterization of clarithromycin nanoparticles as colloidal drug delivery system Colloids and Surfaces B: Biointerfaces 2011 (in press).

[81] Sharma A, Pandey R, Sharma S, Khuller GK. Chemotherapeutic efficacy of poly (d,l-lactide-co-glycolide) nanoparticle encapsulated antitubercular drugs at sub-therapeutic dose against experimental tuberculosis. Int. J. Antimicrob. Agents 2004, 24: 599–604.

[82] Lordonez-Gutierrez, Espada-Fernandez R, Auxiliadora Dea-Ayuela M, Tor-rado J, Bolas-Fernandez F, Alunda JM. *In vitro* effect of new formulations of amphotericin B on amastigote and promastigote forms of Leishmania infantum. Int. J. Antimicrob. Agents 2007, 30: 325–329.

[83] Dillen K, Bridts C, Van D, Cos VP, Vandervoort J, Augustyns K, Stevens W, Ludwig A. Adhesion of PLGA or Eudragit((R))/PLGA nanoparticles to Staphylococc-cus and Pseudomonas. Int. J. Pharm. 2008, 349: 234–240.

[84] Martins D, Costa FTM, Brocchi M, Duran N. Evaluation of the antibacterial activity of poly-(D,l-lactid-co-glycolide) nanoparticles containing violacein. J. Nanopart. Res. 2011, 13:355-363.

[85] Ranjan A, Pothayee N, Seleem M, Jain N, Sriranganathan N, Riffle JS, Kasimanickam R. Drug Delivery using novel nanoplexes against a *Salmonella* mouse infection model. J. Nanopart. Res. 2010, 12:905-914.

[86] Seleem MN, Munusamy P, Ranjan A, Alqublan H, Pickrell G, Sriranganathan N, Silica-antibiotic hybrid nanoparticles for targeting intracellular pathogens. Antimicrobial Agents Chemotherapy. 2009, 53:4270-4274.

[87] Munusamy P, Seleem MN, Alqublan H, Tyler R, Sriranganathan N, Pickrell G. Targeted drug delivery using silica xerogel systems to treat disease due to intracellular pathogens. Materials Sci. Eng. C, 2009, 29:2313-2318.

[88] Kortesuo, P., M. Ahola, S. Karlsson, I. Kangasniemi, J. Kiesvaara, and A. Yli-Urpo.. Sol-gel-processed sintered silica xerogel as a carrier in controlled drug delivery. J. Biomed. Mater. Res. 1999, 44:162–167.

[89] Slowing II, Vivero-Escoto JL, Wu CW, Lin VS. Meso- porous silica nanoparticles as controlled release drug delivery and gene transfection carriers. Adv. Drug Deliv. Rev. . 2008, 60:1278–1288.

[90] Trewyn BG, Slowing II, Giri S, Chen HT, lin VSY. Synthesis and functionalization of a mesoporous silica nanoparticle based on the sol-gel process and applications in controlled release. Acc. Chem. Res.. 2007, 40: 846-853.

[91] Zhu, Y., and J. Shi. 2007. A mesoporous core-shell structure for pH-controlled storage and release of water-soluble drug. Microporous Mesoporous Mater. 103:243–249.

[92] Zhu, Y., J. Shi, Y. Li, H. Chen, W. Shen, and X. Dong. 2005. Hollow mesoporous spheres with cubic pore network as a potential carrier for drug storage and its *in vitro* release kinetics. J. Mater. Res. 20:54–61.

[93] Ochoa J, Irache JM, Tamayo I, Walz A, DelVecchio VG, Gamazo C. Protective immunity of biodegradable nanoparticle-based vaccine against an experimental challenge with *Salmonella Enteritidis* in mice. Vaccine 2007, 25: 4410-4419.

[94] QingShan W, Jian J, JinHong F, JiaCong S, Norvancomycin-capped silver nanoparticles: synthesis and antibacterial activities against *E. coli*. Sci. China B Chem, 2007, 50:418-424.

[95] Oloffs A., Crosse-Siestrup C., Bisson S., Rinck M., Rudolvh,R, Gross U. Biocompatibility of silver-coated polyurethane catheters and silver-coated Dacron R _ material. Biomaterials, 1994, 15: 753-758.

[96] Tokumaru T, Shimizu Y, Fox C. Antiviral activities of silver sulfadiazine and ocular infection. Res. Comm. Chem. Pathol. Pharmacol. 1984, 8:151-158.

[97] Patil SS, Dhumal RS, Varghese MV, Paradkar AR, Khanna PK, Synthesis and antibacterial studies of chloramphenicol loaded nano-silver against *Salmonella typhi*. Synthesis Reactivity in Inorganic, Metal-Organic, and Nano-metal Chemistry, 2009, 39:65-72.

[98] Rai M, Yadav A, Gade A. Silver nanoparticles as a new generation of antimicrobials. Biotechnology Advances 2009, 27: 76-83.

[99] Pal S, Tak Y, Song J, Does the antibacterial activity of silver nanoparticles depend on the shape of the nanoparticles? A study of the Gram-negative bacterium *Escherichia coli*. Appl. Environ. Microbial. 2007, 73: 1712-1720.

[100] Chen X, Schluesener H. Nanosilver: a nanoproduct in medical application. Toxicol. Lett. 2008, 176: 1-12.

[101] Durán N, Marcato PD, De Conti R, Alves OL, Costa FTM, Brocchi M. Potential Use of Silver Nanoparticles on Pathogenic Bacteria, their Toxicity and Possible Mechanisms of Action. J. Braz. Chem. Soc. 2010; 21: 949-959.

[102] Durán N, Marcato PD, Alves OL, De Souza GIH, Esposito E. Mechanistic aspects of biosynthesis of silver nanoparticles by several Fusarium oxysporum strains. J. Nanobiotechnol. 2005, 3:1-10.

[103] Durán N, Marcato PD, Alves OL, De Souza GIH, Espósito E. Antibacterial effect of silver nanoparticles produced by fungal process on textile fabrics and their effluent treatment J. Biomed. Nanotechnol. 2007, 3:203-208.

[104] Kalishwaralal K, Deepak V, Ramkumarpandian S, Nellaiah H, Sangiliyand G. Extracellular biosynthesis of silver nanoparticles by the culture supernatant of Bacillus licheniformis Mater. Lett. 2008, 62: 4411-4413.

[105] Mohanpuria P, Rana NK, Yadav SK. Biosynthesis of nanoparticles: technological concepts and future applications. J. Nanopart. Res. 2008, 10:507-517.

[106] Mukherjee P, Roy M, Mandal BP, Dey GK, Mukherjee PK, Ghatak J, Tyagi AK,Kale SP. Green synthesis of highly stabilized nanocrystalline silver particles by a non-pathogenic and agriculturally important fungus T-asperellum Nanotechnology 2008, 19: 075103.

[107] Sadowski Z, Maliszewska IH, Grochowalska B, Polowczyk I, Koźlecki T. Synthesis of silver nanoparticles using microorganisms Mater. Sci., Poland 2008, 26: 419-424.

[108] Maliszewska I, Sadowski Z. Synthesis and antibacterial activity of of silver nanoparticles J. Phys. Conf. Ser. 2009, 146: 012024.

[109] Ingle A, Gade A, Pierrat S, Sönnichsen C, Rai M. Mycosynthesis of silver nanoparticles using the fungus Fusarium acuminatum and its activity against some human pathogenic bacteria Curr. Nanosci. 2008, 4:141.

[110] Shankar, S.S, Rai A, Ahmad A, Sastry M. Rapid synthesis of Au, Ag, and bimetallic Au core-Ag shell nanoparticles using Neem (Azadirachta indica) leaf broth J. Colloid Interface Sci. 2004, 275: 496-502.

[111] Nomiya K, Yoshizawa A, Tsukagoshi K, Kasuga NC, Hirakawa S, Watanable J. Synthesis and structural characterization of silver(I), aluminium(III) and cobalt(II) complexes with 4-isopropyltropolone (hinokitiol) showing noteworthy biological activities. Action of silver(I)–oxygen bonding complexes on the antimicrobial activities J. Inorg. Biochem. 2004, 98: 46-60.

[112] Gupta A, Silver S. Silver as a biocide: will resistance become a problem? Nat. Biotechnol. 1998, 16: 888.

[113] Feng QL, Wu J, Chen G Q, Cui FZ, Kim TN, Kim JO. A mechanistic study of the antibacterial effect of silver ions on Escherichia coli and Staphylococcus aureus J. Biomed. Mater. Res. 2000, 52: 662–668

[114] Panáček A, Kvitek L, Prucek R, Kolář M, Večeřová R, Piźurová N, Sharma V K,Něěcná T, Zbořil R. Silver colloid nanoparticles: synthesis, characterization, and their antimicrobial activity J. Phys. Chem. B. 2006, 110:16248–16253.

[115] Chopra I. The increasing use of silver-based products as antimicrobial agents; a useful development or a concern J. Antimicrob. Chemother. 2007, 59: 587–590.

[116] Asharani PV, Wu YL, Gong ZY, Valiyaveettil S. Toxicity of silver nanoparticles in zebrafish models Nanotechnology 2008, 19: 255102.

[117] Drake PL, Hazelwood K J. Exposure-related health effects of silver and silver compounds: a review Ann. Occup. Hyg. 2005, 49:575–585.

[118] Inbaraj BS, Kao TH, Tsai TY, Chiu CP, Kumar R, Chen BH. The synthesis and characterization of poly(γ -glutamic acid)-coated magnetite nanoparticles and their effects on antibacterial activity and cytotoxicity. Nanotechnology 2011, 22: 075101 (9pp).

[119] Khanna PK, Gokhale R, Subbarao V. Poly(vinyl pyrolidone) coated silver nano powder via displacement reaction. J. Mater. Sci. 2004, 39: 3773–3776.

[120] Lim P, Liu R, She P, Hung C, ShihH. Synthesis ofAg nanospheres particles in ethylene glycol by electrochemical-assisted polyol process. Chem. Phys. Lett., 2006, 420: 304–308.

[121] Sergeev B, Kiryukhin M, Rubtsova M, Prusov A. Synthesis of protein A conjugates with silver nanoparticles. Colloid J. 2003, 65: 636–638.

[122] Yakutik I, Shevchenko G. Self-organization of silver nanoparticles forming on chemical reduction to give monodisperse spheres. Surface Sci. 2004, 566: 414–418.

[123] Kasuga N, Sato M, Amano A, Hara A, Tsuruta S, Sugie A, Nomiya K. Light-stable and antimicrobial active silver(I) complexes composed of triphenylphosphine and amino acid ligands: synthesis, crystal structure, and antimicrobial activity of silver(I) complexes constructed with hard and soft donor atoms (n∞{[Ag(L)(PPh3)]2} with L= α-ala−or asn−and n = 1 or 2). Inorg. Chim. Acta, 2008, 361: 1267–1273.

[124] Temgire M. Joshi S. Optical and structural studies of silver nanoparticles. Rad. Phys. Chem. 2004, 71:1039–1044.

[125] Khanna PK, Singh N, Charan S, Subbarao V, Gokhale R, Mulik U. Synthesis and characterization of Ag/PVA nanocomposite by chemical reduction method. Mater. Chem. Phys., 2005, 93: 117–121.

[126] Vaseashta A, Dimova-Malinovska D. Nanostructured and nanoscale devices, sensors and detectors. Sci Technol Adv Mater 2005, 6:312–318.

[127] Comini E. Metal oxide nano-crystals for gas sensing. Anal Chim Acta 2006, 568:28–40.

[128] Raveh A, Zukerman I, Shneck R, Avni R, Fried I. Thermal stability of nanostructured superhard coatings: a review. Surf Coat Technol 2007, 201: 6136–6142.

[129] Long TC, Saleh N, Tilton RD, Lowry GV, Veronesi B. Titanium dioxide (P25) produces reactive oxygen species in immortalized brain microglia (BV2): implications for nanoparticle neurotoxicity. Environ Sci Technol 2006, 40: 4346–4352.

[130] Porter AE, Muller K, Skepper J, Midgley P, Welland M. Uptake of C60 by human monocyte macrophages, its localization and implications for toxicity: studied by high resolution electron microscopy and electron tomography. Acta Biomater 2006, 2: 409–419.

[131] Ruparelia JP, Chatterjee AK, Duttagupta SP, Mukherji S. Strain specificity in antimicrobial activity of silver and copper nanoparticles. Acta Biomaterialia 2008, 4: 707–716.

[132] Sawosz E, Chwalibog A, Szeliga J, Sawosz F, Grodzik M, Rupiewicz M, Niemiec T, Kacprzyk K, Visualization of gold and platinum nanoparticles interacting with Salmonella Enteritidis and Listeria monocytogenes. Int. J. Nanomed. 2010, 5: 631-637.

[133] Wang S, Lawson R, Ray PC, Yu H, Toxic effects of gold nanoparticles on Salmonella typhimurium bacteria. Toxicology Ind. Health, 2011, 27:547-554.

[134] Goodman C, McCusker C, Yilmaz T, Rotello V, Toxicity of gold nanoparticles functionalized with cationic and anionic side chains. Bioconjugate Chemistry, 2004, 15:897-900.

[135] Patra HK, Banerjee S, Chaudhuri U, Lahiri P, Dasgupta AK, Cell-selective response to gold nanoparticles. Nanomedicine, 2007, 3:111-119.

[136] Bhat GK, Suman E, Shetty A, Hegde BM. A Study on the ASAP Nano-silver Solution on Pathogenic Bacteria and Candida. JIACM 2009, 10: 15-17.

[137] Rastogi SK, Rutledge VJ, Gibson C, Newcombe DA, Branen JR, Branen AL. Ag colloids and Ag clusters over EDAPTMS-coated silica nanoparticles: synthesis, characterization, and antibacterial activity against Escherichia coli. Nanomedicine: Nanotechnology, Biology, and Medicine 2011; 7: 305–314.

[138] Wahab R, Mishra A, Yun S, Kim YS, Shin HS. Antibacterial activity of ZnO nanoparticles prepared via non-hydrolytic solution route. Appl Microbiol Biotechnol 2010, 87:1917–1925.

[139] Irwin P, Martin J, Nguyen LH, He Y, Gehring A, Chen CY. Antimicrobial activity of spherical silver nanoparticles prepared using a biocompatible macromolecular capping agent: evidence for induction of a greatly prolonged bacterial lag phase. J. Nanobiotechnology 2010, 8:34.

[140] Jin T, Sun D, Su JY, Zhang H, Sue HJ. Antimicrobial Efficacy of Zinc Oxide Quantum Dots against *Listeria monocytogenes, Salmonella Enteritidis*, and *Escherichia coli* O157:H7. J. Food Sci. 2009, 74: 46-52.

[141] Mohammed Fayaz A, Balaji K, Girilal M, Yadav R, Kalaichelvan PT, Venketesan R. Biogenic synthesis of silver nanoparticles and their synergistic effect with antibiotics: a study against gram-positive and gram-negative bacteria. Nanomedicine: Nanotechnology, Biology, and Medicine 2010, 6: 103–109.

[142] Sadhasivam S, Shanmugam P, Yun KS. Biosynthesis of silver nanoparticles by Streptomyces hygroscopicus and antimicrobial activity against medically important pathogenic microorganisms. Colloids and Surfaces B: Biointerfaces 2010, 81: 358–362.

[143] Nanda A, Saravanan M. Biosynthesis of silver nanoparticles from Staphylococcus aureus and its antimicrobial activity against MRSA and MRSE. Nanomedicine: Nanotechnology, Biology, and Medicine 2009, 5: 452–456.

[144] Thirumurugan A, Ganesh RJ, Akila S, Tomy NA, Meruvu H. Biosynthesis of silver nanoparticles using Psuedomonos aerogenosa and its Effect on the antibacterial activity of various antibiotics against clinically isolated organism. Journal of Pharmacy Research 2010, 3: 2510-2511.

[145] Kumar A, Pandey AK, Singh SS, Shanker R, Dhawan A. Cellular uptake and mutagenic potential of metal oxide nanoparticles in bacterial cells. Chemosphere 2011; 83: 1124–1132.

[146] Shrivastava S, Bera T, Roy A, Singh G, Ramachandrarao P, Dash D. Characterization of enhanced antibacterial effects of novel silver nanoparticles. Nanotechnology 2007, 18: 225103: 9pp.

[147] Mahapatra O, Ramaswamy S, Kumar Nune SV, Yadavalli T, Gopalakrishnan C. Corn flake-like morphology of iron nanoparticles and its antibacterial property. J. Gen. Appl. Microbiol. 2011, 57: 59-62.

[148] Salem HF, Eid KAM, Sharaf MA. Formulation and evaluation of silver nanoparticles as antibacterial and antifungal agents with a minimal cytotoxic effect. Int. J. Drug Delivery 2011, 3: 293-304.

[149] Shirokov LN, Alexandrov VA, Egorov EM, Vihorev GA. Macromolecular Systems and Bactericidal Films Based on Chitin Derivatives and Silver Nanoparticles. Applied Biochemistry and Microbiology 2009, 45: 380–383.

[150] Ingle A, Gade A, Pierrat S, Sönnichsen C, Rai M. Mycosynthesis of Silver Nanoparticles Using the Fungus *Fusarium acuminatum* and its Activity Against Some Human Pathogenic Bacteria. Current Nanoscience, 2008,4: 141-144.

[151] Oha SD, Lee S, Choi SH, Lee IS, Lee YM, Chun JH, Park HJ. Synthesis of Ag and Ag–SiO2 nanoparticles by _-irradiation and their antibacterial and antifungal efficiency against *Salmonella enteric* serovar *Typhimurium* and *Botrytis* cinerea. Colloids and Surfaces A: Physicochem. Eng. Aspects 2006, 275: 228–233.

[152] Suwanboon S, Amornpitoksuk P, Bangrak P. Synthesis, characterization and optical properties of $Zn_{1-x}Ti_xO$ nanoparticles prepared via a high-energy ball milling technique. Ceramics Int. 2011, 37: 333–340.

[153] Sawosz E, Chwalibog A, Szeliga J, Sawosz F, Grodzik M, Rupiewicz M, Niemiec T, Kacprzyk K. Visualization of gold and platinum nanoparticles interacting with

Salmonella Enteritidis and *Listeria monocytogenes*. International Journal of Nanomedicine 2010, 5: 631–637.

[154] Mahapatra O, Bhagat M, Gopalakrishnan C, Arunachalam KD. Ultrafine dispersed CuO nanoparticles and their antibacterial activity. Journal of Experimental Nanoscience 2008, 3: 185-193.

[155] Donaldson K, Tran CL, MacNee, W. Deposition and effects of fine and ultrafine particles in the respiratory tract. Eur. Respir. Mon. 2002, 7:77–92.

[156] Kruter J, Shamenkov D, Petrov V.. Apolipoprotein-mediated transport of nanoparticle-bound drugs across the blood-brain barrier. J. Drug Target. 2002, 10:317–325.

[157] Oberdörster G, Sharp Z, Elder AP.. Translocation of inhaled ultrafine particles to the brain. Inhal Toxicol. 2004, 16:437–445.

[158] Park S, Lee YK, Jung M, Kim KH, Chung N, Ahn EK, Lim Y, Lee KH. Cellular toxicity of various inhalable metal nanoparticles on human alveolar epithelial cells. Inhal. Toxicol. 2007;19 Suppl 1: 59-65.

8

Mexican Plants Used in the Salmonellosis Treatment

Daniel Arrieta-Baez[1], Rosario Ruiz de Esparza[2]
and Manuel Jiménez-Estrada[2]
[1]Escuela Nacional de Ciencias Biológicas-IPN
Prolongación de Carpio y Plan de Ayala S/N
[2]Instituto de Química, UNAM. Cd. Universitaria, Coyoacán
México

1. Introduction

Diarrhoeal diseases constitute a major public health problem, particularly in developing countries, where the rate of mortality and morbidity is very high (Bern et al., 1992). The World Health Organization (WHO) has estimated that 1.5 billion episodes of diarrhea occur every year in these countries, resulting in 3 million deaths (Alper 2003). In Mexico, intestinal infectious diseases are the second cause of morbidity, for which children younger than 5 years of age and adults over 65 years are the most affected (Secretaría de Salud, 2005). These diseases are the 18th leading cause of mortality in the general population and, more importantly, the primary cause of death in children (1 to 4 years) and the fourth for infants (1 to 12 months) (Secretaría de Salud, 2005). In addition, certain social groups are more prone to suffer intestinal infections. These diseases are the second cause of mortality among the Mexican rural population (Tapia-Conyer, 1994).

One study suggests that at least 37% of the Mexican population uses medicinal plants. From the 3034 plants used as medicine in Mexico, 1024 (34%) are used to treat diseases in the digestive tract (Argueta et al., 1994). The high percentage of species used to treat gastrointestinal diseases may reflect the high level of incidence of these illnesses among rural and native communities where the plants are their principal, if not only, available health resource (Tapia-Conyer, 1994; Waller, 1993).

The etiological agents of diarrhea described in epidemiological studies are transmitted as waterborne and foodborne. Some foodborne pathogens have been recently considered as emerging diseases (WHO Media Centre, 2007), despite the fact they have been known since a long time ago. For example, outbreaks of salmonellosis have been described for many decades, and yet their incidence has increased over the last 25 years. Diarrhoeal infections can be caused by many etiological agents, but mainly by enterobacteria such as *Escherichia coli*, *Salmonella spp.*, *Shigella spp.*, *Campylobacter jejuni* and *Vibrio cholerae*; as well as parasites such as *Entamoeba histolytica* and *Giardia intestinalis*, and some rotaviruses are also important agents (Guerrant et al., 1990). *Salmonella spp.* is a facultative, gram negative, flagellated member of the *Enterobacteriaceae* family. The most extensive accepted classification of

Salmonella strains is based on the diversity of two differentially expressed H flagellar antigens: flagellin phase I and phase II antigens (codified by fliC and fljB genes), and the O antigens of the bacterial lipopolysaccharide, both determined by serotyping. Until now, 2501 serotypes have been described; which turns Salmonella classification into a complex and laborious process in the clinical laboratory; therefore, several PCR based methods have recently been developed, and were reported to be a simple, highly sensitive, fast and reliable alternative when compared to traditional clinical laboratory methods.

As a part of our contribution in this area, we have studied an important plant used in this public health problem. *Piqueria trinervia* Cav. is a perennial herb of the family *Asteraceae* that grows commonly in open areas of the pine-oak forests of the mountains throughout Mexico and Central America. It is employed to treat diarrhea, dysentery, "empachos" (a cultural disease with manifestations of various gastrointestinal disorders), intestinal infections, stomach pain, and typhoid fever (Argueta et al., 1994). To cure diarrhea or intestinal infections, an infusion or decoction is prepared with 10 g of the aerial parts in 0.5L of water; the tea is consumed throughout the day until the sickness disappears (Linares, 1991). The biological and chemical study of this plant would be described in the present chapter.

2. Traditional medicine in the treatment of Salmonellosis

The use of medicinal plants has occurred in Mexico since pre-Hispanic times. Aztecs had traditionally viewed the disease and health as dynamic changes in the body. According with this, health is maintained when the body is balanced, and the disease comes when this balance is lost (Ortiz de Montellano 1987; Viesca Trevino 1986). The most common causes of illness were the "bad" spirits called "ehecatl". This concept still remains until our days in the Nahua region. These spirits can get into the body of a person and bring bad luck, illness or even death. This concept can be easily associated etiologic with the popular cause of "mal aire" (bad air: offensive spirits in the air) found in different areas of Mesoamerica. Treatment for the disease among the Nahua is a series of techniques to get the offensive spirit of the patient's body, including complex rituals that use a combination of prayers, invocations, and medicinal plants. An important part of the ritual is the "limpia" or the ritual of the cleanliness, through which the body is cleaned of the spirits that cause disease. The specialist in healing restores the balance by removing the spirit from the body of the person, and then he use different formulations based on medicinal plants to complete the treatment of the patient. Aztec/Nahua healing specialists are very knowledgeable about plants that can be used medicinally and they have passed this knowledge from generation to generation.

The Yucatán peninsula is a region particularly recognized for its culture, endemic elements and ancient Mayan buildings. Furthermore, although less known, there is an important ancestral knowledge about the medicinal properties of plants, which is widely used by the general population (Ankli et al., 1999). The Mayan traditional healers keep the knowledge of the different properties of the plants, and how to make medicinal mixtures. These formulations (also known as potions) vary according to the disease to be treated and are used since ancestral times (Leonti et al., 2003) to cure some of the most common health problems in the population, such as diarrheal, cutaneous diseases and those from the respiratory tract as documented by Roys (1931). Among the infectious bowel diseases in the Mayan traditional medicine, there is a group known as dysenteries classified as white and red. The main symptom associated with the red dysentery is bloody diarrhea, and the main symptom associated with the white dysentery is foamy diarrhea (Roys, 1931); other symptoms include abdominal cramps, fever or severe pain during defecation (Waldman et al., 1994). Since the traditional diseases are not easy to correlate with the causal agents of infectious bowel diseases, it has taken into account the most characteristic symptoms: bloody and/or mucous diarrhea, foamy diarrhea, and abdominal pain to select the microorganisms to be tested

With the presence of nearly 10,000,000 indigenous people speaking nearly 85 different languages, who still depend upon plants for primary therapy from the diverse flora (almost 5,000 medicinal plants), Mexico represent a good area to perform ethnopharmacological studies. Studying the biological diversity of plants related to their traditional use as medicines can lead us to understand how they act and to assure the rational exploitation of the resources and their further development as phytomedicines. Because medicinal plants continue to be culturally suitable as treatments for several illnesses, it is important to document their uses and perform studies about their pharmacological activities to assure their efficacy and safety. Despite the vast literature that exists in Mexico (in Spanish) about ethnobotanical studies, only a few efforts to publish these data in international journals have been done. However, Heinrich and his group published more than 18 works (i.e. Weimann and Heinrich, 1997; Heinrich et al., 1998) with a combination of adequate field work and appropriate interpretation of the data.

The people in Mexico still depend upon the use of medicinal plants to treat simple health problems, including those who live in regions where it is still possible to find people who speak the pre-Hispanic Nahua language. The cultural knowledge about the use of medicinal plants converges with the richness in the surrounding flora making possible to select different regions in Mexico to study selected traditionally used medicinal plants.

3. Mexican medicinal plants used in Intestinal diseases

Today in Mexico there is a strong attachment to the use of plants to cure various diseases. For large sectors of Mexican society, particularly for the indigenous, traditional medicine is the main or only source to address the health problems (Tapia-Conyer, 1994).

The presence of traditional medicine is clearly observed in the southern areas of Mexico such as Chiapas and Oaxaca, where native culture prevails (Lozoya, 1990). Although there are medical services in some rural communities, people prefer traditional medicine. In addition to the cultural, traditional medicine is more accessible to people living in rural communities, either because health facilities are far from the locality or because it is more expensive medical treatment with allopathic (Frei et al., 1998).

In a study from 1983 to 1985 in rural areas of Mexico, through surveys of the program IMSS-COPLAMAR clinics to traditional doctors, they enlisted the 1.950 most used medicinal species in the country. 140 plant species were highlighted for the frequency of use, of which 38.0% were used to cure digestive disorders (deworming, antispasmodics, laxatives, anti-diarrhea and cholera), 13.6% were used to cure diseases of the respiratory system and 13.5% for the treatment of skin lesions. (Lozoya et al., 1987).

Between 1994 and 1995, the most common diseases treated by traditional healers in rural areas of the country were digestive (44.0%), respiratory (11.0%) and injuries (9.0%) (Lozoya,

1990). Most of the medicinal plants used in three indigenous communities of Oaxaca (Maya, Nahua and Zapotec) were used to treat gastrointestinal, skin and respiratory diseases (Heinrich et al., 1992). Zapotec Healers from the Tehuantepec Isthmus used 205 plant species, of which 46.1% were used to treat skin problems and 39.6% to treat gastrointestinal diseases (Frei et al., 1998).

According to traditional Mexican medicine, there are diverse healing strategies, as well as different preferences for the plant parts used for various diseases. It is generally accepted that the beneficial effects of medicinal plants can be obtained from active constituents present in the whole plants, parts of plant (as flowers, fruits, roots or leaves), or plant materials or combinations thereof, whether in crude or processed state.

According with our etnobotanical research, the next plant are the most used in treatment of intestinal diseases: Buddleia scordioides, Byrsonima crassifolia, Geranium seemannii, Guazuma ulmifolia, Larrea tridentate, Persea americana, Piqueria trinervia, Psidium guajava.

3.1 Escobilla (*Buddleia scordioides* H. B. K)

Buddleja scordioides HBK (Buddlejaceae) is a shrub which grows in the Chihuahuan desert and in the state of Coahuila, Nuevo León, Tamaulipas, Durango, Zacatecas, Aguas Calientes, San Luis Potosí, Guanajuato, Queretaro, Hidalgo, Jalisco, México and Distrito Federal (Avila et al., 2005). *Buddleja scordioides* HBK (KUNTH), *Loganiaceae*, is commonly known as escobilla, butterfly-bush, mato, salvia real and salvilla (Martínez, 1979).

Its medicinal use includes the treatment of eating disorders, especially stomach aches and diarrhea. Also used as eupeptic, it is recommended to prepare branch or root infusions, these parts are boiled in milk and given to children when they have colic (Argueta, 1994).This plant is widely used for the treatment of diarrhea, stomacha che (colic) and gastrointestinal disorders.

It was reported that the amebicide activity of *B. cordata* is caused by linarin. This plant can be a source for this compound since it constitutes 24% of the methanol extract of aerial parts.

It was found that extracts of the aerial parts of *B. scordioides* and *B. perfoliata* showed antispasmodic activity and had a relaxing effect on rabbit jejunum and ileum of guinea pigs. Such activities may be the cause of its use in traditional medicine in the treatment of gastrointestinal pain, spasms and cramps (Cortés et al., 2006). Decoctions of this plant are commonly used orally or topically for treating several illnesses such as diarrhea, headache, and hurts (Avila et al., 2005).

In addition, a verbascoside with antibacterial activity, triterpenoid saponins and other glycosides, have been extracted from this species. The presence of some flavonoids such as rutine, quercetin and quercitrin has also been reported (Cortés 2006)

3.2 Nanche (*Byrsonima crassifolia* (L.) Kunth)

Byrsonima crassifolia (Malpighiaceae) is a tropical tree widely distributed in Mexico, Central and South America. The pharmacological activities of *B. crassifolia* extracts as a bactericide, fungicide, leishmanicide, and as a topical anti-inflammatory (Maldini et al. 2009) have been described. *B. crassifolia* is popularly known as "nanche" and it has been used medicinally since prehispanic times, mainly to treat gastrointestinal afflictions and gynecological inflammation (Heinrich et al. 1998). The most often medicinal use of *B. Crassifolia* has been as an antidiarrheal, but has been also indicated to treat other disorders of the digestive system such as dysentery, stomach pain, indigestion and poor digestion. A decoction of the bark is usually used to treat these conditions (Martinez, 1959; Argueta, 1994). Some other reports indicate that *B. crassifolia* has been employed in the treatment of nervous excitement and to induce a pleasant dizziness (Maldonado 2008).

Different phytochemical studies have been carried out to isolate the main active compounds. The presence of terpenes, saponins, flavonoids and glycosides has been reported from the root of *B. crassifolia* and saponins, flavonoids, glycosides and tannins have been isolated from the stem as well.

3.3 Pata de León (*Geranium seemannii* Peyr)

Geranium seemanni Peyr is a perennial herbaceous plant, with flowers with different colours varying from purple to white (Rzedowski Rzedowski, 1995). It has been located in the states of Sinaloa, Chihuahua, Coahuila, Nuevo Leon, Tamaulipas, Durango, Zacatecas, San Luis Potosi, Aguascalientes, Guanajuato, Mexico City, Queretaro, Hidalgo, Guerrero, Jalisco, Michoacan, Morelos, Oaxaca , Puebla, Tlaxcala, Veracruz and Chiapas (Sanchez, 1979; Rzedowski and Rzedowski, 1995).

The Codigo Florentino, one of the most important books written in ethnobotany in the century XVI, mention that the ground plant applied in the face can remove stains on the skin face. Francisco Hernandez, in the same century says: "is astringent, cures dysentery and other flows, eye inflammation, hemorrhoids and indigestion, and cools down some fevers, among other diseases."

Currently the most common use of this plant is for digestive disorders such as vomiting and diarrhea. The decoction of the plant can be used as antigastralgic and the infusion of the leaves as a purgative. The decoction of the stem, leaves and flowers is usually used to relieve the itch (Argueta, 1994).

This specie has not been studied to analyze its bioactive compounds. However, different species of the *Geranium sp*, used with therapeutic purposes in Mexico, are reported in the literature (Calzada et al., 1999; Serkedjieva e Ivancheva, 1999; Akdemir et al., 2001). The geranin is the most abundant tannin founded in the *Geranium* genre. According with Okuda et al., in 1980, a 9.8 to 12% of this compound is present in dry leaves of the plant. (-)-Epicatechin, (+)-catechin, β-sitosterol-3-O-β-glucoside, tiramine y saccharose were isolated from the roots of the *G. mexicanum* (Calzada et al., 2005).

The aqueous and methanolic extracts prepared from the aerial part and root of *G. mexicanum* showed antimicrobial activity against strains of *E. coli, Shigella sonnei, S. flexneri* and *Salmonella sp.* (Alanis et al., 2005).

3.4 Guacimo (*Guazuma ulmifolia* Lam.)

Trees of *Guazuma ulmifolia* Lam. (*Sterculiaceae*), commonly known as guácimo, caulote, tapaculo, or aquiche, occupy dry lowlands from Peru, north and east to Venezuela and to northern México. The species is common in pastures and fencerows and its foliage and fruits are valuable cattle and horse fodder (Seigler et al., 2005).

This plant is used in the Mexican traditional medicine to treat various diseases. In Guerrero, Puebla and Veracruz, is used to treat gastrointestinal disorders. The decoction of the bark, leaves or buds, are used to treat diarrhea. A tea prepared with guacimo shoots, shoots of guayaba (*Psidium guajava*), the stem of nanche (*Byrsonimia crassifolia*) and oak (Quercus sp.) is used in children suffering intestinal infection with diarhea (Argueta, 1994). *G. ulmifolia* is used in some Mixe communities of Oaxaca and Veracruz to treat diarrhea (Leonti et al., 2003). The species is widely used by the Zapotec of Oaxaca to treat gastrointestinal diseases. Indigenous healers recognize plants with astringent properties (high levels of tannins) as useful in the treatment of gastrointestinal disorders (Frei, 1998).

Tannins are the main components of *Guazuma sp.* The variation of these compounds depends on the part of the plant, thus the leaves has 0.145 mg/g, in the leaves with stems 0.115 mg/g, in the stems 0.087 mg/g, and were not detected in fruit (Ortega et al., 1998).

The following compounds were isolated from the ethanolic extract of the stem bark, tannin acid, (-)-epicatechin-[4β→8]-(-)-epicatechin-4β-benzilthioether, (-)-epicatechin-[4β→6]-(-)-epicatechin-4β-benzilthioether; (-)-epicatechin; the dimers procyanidin B2 and B5; the trimers procyanidin C1, (-)-epicatechin-[4β→6]-(-)-epicatechin-[4β→8]-(-)-epicatechin and (-)-epicatechin-[4β→8]-(-)-epicatechin-[4β→6]-(-)-epicatechin and the tetramer (-)-epicatechin-[4β→8]-(-)-epicatechin-[4β→8]-(-)-epicatechin-[4β→8]-(-)-epicatechin (Hörr et al., 1996).

3.5 Gobernadora (*Larrea tridentata* (DC) Cav.)

Larrea tridentata (Sesse and Moc. Ex DC, Zygophyllaceae) also known as gobernadora, coville, larrea, chaparral, or creosote bush, is a shrubby plant belonging to the family Zygophyllaceae. *L. tridentada* is a common shrub of North American warm deserts. Its dominance has increased within 19 million ha of lands previously considered desert grasslands in response to disturbances such as grazing (Arteaga et al., 2005). While often viewed as an indicator of desertified conditions and the focus of extensive control efforts it is also an important plant with a long history of medicinal use (Arteaga et al., 2005).

Tea brewed from the leaves of *L. tridentata* has been used in traditional medicine to treat digestive disorders, rheumatism, venereal disease, sores, bronchitis, chicken pox, and the common cold (Sinnott et al., 1998). This plant is often used to treat gynecological problems. In cases of infertility, the decoction of the leaves is used in vaginal washings or taken as a tea for nine days after the period (for three consecutive months)(Argueta, 1994). Among the proposed medicinal properties of creosote bush, the most prominent is its antioxidant effects (Sheikh et al., 1997).

Phytochemical studies carried out on *L. tridentate* showed that it contains a series of lignans, flavonoids, condensed tannins, triterpene saponins, and naphthoquinones (Abou-Gazar et al., 2004). The extracts or constituents of *L. tridentata* have been reported to possess antioxidant, anti-HIV, antimicrobial, enzyme inhibitory, anti-tumor, and anti-hyperglycemic (Abou-Gazar et al., 2004) activities. The plant contains the powerful antioxidant, nordihydroguaiaretic acid (NDGA) which is suspected to contribute to the toxic effects associated with the consumption of chaparral products (Sinnott et al., 1998).

3.6 Aguacate (*Persea americana* Miller)

Persea americana mill (lauraceae) is a tree plant also called avocado or alligator pear. It is chiefly grown in temperate regions and sparsely grown in tropical regions of the world. Since ancient times this plant has been valued for its nutritional and medicinal properties. The fruit is highly prized for its aroma and exquisite taste (Lozoya and Lozoya, 1982). Avocado is used in 25 states in Mexico to treat various digestive disorders (Lozoya et al., 1987) as dysentery, stomach pain, constipation, stomach gas, vomiting, among others (Argueta, 1994). Peel of the avocado is used in infusion for treating intestinal parasites (Martinez 1959; Lozoya and Lozoya, 1982; Argueta, 1994). It is recommended for anemia, exhaustion, hypercholesterolemia, hypertension, gastritis, and gastroduodenal ulcer. The infusion prepared from the leaves is used in the treatment of diarrhea and some cases of indigestion. The leaves have also been reported as an effective antitussive, antidiabetic, and relief for arthritis pain by traditional medicine practitioners. Analgesic and antiinflammatory properties of the leaves have been reported (Adeyemi et al., 2002).

The result of the phytochemical screening of the aqueous leaf extract of *Persea americana* revealed that the extract contained various pharmacologically active compounds such as saponins, tannins, phlobatannins, flavonoids, alkaloids, and polysaccharides.

From the aqueous extract of the avocado leaves, two new monoglycosyl flavonols were isolated, 3-O-α-D-of kaempferol and quercetin arabinopyranoside, along with the 3-O-α-L-ramnopiranosido-kaempferol (Afzeliana), 3-O-α-L-ramnopiranosido-quercetin (quercitrin), 3-O-β-gluco-pyranoside-quercetin and quercetin 3-O-β-galactopyranoside, quercetin (Almeida et al., 1998).

3.7 Guayaba (*Psidium guajava* L.)

Psidium guajava, a tropical fruit guava of the family *Myrtaceae*, is widely recognized as a plant of many herbal medicines. The leaf, root, and bark of P. *guajava* are used in indigenous herbal medicine for the treatment of various ailments including those that are bowel related. In 23 states of Mexico is used to treat gastrointestinal diseases (Lozoya et al., 1987), most notably diarrhea. Generally, a decoction or infusion prepared with the leaves of guava tree is taken against gastrointestinal diseases (Argueta, 1994). The Tzotzil prepare an infusion to relieve severe diarrhea, weakness, vomiting, stomach pain, when present watery or bloody stools that can last days (Argueta, 1994).

The decoction of the leaves has showed *in vitro* antimicrobial activity against *Escherichia coli*, *Salmonella typhi*, *Staphylococcus aureus*, *Proteus mirabilis*, and *Shigella dysenteria*. Another paper showed the effectiveness of the leaf extract against *Staphylococcus aureus* (Gnan and Demello, 1999). It was shown to antibacterial in another study and in addition to *Staphylococcus aureus* was also useful against *Streptococcus spp*. The leaves are rich in tannin, and have antiseptic properties. Modern proof of the traditional use can be found in modern studies. The methanolic extract of P. *guajava* (leaves) showed significant inhibitory activities against the growths of 2 strains of *Salmonella*, *Shigella spp*. (*Shigella flexneri*, *Shigella virchow* and *Shigella dysenteriae*) and 2 strains of the enteropathogenic *Escherichia coli*. The results have confirmed the effectiveness of this medicinal plant as an antidiarrheal agent. Guava sprout extracts (P. guajava) by 50% diluted ethanol showed the most effective inhibition of *E. coli*, while those in 50% acetone were less effective. It is concluded that guava sprout extracts constitute a treatment option for diarrhoea caused by *E. coli* or by *S. aureus* produced toxins, due to their quick therapeutic action, easy availability in tropical countries and low cost.

The leaves contains essential oil with the main components being α-pinene, β-pinene, limonene, menthol, terpenyl acetate, isopropyl alcohol, longicyclene, caryophyllene, β-bisabolene, caryophyllene oxide, β-copanene, farnesene, humulene, selinene, cardinene and curcumene. The essential oil from the leaves has been shown to contain, nerolidiol, β-sitosterol, ursolic, crategolic, and guayavolic acids have also been identified. In addition, the leaves contain an essential oil rich in cineoland four triterpenic acids as well as three flavonoids; quercetin, its 3-L-4-4-arabinofuranoside (avicularin) and its 3-L-4-pyranoside with strong antibacterial action.

4. The *Asteraceae* (*Compositae*) family

The family *Asteraceae* (*Compositae*) is one of the largest families of flowering plants with about 1100 currently accepted genera and 25000 species (Heywood, 1977). It is of worldwide distribution particularly in semiarid region of the tropics and subtropics. The most members are evergreen shrubs or subshrubs or perennial rhizomatous herbs; biennial and annual herbs are also frequent. It is generally accepted that *Compositae* are a "natural" family with well established limits and a basic uniformity of floral structure imposed on all members by the common possession of charaters such as the aggregation of the flowers into capitula and the special features of the stamensand corolla.

Plants in *Asteraceae* are medically important in areas that don't have access to Western medicine. They are also commonly featured in medical and phytochemical journals because the sesquiterpene lactone compounds contained within them are an important cause of allergic contact dermatitis. *Asteraceae* (*Compositae*) are useful for therapeutic application due to their antihepatoxic, choleretic, spasmolytic, antihelminthic, antiphlogistic, antibiotic or antimicrobial activities. Some of them possess remarkable bacteriostatic and fungistatic properties and they probably participate in the pharmaceutical activity of some drugs and hence the elucidation of the structure of some members of the family. Some preliminary studies of *Piqueria trinervia* have demonstrated that active substances are present in these plants.

4.1 Piqueria trinervia Cav from the work "antibacterial activity of *Piqueria trinervia*, a Mexican medicinal plant used to treat diarrhea", Pharmaceutical Biology, 2007, vol. 45, No. 6, pp. 446–452

Piqueria trinervia Cav. is a perennial herb of the family *Asteraceae* that grows commonly in open areas of the pine-oak forests of the mountains throughout Mexico and Central America. It is usually called "hierba de San Nicolás" or "hierba de tabardillo." During the 16th century, it was known by its Nahuatl name as "cuapopolchi" and was used as a febrifuge as well as in the treatment of various gastrointestinal ailments including diarrhea. Today, it is employed to treat diarrhea, dysentery, "empachos" (a cultural disease with manifestations of various gastrointestinal disorders), intestinal infections, stomach pain, and typhoid fever (Argueta et al., 1994). To cure diarrhea or intestinal infections, an infusion or decoction is prepared with 10 g of the aerial parts in 0.5L of water; the tea is consumed throughout the day until the sickness disappears. This preparation is used as an enema for 2 days, once in the morning and once at night (Torres, 1984). To cure "empacho," people drink a half glass of the root decoction of the tea on an empty stomach for 2 days or until the sickness disappears.

4.2 Biological activity in salmonellosis

Because of its long history in traditional medicine in Mexico, the antimicrobial activity of *Piqueria trinervia* was investigated. Previous studies reported antibacterial activity only from the aerial parts. In our study, antimicrobial activity of extracts derived from both above-ground and root portions of "hierba de tabardillo" was evaluated using bacteria that are common to intestinal infections.

4.3 Plant material

The plant material used in this study was collected in the Ajusco zone, Distrito Federal, Mexico, during May and June. The voucher specimen is deposited in the National Herbarium (MEXU) located at the Instituto de Biología, UNAM. The plant was divided in three parts, thick (primary) roots, thin (secondary, 5mm or less) roots, and aerial parts (stem, leaf, flower, and fruit), and dried at room temperature. Each part was ground separately using a mill with rotary knives. The roots were extracted sequentially at room temperature with hexane, ethyl acetate, dichloromethane, and methanol.

4.4 Microbiological test

Test microorganisms. The microorganisms tested were strains of *Escherichia coli*, *E. coli* multidrug resistant (MDR), *Salmonella typhi*, *Shigella boydii*, *Staphylococcus aureus*, *Staphylococcus epidermidis*, *Yersinia enterocolitica*, nontoxic *Vibrio cholerae*, *Bacillus subtilis*, *Enterobacter aerogenes*, and *Enterobacter agglomerans*. The strains used for this study were provided by Dr. José Guillermo Avila of the Laboratory of Phytochemistry, UBIPRO, Facultad de Estudios Superiores, Iztacala, UNAM.

Bioassay. *In vitro* antibacterial activity was evaluated using the agar disk diffusion method in which the minimum inhibitory concentration (MIC) and minimum bactericidal concentration (MBC) were determined by microdilution assay in 96-well plates. Disks of 6-mm-diameter sterilized filter paper with concentrations of the extract 2 and 3 mg/disk were prepared as follows: (1) the extracts and fractions were dissolved in dimethyl sulfoxide (DMSO) and diluted with distilled water to obtain concentrations of 133 and 200 mg/mL; (2) then, 15 mL was applied to each disk. The methodology proposed by Bauer et al. (1966) and

Cáceres et al. (1990) was followed. A pure culture of each of the bacteria tested was incubated at 37 °C for 24 h in 5 mL of 1% peptone water (pH 7.2). Afterward, the culture was adjusted to a MacFarland value of 0.5 and diluted in a proportion of 1:20 with 0.15 M saline solution. An aliquote of 100 mL of the bacteria solution was spread over Müller-Hinton agar in the Petri dish, and the disks with the extracts were placed equidistant over the agar surface. After maintaining the Petri dish for 25 min at room temperature, incubation was followed at 37 °C for 24 h. The diameter of each inhibition zone is expressed in millimeters.

4.5 Results and discussion

Activity of organic extracts obtained by liquid-liquid separation. The partition of the aqueous extract of the aerial parts with dichloromethane and microbiological evaluation of the fractions showed activity only in the organic fraction. At a concentration of 3 mg/disk, this fraction was active against all the tested strains with an inhibition halo that ranged from

9 to 20 mm. *Y. enterocolitica* and *S. typhi* were the most sensitive bacteria with halos of 20 and 18 mm, respectively. Also, antibacterial activity was detected in the organic fraction of the aqueous root extract and inhibited the growth of *Y. enterocolitica*, *B. subtilis*, all the tested *Salmonella* strains, *Staphylococcus aureus*, and *Staphylococcus epidermidis* with halos ranging from 7 to 11mm. *Escherichia coli*, *Enterobecter aerogenes*, and *Enterobecter agglomerans* were resistant to the root extracts. These extracts produced bactericidal effects on *Salmonella dublin* (MIC= 3.0 mg/mL) and *Salmonella gallinarum* that produced diarrhea in chickens as well as on *Salmonella typhi* and *Staphylococcus aureus* with MBC ranging from 5 to 7.5 mg/mL.

Antibacterial activity of the organic extracts from the roots. The hexane extract from thick roots (TkR) at a concentration of 3 mg/disk produced antibacterial activity against all 11 strains assayed; meanwhile, the hexane extract obtained from the thin roots (TnR) was active only against *B. subtilis*, *E. agglomerans*, *S. aureus*, *S. epidermidis*, and *S. boydii* at the two tested concentrations (Tables 1 and 2). This TkR extract inhibited the growth of *V. cholerae* with a halo of 32.5mm as well as that of *Y. enterocolitica*, *S. boydii*, *E. aerogenes*, *S. epidermidis*, and *S. typhi* with inhibition halos between 24.0 and 11.5 mm. The activity of the extract at a concentration of 2mg/disk ranged from 14.0 to 11.5 mm, and the most sensitive strains were *S. epidermidis*, *S. aureus*, *B. subtilis*, *S. boydii*, *S. typhi*, and *E. Coli* (Table 1). In general, the TnR extracts had less antibacterial activity than the TkR extracts (Table 2). Nonetheless, *S. aureus* was highly sensitive to TkR and TnR hexane extracts with halos of 17.5 and 12 mm, respectively. Even at a higher dose of TnR hexane extract, the activity maintained the same order (Table 2). Dichloromethane extracts were less active than the hexane extracts (Table 2). In comparison with TnR, TkR dichloromethane showed the broadest antimicrobial action. At a concentration of 3mg/disk, the bacteria *V. cholerae*, *S. aureus*, *S. boydii*, *Y. enterocolitica*, *B. subtillis*, and *S. epidermidis* were sensitive to TkR, whereas TnR inhibited only *B. subtilis* and *E. agglomerans*.

Strains	Inhibition zone (mm)						
	Organic Extract						
		2 mg/disk			3 mg/disk		
	H	DM	EA	M	H	DM	EA
B. subtilis	13.0	14.5	12.5	9.5	12.0	10.0	12.0
E. coli	9.0	-	-	-	11.5	-	-
E. coli MDR	-	-	-	-	10.5	-	-
E. aerogenes	-	-	7.5	-	12.5	-	7.5
E. agglomerans	-	-	-	-	10.5	-	-
S. aureus	13.5	11.0	11.0	-	17.5	16.5	11.0
S. epidermis	14.0	12.0	13.0	-	11.5	10.5	8.0
S. boydii	11.5	12.0	11.0	-	17.0	16.5	11.5
S. typhi	11.0	-	-	-	11.5	-	8.5
Y. enterocolitica	NT	NT	NT	-	24.0	13.5	17.0
V. cholerae	NT	NT	NT	-	32.5	30.0	26.5

H, hexane; DM, dichloromethane; EA, ethyl acetate; M, methanol; -, no inhibition, NT, not tested.

Table 1. Antibacterial activity of the organic extracts of the thick roots of *Piqueria trinervia*.

The halo diameters for the effective dichloromethane extract of TkR at the concentration of 3mg/disk ranged from 16.5 to 10 mm, with the exception of the 30mm halo produced by V. cholerae. The ethyl acetate extract from TkR showed an inhibitory effect with halos between 17.0 and 8.0 mm. This extract was effective against most of the tested strains. TkR was inactive against E. coli, E. coli MDR, and E. agglomerans, and TnR was inactive against E. Coli MDR and E. agglomerans. Methanol extracts from TkR and TnR were generally inactive with the exception of the inhibition of B. subtilis, which produced a 9-mm halo (Tables 1 and 2).

In the majority of the inhibition zones, no bacterial growth was observed, but in some inhibitions zones we detected traces of bacterial growth. The inhibition zones without growth are indicated in the tables with bold numbers.

	Inhibition zone (mm)						
	Organic Extract						
	2 mg/disk				3 mg/disk		
Strains	H	DM	EA	M	H	DM	EA
B. subtilis	12.5	8.5	12.5	9.0	11.5	9.5	10.0
E. coli	-	-	-	-	-	-	7.0
E. coli MDR	-	-	-	-	-	-	-
E. aerogenes	-	-	-	-	-	-	8.5
E. agglomerans	8.7	-	-	-	9.0	8.0	-
S. aureus	12.0	-	11.0	-	12.0	-	9.5
S. epidermis	11.5	10.5	11.0	-	10.5	-	8.0
S. boydii	9.5	-	9.0	-	10.5	-	9.0
S. typhi	9.5	-	-	-	-	-	8.5
Y. enterocolitica	NT	NT	NT	-	NT	NT	12.5
V. cholerae	NT	NT	NT	-	NT	NT	27.0

H, hexane; DM, dichloromethane; EA, ethyl acetate; M, methanol; –, no inhibition; NT, not tested.

Table 2. Antibacterial activity of the organic extracts of the thin roots of *Piqueria trinervia*.

Activity of TkR hexane fractions. Nine fractions were obtained by TLC chromatography (hexane-ethyl acetate, 8:2) of the hexane TkR extract. B. subtilis and V. cholerae were highly sensitive to most of the isolated fractions, and the halos of the effect strains ranged from 8 to 18 mm. The greatest growth inhibition of B. subtilis was produced by F2, while that of V. cholerae (30 mm) came from the polar residue at the application zone. A moderate activity was found with fractions F1–F3 against S. aureus, S. epidermidis, and S. boydii (8 to 12 mm). E. aerogenes, E. agglomerans, and S. typhi were resistant to all the fractions. E. coli and E. coli MDR were inhibited by F1 and F2.

Chemical analyses. The organic fraction (dichloromethane) of the aerial part is a light oil, and the hexane extract of the roots is a dark yellow oil. Both oils were submitted for GC-MS study. According to the electronic database of the equipment, it was possible to identify 20 compounds besides Piquerol A (Fig. 1). The numbers are the proportion of each compound in the oil. The compounds present in both oils are the phenol, which was obtained also by Jiménez-Estrada et al. (1996) by chemical reaction of the piquerol; the other compound is the carquejol. The antibacterial activity is attributed to these compounds, as the antibacterial activity of the phenols have been reported. The carquejol and the piquerol A (Fig. 1) monoterpenes have a biogenetic structural arrangement characteristic of this plant; the destitution of the group on the six member ring is in the vicinal positions 5, 6. Thus, we assigned aromatic compounds the same substitution to the aromatic compounds.

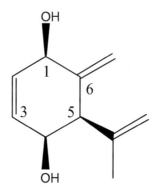

Fig. 1. Compound I: Piquerol A.

5. Conclusion

The organic portion of the aqueous extract prepared with the aerial part has highly activity and favors sustainable harvesting practices that would not damage the roots of plants in natural or cultivated populations. The extracts prepared with the thick root showed more activity with the hexane extract being the most active, followed by the ethyl acetate, dichloromethane, and methanol extracts, respectively. When the hexane extract of the thick root was subfractionated, the greatest antibacterial activity was retained in the residue at the application point and in the three most polar fractions. Based on these results, we conclude that P. trinervia exhibited antimicrobial activity.

6. Acknowledgments

This work was funded by financial aid from CONACYT, Mexico (project CONACYT 38608-M). We thank Dra. Gloria Gutierrez for the facilities provided for antimicrobial testing.

7. References

Abou-Gazar, H.; Bedir, E.; Takamatsu, S.; Ferreira, D.; Khan, I.A. (2004) Antioxidant lignans from *Larrea tridentata*. *Phytochemistry* Vol.65, pp. 2499–2505, ISSN 0378-8741.

Adeyemi, O.O.; Okpo, S.O.; Ogunti, O.O. (2002). Analgesic and anti-inflammatory effects of some aqueous extracts of leaves of *Persea americana* Mill (Lauraceae). *Fitoterapia*. Vol.73, pp. 375-80, ISSN: 0367-326X.

Akdemir Z. Ş., Tatli İ. İ., Saracoğlu İ., İsmailoğlu U B., Şahin-Erdemli İ. and Çaliş, (2001). Polyphenolic compounds from *Geranium* pratense and their free radical scavenging activities. *Phytochemistry*. Vol.56, pp. 189-193, ISSN 0378-8741.

Alanís A. D., Calzada F., Cervantes J. A., Torres J. and Ceballos G. M., (2005). Anti-bacterial properties of some plants used in Mexican traditional medicine for the treatment of gastrointestinal disorders. *Journal of Ethnopharmacology*. Vol.100, pp. 153-157, ISSN 0378-8741.

Almeida de A. P., Miranda M. M. F. S., Simoni I. C., Wigg M. D. and Lagrota M. H. C. and Costa S. S., (1998). Flavonol monoglycosides isolated from the antiviral fractions of *Persea americana* (Lauraceae) leaf infusion. *Phytotherapy research*. Vol.12, pp. 562-567, ISSN 0951-418X.

Alper, J. (2003). Date gaps need bridging to assess infectious gastrointestinal diseases. *ASM News* Vol.69, pp. 65-68, ISSN: 0044-7897.

Ankli, A., Heinrich, M., Bork, P., Wofram, L., Bauerfeind, P., Brun, R., Schmid, C., Weiss, C., Bruggisser, R., Gertsch, J., Wasescha, M., Sticher, O., (2002). Yucatec Mayan medicinal plants: evaluation based on indigenous uses. *Journal of Ethnopharmacology*. Vol.79, pp. 43–52, ISSN: 0378-8741.

Argueta, V.A.; Cano, A.L.M.; Rodarte, M.E. (1994): Atlas de las Plantas de la Medicina Tradicional Mexicana, Vol. 2. México, Instituto Nacional Indigenista, pp. 747–748. ISBN: 968-29-7324-4

Arteaga, S.; Andrade-Cetto, A.; Cárdenas, R. (2005). *Larrea tridentata* (Creosote bush), an abundant plant of Mexican and US-American deserts and its metabolite nordihydroguaiaretic acid. *Journal of Ethnopharmacology*. Vol.98, pp. 231–239, ISSN 0378-8741

Avila A. J.G.; M. C. Castañeda, J. C. Benítez F., R. Durán V.; G. Martínez C., J. L. Muñoz L., A. Martínez C., A. Romo de Vivar. (2005). Photoprotective Activity of Buddleja scordioides. Fitoterapia Vol.76, pp. 301-309, ISSN 0367-326X.

Bern, C.; Martines, J.; de Zoysa, I.; Glass, R.I. (1992). The magnitude of the global problem of diarrhoeal disease: a ten-year update. *Bull WHO* Vol.70, pp 705-714, ISSN 0042-9686

Calzada F., Cerda-García-Rojas C. M., Meckes M., Cedillo-Rivera R., Bye R. and Mata R., (1999). Geranins A and B, new antiprotozoal a type proanthocyani-dins from Geranium niveum. *Journal of Natural Products*. Vol.62, 705-709, ISSN 0163-3864.

Calzada F., Cervantes-Martínez J. A. and Yépez-Mulia L., 2005. In vitro antiproto-zoal activity from the roots of *Geranium mexicanum* and its constituents on *Entamoeba histolytica* and *Giardia lamblia*. *Journal of Ethnopharmacology*. Vol.98, pp. 191-193, ISSN 0378-8741.

Cortés A. R., Delgadillo A. J., Hurtado M., Domínguez-Ramírez A. M., Medina J. R. and Aoki K., (2006). The antispasmodic activity of Buddleja scordioides and Buddleja perfoliata on isolated intestinal preparations. *Biological and Pharmaceutical Bulletin*. Vol.29, pp. 1186-1190, ISSN 0918-6158.

Farias H (1886): Ligero estudio sobre la Yerba del Tabardillo. Bachelor Thesis, Escuela Nacional de Medicina y Farmacia, Facultad de Medicina, México, pp. 1–26.

Frei B., Baltisberger M., Sticher O. and Heinrich M. (1998). Medical ethnobotany of the Zapotecs of the Isthmus Sierra (Oaxaca, Mexico): Documentation and assessment of indigenous uses. *Journal of Ethnopharmacology*. Vol.62, pp. 149-165, ISSN: 0378-8741.

Gnan, S. O., Demello, M. T. (1999). Inhibition of *Staphylococcus aureus* by aqueous Goiaba extracts. *Journal of Ethnopharmacology*. Vol.68, pp. 103-108, ISSN: 0378-8741.

Goldhaber Pasillas GD (2004). Actividad Antimicrobiana de Piqueria Trinervia Cav. Sobre Algunas Bacterias Enterop patogenas. Bachelor Thesis, Facultad de Estudios Superiores Iztacala, UNAM, p. 152.

Gómez Salazar LC, Chong de la Cruz I (1985). Conocimiento y Usos Medicinales de La Flora de Amatlán, Municipio de Tepozotlán, Morelos. Bachelor Thesis, Facultad de Ciencias, Universidad Nacional Autónoma de México, p. 83.

Guerrant RL, Hughes JM, Lima NL, Crane J. (1990) Diarrhoea in developed and underdeveloping countries: magnitude, special settings, and etiologies. *Rev Infect Dis*, 12:s41-s50, ISSN: 0162-0886.

Heinrich M., Rimpler H. and Barrera A., (1992). Indigenous phytotherapy of gastrointestinal disorders in a lowland Mixe community (Oaxaca, Mexico): Ethnopharmacologic evaluation. *Journal of Ethnopharmacology*. Vol.36, pp. 63-80, ISSN: 0378-8741

Heinrich, M. (1998). Indigenous concepts of medicinal plants in Oaxaca, Mexico: Lowland Mixe plant classification based on organoleptic characteristics. Journal of Applied Botany. Vol.7, pp. 75-81, ISSN 0949-5460.

Heinrich, M., Ankli, A., Frei, B., Weimann, C., Sticher, O., (1998). Medicinal plants in Mexico: healers' consensus and cultural importance. *Soc. Sci. Med.* Vol.47, pp. 1859-1871, ISSN 02779536.

Heywood, V.H. et al., (eds.), (1977). The Bio. and Chem.of Compositae, 2:141–65. London

Hörr M., Heinrich M. and Rimpler H., (1996). Proanthocyanidin polymers with antisecretory activity and proanthocyanidin oligomers from *Guazuma ulmifolia* bark *Phytochemistry*. Vol.42, pp. 109-119, ISSN: 0031-9422.

Leonti, M., Sticher, O., Heinrich, M., (2003). Antiquity of medicinal plant usage in two macro-Mayan ethnic groups (México). *Journal of Ethnopharmacology*. Vol. 88, pp. 119-124, ISSN: 0378-8741

Linares Altamirano MME (1991): Flora útil de dos localidades del municipio de Tecali de Herrera Puebla. Bachelor Thesis, Facultad de Ciencias, Universidad Nacional Autónoma de México, pp. 91–92.

Lozano MascarruaGI (1996) Plantas medicinales utilizadas por los mazahuas del municipio de San Felipe de Progreso, Estado de México. Bachelor Thesis, Facultad de Ciencias, UniversidadNacionalAutónoma deMéxico, pp. 113–203.

Lozoya, X. y Lozoya M., (1982). Flora Medicinal de México. Primera parte: Plantas Indígenas. Instituto Mexicano del Seguro Social, México. pp. 17-29

Lozoya X., Aguilar A., y Camacho J. R. (1987). Encuesta sobre el uso acutual de plantas en la Medicina Tradicional Mexicana. *Revista Médica IMSS* (Mex) Vol.25, pp. 283-291

Lozoya, X., (1990). An Overview of the System of Traditional Medicine Currently Practised in Mexico. In: Economic and Medicinal Plant Research, Vol. 4, Plants and Traditional Medicine., Ed. by H. Wagner and N. R. Fransworth. Academic Press, pp. 71-93, ISBN: 9781741759242.

Maldini, M., Sosa, S., Montoro, P., Giangaspero, A., Balick, M.J., Pizza, C., Della Loggia, R., (2009). Screening of the topical anti-inflammatory activity of the bark of Acacia cornigera Willdenow, Byrsonima crassifolia Kunth. Sweetia panamensis Yakovlev and the leaves of Sphagneticola trilobata Hitchcock. *Journal of Ethnopharmacology.* Vol.122, pp. 430–433, ISSN: 0378-8741

Maldonado Almanza B. J., 1997. Aprovechamiento de los recursos florísticos de la Sierra de Huautla Morelos, México. Tesis Maestría en Ciencias (Biología), Facultad de Ciencias, Universidad Nacional Autónoma de México. 149 p

Martínez M., 1959. Las Plantas Medicinales de México. Ediciones Botas, 4ª Edición, México. 657 p.

Martínez M., 1979. Catálogo de nombres vulgares y científicos de plantas mexicanas, 1ª Ed. Fondo de Cultura Económica, México. 1220 p. ISBN: 9789681600112

Ortega M. E., Carrasco M. E. Mendoza G. and Castro G., (1998). Chemical composition of *Guazuma ulmifolia* Lam and its potential for rumiant feeding. *Cuban Journal of Agricultural Science.* Vol.32, pp. 383-386, ISSN 0864-0408.

Ortiz de Montellano, B. (1987). Caída de Mollera: Aztec Sources for a Mesoamerican Disease of Alleged Spanish Origin. *Ethnohistory* Vol.34, No.4, pp. 381-399, ISSN: 0014-1801.

Paray L (1953): Las Compuestas del Valle de México. Boletín de la Sociedad Botánica de México 15: 1–12.

Rzedowski J. y Rzedowski C de, G., (1995). Flora del Bajio y de Regiones adyacentes, fascículo 40, Instituto de Ecología A. C., México.

Roys, R.L., (1931). The Ethnobotany of the Maya, second ed. The Department of Middle American Research, The Tulane University of Louisiana, New Orleans.

Sánchez S., O. (1979). Flora del Valle de México, pp. 402

Secretaría de Salud, Mexico. (2005). Estadísticas de mortalidad en México: muertes registradas en el año 2003. Salud Pública Méx, 47:171-187.

Seigler, D.S.; Pauli, G.F. ; Fröhlich, R.; Wegelius, E. (2005). Cyanogenic glycosides and menisdaurin from *Guazuma ulmifolia, Ostrya virginiana, Tiquilia plicata,* and *Tiquilia canescens. Phytochemistry.* Vol.66, pp. 1567–1580, ISSN: 0031-9422.

Sheikh, N., Philen, R., Love, L., (1997). Chaparral-associated hepatotoxicity. *Archives of International Medicine.* Vol.157, pp. 913–919, ISSN 1538-3679.

Sinnott, R.A., Clark, D.W., De Boer, K.F., (1998). Non-toxic therapeutic extract of *Larrea tridentata.* PCT Int. Appl. WO 97- 4518103, 27pp.

Serkedjieva, J. and Ivancheva, S., (1999). Antiherpes virus activity of extracts from the medicinal plant *Geranium sanguineum* L. *Journal of Ethnopharmacology.* Vol.65, pp. 59-68, ISSN: 0378-8741.

Tapia-Conyer, R. (1994). La salud de los pueblos indígenas en México in : *Cuadernos de Salud.* Sepúlveda AJ (Ed.). México, Secretaría de Salud, Instituto Nacional Indigenista, p. 60, México, DF.

Torres González J (1984). Etnobotánica Mexicana Plantas Tradicionalmente Utilizadas Para el Tratamiento de Enfermedades Gastrointestinales en el Estado de Michoacán. Bachelor Thesis, Facultad de Ciencias, Universidad Nacional Autónoma de México, pp. 53–54.

Viesca T.C. (1986) *Medicina Prehispánica de México: El Conocimiento Médico de los Nahuas.* México: Panorama Editorial. ISBN 9683801609, 9789683801609

Waldman, R., Fontaine, O., Richards, L., 1994. Epidemic dysentery. In: A Supplement to Dialogue on Diarrhoea, vol. 55. CDR, WHO, CH-121, Geneva 27, Switzerland, pp. 1–2.

Waller, D.P. (1993). Methods in ethnopharmacology. *J. Ethnopharmacol.* Vol.38, pp. 89–195, ISSN: 0378-8741

Weimann, C. and Heinrich, M. (1997) Indigenous medicinal plants in Mexico: the example of the Nahua (Sierra de Zonglicia). *Botanica Acta.* Vol.110, pp. 62-72, ISSN: 0932-8629.

WHO Media centre. (2007). Food safety and foodborne illness in
http://www.who.int/mediacentre/factsheets/fs237/en/index.html

9

Xylariaceae Endophytic Fungi Metabolites Against *Salmonella*

Fernanda Pinheiro de Carvalho Ribeiro[1], Fernanda Carolina Sousa Fonseca[1],
Isabella Alves Reis[1], Isabella Santos Araújo[1], Hélio Mitoshi Kamida[1],
Alexsandro Branco[1] and Ana Paula Trovatti Uetanabaro[2]
[1]State University of Feira de Santana
[2]State University of Santa Cruz
Brazil

1. Introduction

For several years, natural products have been used directly as drugs or have provided the basic chemical architecture for deriving such drugs. Natural products are naturally derived metabolites and/or byproducts from plants, animals or microorganisms. These products have been exploited for human use for thousands of years and plants have been chief source of compounds used for medicine. Besides plants, microorganisms constitute a major source of natural products with desirable bioactive properties. The ultimate purpose of the researchers is looking for new sources of metabolites. The marine organisms, for example, have been studied with more attention in the recent decades. Endophytic fungi appear to be another interesting source of research. Because of what appears to be their contribution to the host plant, the endophytes may produce a plethora of substances that may have potential use to modern medicine, agriculture and industry.

Infections caused by pathogenic microorganisms are responsible for high rates of morbidity and mortality in world (Coelho et al., 2007; Souza et al., 2007). These infections can occur in invasive form, and are an increasing problem due to the increase of their incidence in hospitals, especially in patients who are undergoing cancer treatment, transplantation or are immunosuppressed for other reasons (Oliveira et al., 2001).

The genus *Salmonella* is extremely heterogeneous, comprising almost 2000 serotypes, of which only a few are major human pathogens (Kaufmann et al, 2001). However, despite this apparent complexity, *Salmonella* species are actually quite similar genetically, with the serotype differences based on surface antigen differences such as LPS and flagella (Kaufmann et al, 2001). The two main symptoms of salmonellosis are typhoid or typhoid-like fever and gastroenteritis (Tahergorabi et al, 2011). Salmonella spp. are a leading cause of acute gastroenteritis in several countries, and salmonellosis remains an important public health problem worldwide, particularly in the developing countries (Mead et al., 1999). Food is a transmission vehicle for *Salmonella* that causes about 96% of all salmonellosis cases (Tahergorabi et al, 2011). Gastroenteritis-causing pathogens are the second leading cause of morbidity and mortality worldwide, with children under the age of 5 years at greatest risk

(Guerrant et al, 2001). Such serious infections are most common in children and the elderly (Rotimi et al, 2008).

S. typhi remains an important health threat for humankind, with more than 16 million cases and 600000 deaths annually, worldwide (Kaufmann et al, 2001). Ironically, typhoid fever is declining worldwide, but non-typhoidal *Salmonella* infections are increasing rapidly, due to increased automation in food processing and other factors (Kaufmann et al, 2001). Non-typhoidal species also cause serious disease in immunocompromised individuals. (Rotimi et al., 2008). Up to a decade ago, in many countries, conventional 1st-line antimicrobial agents, such as ampicillin, chloramphenicol, and trimethoprim–sulfamethoxazole, were the drugs of choice for the treatment of lifethreatening salmonella infections (Rotimi et al 2008). However in the past two decades, these species are also becoming increasingly resistant to most antibiotics, which has significantly increased the concern about these food and water-borne pathogens (Threlfall et al., 1997; Therefall, 2002; Wedel et al., 2005).

In recent years, several powerful new technologies have been developed that have significantly enhanced our knowledge of *Salmonella* pathogenesis (Beuzon & Holden, 2001; Kaufmann et al., 2001). Novel approaches for development of new antibiotics have been pursued, such as combinatory chemistry tools but only a few new antibiotics are produced by the pharmaceutical industry nowadays (Coates & Hu, 2007) Despite of the huge expectative on synthetic molecules with effective antimicrobial properties, natural products are still a worth promise.

Thus, the search for new compounds with antimicrobial activity from plants and fungi has been the subject of intense research in recent years (Harvey, 2007; Lee et al., 2007; Hostettmann et al., 2003). This is due mainly to the fact that the plants are widely used in folk medicine to combat various diseases in humans caused by bacteria and fungi (Stefanello et al., 2006; Duarte et al., 2004; Cruz et al., 2007). In this sense, many researchers are aiming to scientifically prove the use of plant extracts as an effective control of infections of the skin (Weckesser et al., 2007), the mouth (More et al., 2008) and other infections caused by a range of Gram-positive and Gram-negative bacteria (Vuuren, 2008; Lee et al., 2007; Chauhan et al.,2007).

2. Endophytic fungi

Endophyte is one which resides in the tissues beneath the epidermal cell layers without causing any immediate, overt negative effects (Stone et al., 2000). It is worthy to note that studies have shown that, nearly 300000 plant species that exist on earth, each individual plant is the host to one or more endophytes, the population of a given endophytic species varies from several to a few hundreds strains (Strobel & Daisy, 2003; Huang et al., 2007; Yu et al., 2010). The endophytes may live in plants air parts, especially in leaves, but can also be found living in intracellular gaps of roots, that is one of the main entrance door for these microorganisms (Azevedo et al., 2001). The endophytic colonization can be positive to host plants. Both fungi and bacteria are the most common microbes existing as endophytes, but the most frequently isolated are fungi.

The endophytes transmission from one generation to another may occur vertically, among seeds, during plants reproduction, or horizontally, where fungi spores are transmitted by air way, water or insects (Carroll, 1988). The microorganism penetration may occur from natural gaps or insects. The roots are the entrance main door (Kobayashi & Palumbo, 2000).

Of the myriad of ecosystems on earth, those having the greatest biodiversity seem to be the ones also having endophytes with the greatest number and most diversity (Strobel et al., 2005). The observation of Moricca & Ragazzi (2008) indicates that the type of interaction between an endophyte and a plant is controlled by the genes of both organisms and modulated by the environment. The fungi produce secondary metabolites compounds that have various biological activities, and have great bioactive potential (Petrini, 1991). The symbiosis among plants and fungi, mainly endophytic fungi, might be an important source of active pharmacologic compounds.

Beyond production of substances that come from secondary metabolism as the majority of antimicrobial, the symbiosis among plants and endophytic fungi can lead to other benefits to host plants as substances that improve growth and host competitively in nature (Hallman et al., 1997; Azevedo et al., 2000). Besides pharmacological potential, these microorganisms can also be highlighted for their capacity to produce interesting substances for farming, as growth plants regulators or insecticides, acting an important role from the ecological point of view (Souza, 2004). Only a handful of plants have ever been completely studied relative to their endophytic biology, consequently is an opportunity to find novel endophytic microorganisms and antimicrobial metabolites produced by them.

2.1 Biological activity from medicinal plants endophytes

The search for substances with pharmaceutical utility was one of the reasons that boosted the endophytic fungi researches. Beyond the studies about endophytic colonization, the characterization of new metabolites produced from a symbiotic association between fungus and host plant lead to isolation of various compounds with commercial importance. A diversity of biosynthetic classes metabolites were isolated from endophytic fungi and most of them showed interested pharmacological effects (Tan et al., 2001; Gunatilaka, 2006)

The development of new agents with pharmacological proprieties still is a great challenge for science and represents and endless research area. In the last 40 years, plenty of metabolites with different carbon skeleton were isolated from fungi. The isolation of cyclosporine in 1970, as a metabolite from *Cylindrocarpon lucidum* and *Tolypocladium inflation* fungi, represented and important step in the immunosuppressive treatment. (Hanson, 2008). Thought the past of the years, another studies were published describing interesting metabolites isolation. Borges et al. (2006) described the presence of derivatives anthraquinones produced by *Phoma sorghina*, and endophytic fungus associated to a medicinal plant *Tithonia diversifolia* (Asteracea). This plant extracts are used for the treatment of malaria, diarrhea, fever, hepatitis and wounds (Gu et al., 2002, Cos et al., 2002). There are also attributed anti-inflammatory, amebic, antispasmodic, antifungal, antibacterial and antiviral activities (Goffin et al., 2002, Cos et al., 2002).

Lu & co-workers (2000) in a research at Nanjing, China, observed the presence of 11 bioactive metabolites produced by *Colletotrichum* sp. and endophytic fungus isolated from *Artemisia annua* (Asteracea), traditional plant from Chinese medicine. When tested against bacteria, some of there metabolites showed inhibitory activity against gram-negative and positive bacteria, as *Pseudomonas sp.* and *Bacillus subtilis*. Other metabolites were active against pathogenic fungi, as *Candida albicans* and *Aspergillus niger* in the concentration of 200 µg/mL.

2.2 Xylariaceae metabolites biological activity

The Xylariaceae family is considered a great source of a variety of bioactive compounds, showing plenty of chemical structures and biological activity. As an example, can be highlighted the taxol, a diterpene derived that have been used as an effective anti-cancer agent (Stierle et al., 1993). Among fungi that belong to Xylariaceae, the genus *Xylaria* is an important source of new secondary metabolites, with a variety of chemical structures and distinct biological activities.

The chemical investigation of fungi of Xylariaceae family showed as a potential source of biotechnological products, mainly with pharmacological proprieties. The study leadered by Healy et al. (2004) resulted on the isolation of xanthones, compounds isolated from endophytic fungi identified as *Xylaria sp.* The fungus isolated from *Glochidion ferdinandi* plant, and the metabolites extracted showed important pharmacological activities (Peres & Nagem, 1996; Peres et al., 2000) as for example, anti-inflammatory (Lin et al., 1996), antimicrobial (Malet-Cascon et al., 2003), antioxidants (Minami et al., 1994), antifungical (Rocha et al., 1994) e anticancer properties (Ho et al., 2002).

Krohn et al. (2004), developed an similar research, describing the syntesis of xyloketal D, a natural product that belongs to a group of secondary metabolites isolated from *Xylaria* sp. The biotechnological interested for this group is based on inhibitory activity over acethylcholinesterase enzyme.

Based on the abundance of secondary metabolites found in Xylariaceae family, Liu et al. (2008), identified and described the biological activity of 7-amino-4-methylcoumarin, a compound extracted from *Xylaria* sp YX-28 endophytic fungus. The chemical investigation of *Xylaria* (Xylariacea) genus fungus leads to potentials sources of natural products, as Xylarenal A, a terpenoid isolated form *Xylaria persicaria* fermentation (Smith et al., 2002) e xylactam, a nitrogened compound obtained from *Xylaria euglossa* ascomycete (Wang et al, 2005).

The *Hypoxylon, Nodulisporium* and *Daldinia* constitute one of the largest and most important genus of Xylariaceae, and they show a great diversity and production of secondary metabolites (Laessoe et al., 2010; Stadler et al., 2001; Kamisuki et al., 2007). The figure 1 shows some chemical structures isolated from genus *Hypoxylon, Nodulisporium* and *Daldinia*.

2.3 Antimicrobial activity of endophytic fungi metabolites

There is a general call for new antibiotics that are highly effective, possess low toxicity and will have minor environmental impact. This search is driven by the development of resistance infectious microorganisms (e.g. *Staphylococcus, Mycobacterium, Streptococcus*) to existing compounds and by the menacing presence of naturally resistant organisms (Strobel et al., 2005). In support of this idea, metabolites of endophytes have been reported to inhibit a number of microorganisms (Petrini, 1991; Gurney & Mantle, 1993). Many important antifungal and antibacterial chemotherapeutics are either microbial metabolites or their semi-synthetic derivates.

Between the years of 1981 to 2006, the Food and Drug Administration (FDA) had approved 1,184 new drugs among about 609 (51.4%) were natural products related: 55 were natural products, 270 natural product derived by chemical modification (semi-synthetic), 52 were done by synthesis where the active core came from a natural product, and 232 were synthesized by imitating a natural product (Newman & cragg, 2007).

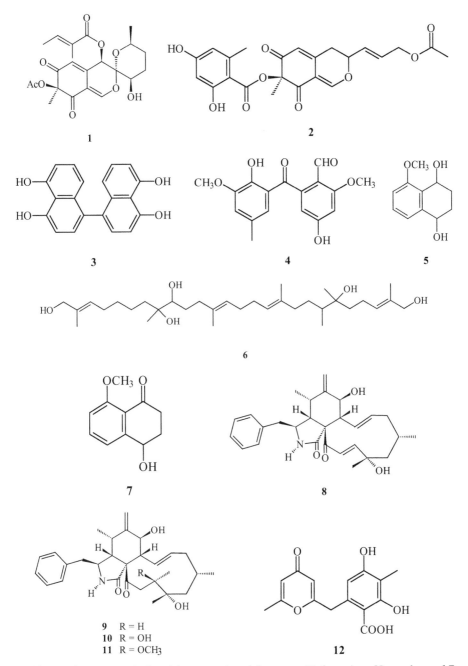

Fig. 1. Chemical structures isolated from species of the genus *Nodosporium, Hypoxylon* and 7 *Daldinia*.1= daldinin C; 2= hypomiltin; 3= BNT; 4= dalninal A; 8= cytochalasin 8 II; 9= cytochalasin I; 10= cytochalasin III; 11= cytochalasin IV; 12= macrocarpon A

Natural products from endophytic microbes have been observed to inhibit or kill a wide variety of harmful microorganisms including, but not limited to phytopathogens, as well as bacteria, fungi, viruses and protozoan that affect humans and animals (Strobel et al., 2005). Taechowisan et al. (2008), described the biological activity of chemical compounds of *Streptomyces* sp., an endophytic fungus isolated from Alpinia galanga (Zingiberaceae) roots, medicinal plant collected in Nakom Pathom, Tailândia, surrounds. According to the authors, the isolated substances showed antimicrobial activity against the following microorganism: Staphylococcus aureus ATCC25932, *Bacillus subtilis* ATCC6633, *Escherichia coli* ATCC10536, *Pseudomonas aeruginosa* ATCC27853, *Candida albicans* ATCC90028 e *Colletrotrichum musae*.Taechowisan et al. (2008), described the biological activity of chemical compounds of *Streptomyces* sp., an endophytic fungus isolated from Alpinia galanga (Zingiberaceae) roots, medicinal plant collected in Nakom Pathom, Tailândia, surrounds. According to the authors, the isolated substances showed antimicrobial activity against the following microorganism: Staphylococcus aureus ATCC25932, *Bacillus subtilis* ATCC6633, *Escherichia coli* ATCC10536, *Pseudomonas aeruginosa* ATCC27853, *Candida albicans* ATCC90028 e *Colletrotrichum musae*. Rocha et al. (2010) observed antagonistic activity of endophytic fungi against to conidia of *Microcyclus ulei*, the agent of South American Leaf Blight responsible for the weak development of rubber plantations in Latin America. Endophytic fungi were isolated from *Hevea brasiliensis* (the rubber tree) leaves, cultivars FX3864, CDC312, MDF180, exhibiting distinct resistance levels to the attack by *M. ulei*. Lyophilized culture filtrates obtained from fungal isolates, grown in liquid malt extract medium, were tested *in vitro* and showed activity against germination of *M. ulei* and exhibited marked inhibitory activity on *M. ulei* conidia germination *in vitro*. The lyophilized culture filtrate of eleven foliar endophytic isolates achieved high inhibitory activity on *Microcyclus ulei* conidia germination and belong to seven genera: *Fusarium* sp., *Gibberella* sp., *Glomerella cingulata*, *Microsphaeropsis* sp., *Myrothecium* sp., *Pestalotiopsis* sp. and *Phomopsis* sp.

Davis et al. (2005) also tested antimicrobial activity of endophytic fungi. After chemical analysis of endophytic fungi cultures, *Eupenicillium* sp. Isolated from an endemic plant in Australia, *Glochidium ferdinandi* (Euphorbiaceae), the authors verified the presence of the following compounds: phomoxin B e C, eupenoxide e phomoxin. The isolated compounds were tested against plenty of microorganism associated to nosocomial infections, including *Staphylococcus aureus* drug multi-resistant (MRSA), *Staphylococcus aureus* (NCCLS 29523), *Escherichia coli* (ATCC 25922), *Enterococcus faecalis* (NCCLS 29212), *Pseudomonnas aeruginosa* (ATCC 27853), *Streptococcus pyogenes* (ATCC 19615), *Acinetobacter anitratus* e *Candida albicans* (ATCC 60193).

The fungus was isolated from a traditional medicinal Chinese plant Ginkgo biloba L. (Ginkogoaceae). The metabolite was identified by NMR and mass spectra. The isolated compound showed activity against Staphylococcus aureus (MIC, 16 µg.mL-1), Escherichia coli (MIC, 10 µg.mL-1), Salmonnela typhia (MIC, 20 µg.mL-1), Salmonnela typhimurium (MIC, 15µg.mL-1), Salmonella enteriditis (MIC, 8,5 µg.mL-1), Aeromonas hydrophila (MIC, 4µg.mL-1), Yerisinia sp. (MIC, 12,5 µg.mL-1), Vibrio anguillarum (MIC, 25 µg.mL-1) Shigella sp. (MIC, 6,3 µg.mL-1), Vibrio parahaemolyticus (MIC, 12,5 µg.mL-1) Candida albicans (MIC, 15 µg.mL-1), Penicilium expansum (MIC, 40 µg.mL-1) e Aspergillus niger (MIC, 25 µg.mL-1).According many researchers (Arnold and Lutzoni 2007; Huang et al., 2008; Tejesvi et al., 2009), endophytes of tropical plants constitute a species-rich ecological assemblage of fungi and should be included in screening programs for novel metabolites (Suryanarayanan et al., 2009).

Metabolites biologically active have been isolated and characterized (Shulz et al., 2002; Tejesvi et al., 2009; Aly et al., 2010), and could be new molecules for many applications. Shulz and Boyle (2005) and Aly et al. (2010) point that these active metabolites belong to the chemical groups, such as phenols, steroids, flavonoids, quinines, terpenoids, xantones, peptides, cytocatalasins, alkaloids, aliphatic compounds, and phenylpropanoids.

3. Xylariaceae metabolites and *Salmonella* strains

The Xylariaceae endophytics fungus were previously isolated by our research group from a cultivate specimem of *Mikania laevigata*, Asteraceae family (Ribeiro, 2011). This plant constituted in a Brazilian medicinal plant more used for respiratory infections. The endophytics fungus species were submitted to sequencing the rDNA ITS region which resulted in the identification of three strains of *Nodulisporium* sp., three strain of *Hypoxylon* sp., one strain of *Daldinia* sp and four strains unknown. The crude extracts above fungi were obtained by its cultivation on potato dextrose agar at 28 °C in Erlenmeyer flasks (3 × 100 m L), for 7 days. The cultures were filtered to separate broth and mycelia. The mycelia were extracted by reflux with methanol (100 mL) to furnish the respective crude extract (Ribeiro et al, unpublished data)

For antimicrobial test four isolates of *Salmonella* were tested: *S. enteritidis* CCMB 522; *S. carrau* CCMB 523 and two isolated from from food (*Salmonella* sp. CCMB 270 and CCMB 281).

3.1 Determination of minimun inhibitory concentration (MIC)

The broth microdilution susceptibility test was used to determinate the minimum inhibitory concentration (MIC) for bacteria as recommended by CLSI (2003). The extract was dissolved in dimethyl sulfoxide (DMSO) and water (50:50) to reduce the inhibitory potential of DMSO and then the extract were sterilized by filtration through cellulose acetate membrane (0.22 mm). Geometric dilutions were prepared using 96-well, flat-bottom microdilution plate which received 90 mL of DMSO extracts diluted in water in lines A1 to A9 containing 90 mL of previously Mueller-Hinton broth two times concentrated. So, the first wells (A1 to A9) contained crude extracts diluted in a concentration from 1 mg.mL^{-1} until 0.008 mg.mL^{-1} (H1 to H9). The suspension of the micro-organism test was adjusted to 1.5 x 10^8 cells mL^{-1} for bacteria in 0.45% sterile saline. After the dilutions were carried out in all wells, each well received 10 µL of the microbial suspension performing a total volume of 100 mL by well (90 µL of the extract and HCM + 10 µL micro-organism). The plates were incubated at 37 ° C for 24 hours. After the incubation period, were added 30 mL of rezasurin (7-hydroxy-3H-fenoxazina-3-one-10-oxide) at final concentration of 0, 01% for quantitative analysis of microbial growth and determining the relative antimicrobial activity of each sample dilution. All tests were performed in triplicate. Dilutions of the antibiotic chloramphenicol (10 mg.mL^{-1}) were used as positive control for comparison of data between independent experiments and as indicators for assessing the relative level of inhibition of the samples tested. Controls of the microbial viability, sterility of the medium, sterility the extract and the potential for inhibition of DMSO on the micro-organisms tested were also carried out. In this work a representative results of MIC values was regarded as equal to or less than 0.5 mg.mL^{-1} of the extracts tested.

3.2 Determination of minimum microbicide concentration (MMC)

After determining the MIC, the minimal microbicidal concentration (MMC) was done. Aliquots of 5 µL of the wells were plated on Mueller Hinton Agar (MHA) and incubated at 37 ° C for 24 hours. The MMC was considered the lowest concentration of the extract which showed no cell growth on the surface of (MHA).

3.3 Xylariaceae extracts against *Salmonella*

Twelve extracts were tested against four strains of *Salmonella*. The results are shown in Table 1.

Extracts		*Salmonella* sp. CCMB 270	*Salmonella* sp. CCMB 281	*Salmonella* Enteritidis CCMB 522	*Salmonella* carrau CCMB 523
Nodulisporium sp. (specie 1)	MIC	0,25	0,5	0,5	0,25
	MMC	0,5	1	-	1
Nodulisporium sp. (specie 2)	MIC	0,5	0,5	0,25	0,5
	MMC	1	1	-	-
Nodulisporium sp. (specie 3)	MIC	0,5	0,5	0,25	0,5
	MMC	1	1	-	-
Hypoxylon sp. (specie 1)	MIC	1	0,5	0,5	0,5
	MMC	1	1	-	-
Hypoxylon sp. (specie 2)	MIC	0,5	0,5	0,5	0,25
	MMC	1	1	-	1
Hypoxylon sp. (specie 3)	MIC	0,5	0,5	0,25	0,5
	MMC	1	1	-	-
Daldinia sp.	MIC	0,5	0,5	N/A	N/A
	MMC	1	1	N/A	N/A
Specie 3 (Unknown)	MIC	0,5	0,5	0,25	0,25
	MMC	1	1	-	-
Specie 8 (Unknown)	MIC	0,5	0,5	0,5	0,25
	MMC	1	1	-	1
Specie 10 (Unknown)	MIC	0,5	0,5	N/A	N/A
	MMC	1	1	N/A	N/A
Specie 12 (Unknown)	MIC	0,25	0,5	0,25	0,5
	MMC	0,5	1	-	-
Specie 14 (Unknown)	MIC	0,5	0,5	0,25	0,5
	MMC	1	1	-	-
CONTROL	CHLORAN	0,31	0,31	0,16	0,31
	DMSO	1	1	1	1

N/A: not applied; CHLORAN: Chloramphenicol; DMSO: dimethylsulfoxide

Table 1. Minimum inhibitory concentration (MIC) and minimal microbicidal concentration (MMC) (mg.mL^{-1}) of Xylariaceae endophytic fungal extracts on *Salmonella* strains

The microdilution technique for determination of minimum inhibitory concentrations (MIC) is often considered as the best methodology for assessing susceptibility or resistance of bacteria to antibiotics (Rivers et al., 1988; Reis, 2006; Alves et al., 2008). According to Ostrosky et al. (2008), MIC has several advantages and one is that this method can be 30 times more sensitive than other methods used in the literature. The DMSO control showed growth inhibition in a dilution corresponding to 1 mg.mL^{-1} extract through MIC determination. Therefore, results were considered representative for MIC values at or below the next lower dilution of the extracts tested (0.5 mg.mL^{-1}). DMSO is a substance that facilitates the diffusion (Vieira, 2005), but it was necessary to control the solvent, since this can enhance the activity of the antimicrobial agent (Herschler, 1970; Ribeiro et al., 2001). All the extracts tested (three strain of *Nodulisporium* sp., three strain of *Hypoxylon* sp., one strain of Daldinia sp. and five strain unknown) samples shown antimicrobial activity less than the representative value for DMSO for MIC (Table 1). Most of samples studied demonstrated MMC= 1.0 mg.mL^{-1} for *Salmonella* sp. CCMB 270 and *Salmonella* sp. CCMB 281. These results show that crude extracts had the same antimicrobial effect on salmonella strains studied. None of the tested extracts showed activity against all microorganisms, because some MIC values obtained were 1 mg.mL^{-1} and were not considered as representative (Table 1). However, the fact of not showing detectable antimicrobial activity does not mean that the fungal extracts evaluated did not possess bioactive compounds against microorganisms tested.

The strain *Salmonella* enteritidis CCMB 523 seems more sensitive compared to the extracts tested, since it was inhibited at 0.25 mg.mL^{-1} by 6 (60%) of the extracts evaluated (Figure 2). Under the same conditions, *Salmonella* carrau CCMB 523 was inhibited by 4 (40%), *Salmonella* sp. CCMB 270 was inhibited by 2 (20%) of the extracts evaluated, while *Salmonella* sp. CCMB 281 was not inhibited at this concentration by none of the tested extracts.

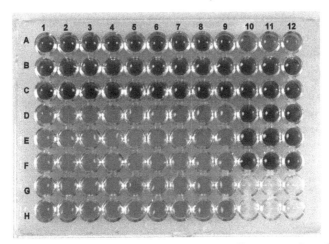

Fig. 2. Determination of minimum inhibitory concentration. Representation of serial dilution of methanolic extracts against *Salmonella enteritidis*. Columns 1-3: **Specie 14 (Unknown)** extract, MIC = 0.25 mg.mL^{-1}, Columns 4-6: *Hypoxylon* **sp. (specie 3)** extract, MIC = 0.25 mg.mL^{-1} and Column 7-9: **Specie 12 (Unknown)** extract, MIC = 0.25 mg.mL^{-1}. Line A 10-11-12: microbial growing control of microorganism tested. Lines B 10-11-12, C 10-11-12 and D 10-11-12: control of extracts and lines F 10-11-12 and G 10-11-12: Control of the sterility of the culture medium (HCM).

The Gram-negative bacteria are reported to possess resistance to several antibiotics. The complexity of Gram-negative bacteria makes them less susceptible to antimicrobial agents (Tadeg et al., 2005).Variations related to determining the MIC of natural extracts can be attributed to several factors. Thus, there is no standardized method for expressing the results of antimicrobial testing of natural products (Fennel et al., 2004; Ostrosky et al., 2008). So the results can be influenced by microorganisms used for testing, the selected method and the solubility characteristics of each substance (Vanden et al., 1991; Valgus et al., 2007).

Through the use of the MIC was possible to demonstrate quantitatively the concentration of extracts that inhibited each microorganism, but this test only indicates the concentration able to cause growth inhibition and does not identify whether the inhibition was bactericidal or bacteriostatic. For this, we used the test Microbicide Minimum Concentration (MMC). Observing the results, the *Nodulisporium* sp. (specie 1) and Specie 3 (Unknown) extracts showed microbicidal activity up to 0.5 mg.mL^{-1} against *Salmonella* sp. CCMB 270.

3.4 HPLC-DAD extracts analysis

Chromatographic techniques are used to separate the constituents from a mixture of substances aiming isolation and identification, being used to chemical investigation of crude extracts and identification of secondary metabolites of interested (Strege, 1999). High Performance Liquid Chromatography (HPLC) is one of methods of choice for determination of secondary metabolites profile to fungi isolated from Xylariaceae family. Ascomycets as *Daldinia, Hypoxylon* and *Xylaria* have been extensively studied using this method for chemical profile determination of majority and minority sample components. The analyses are based on retention time and UV absorption spectra. This technique also facilitates uncolored metabolites detection (Stadler et al., 2004).

The crude extract obtained from methanol extraction and tested to biological activity were analysed and monitored by High Performance Liquid Chromatography with diode array detector (HPLC-DAD), Hitachi, Laechrom Elite model, LiCospher 100 RP18 (5 μm) column, with 150 mm x 04 mm dimensions, Merck, equipped with diodo array detector (DAD). The mobile phase was composed of solvent (A) H_2O/H_3PO_4 0.1% and solvent (B) MeOH. The solvent gradient was composed of A (75-0%) and B (25-100%) for 25 minutes. A flow rate of 1.0mL/min was used, and 20μL of each sample was injected. Chromatographic peaks were monitored at 260 nm and characterized by retention time and UV-vis spectrum (200-600 nm). The HPLC-DAD chromatograms of the methanolic crude extracts from mycelium of *Nodulisporium* sp. (three strains), hypoxylon sp. (three strains) and *Daldinia* sp. were showed in Figure 2. The Figure 3 showed the HPLC-DAD chromatograms of the methanolic crude extracts from mycelium of five unknown species of Xylariaceae.

The *Nodulosporium* sp (specie 2) and *Daldinia* sp demonstrated to contain lack compounds in this analysis conditions when *Nodulosporium* (species 1 and 3) and all species de *Hypoxylon* showed two majoritary compound in the respective chromatograms. In the specific case of unknown species was to verify the presence de several compounds in the respective HPLC-DAD chromatograms (Figure 3). In the crude extract of unknown species 3, 10 and 12 the peaks eluted between 15 to 20 minutes, while in species 6 and 10 the peaks were eluted before 10 minutes, showing two majoritary peaks as the same profile for *Hypoxylon* chromatograms.

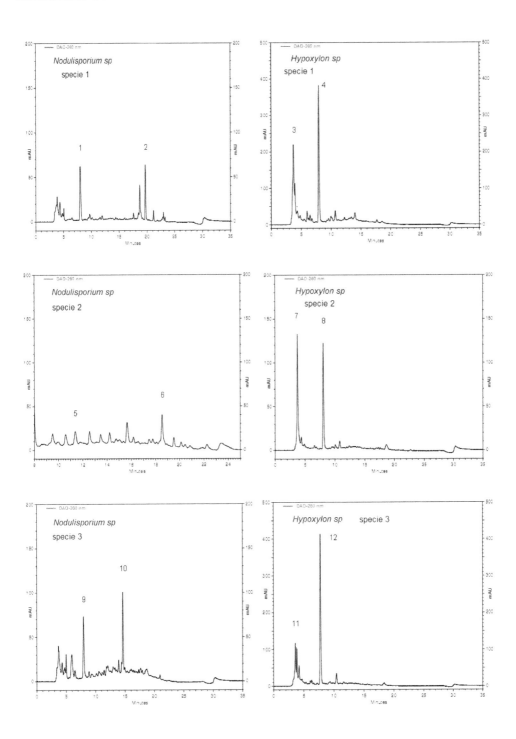

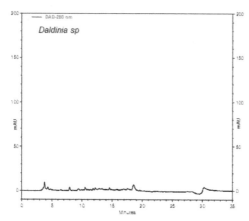

Fig. 3. HPLC-DAD chromatograms of methanolic extracts obtained from *Nodulisporium* sp., *Hypoxylon* sp. and *Daldinia* sp

The *Nodulisporium* chromatograms showed two majoritary compounds eluted in slightly different times but showed similar UV characteristics, suggesting to have the same chromophores (236 and 292 nm). The *Hypoxylon* chromatograms also showed two majoritary compounds eluted in about 3.7 and 7,8 minutes with the same UV profile (260 and 280 nm). On the basis of spectral identification it can be suggest that these compounds might be identified as phenolics.

A possible explanation for the antimicrobial activity of the methanolic extract against salmonella may be the fact that one or some of its constituents caused a significant inhibition of bacterial mobility besides ion permeability alteration on the into bacteria membrane. Antimicrobial activity of phenolic compound toward microorganism, as *salmonella,* is well documented and support this chemical investigation (Orsi et al, 2005; Nohynek et al, 2006). Addition studies are being performed for compounds isolation and identification.

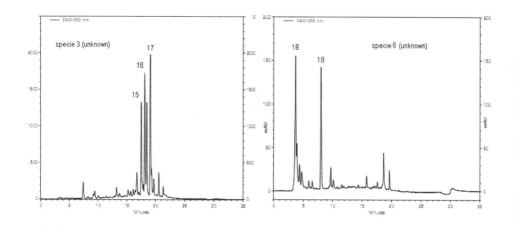

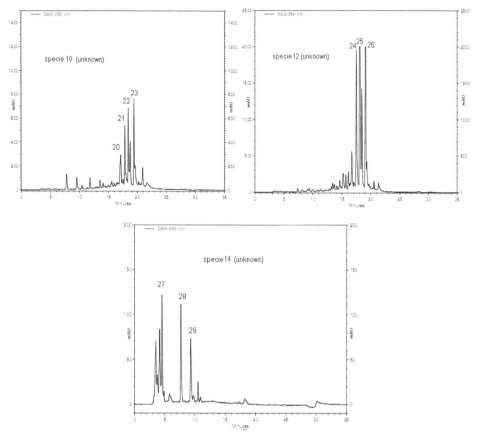

Fig. 4. HPLC-DAD chromatograms of methanolic extracts obtained from five unknown species

4. Conclusion

This work evaluated the antimicrobial activity against four *Salmonella* species of the crude extracts of endophytic fungi. Eleven fungi were isolated from *Mikania laevigata* (Asteraceae), a Brazilian medicinal plant and identified by our group and identified as being from Xylariaceae family. All the extracts of *Nodulisporium* sp., *Hypoxylon* sp., *Daldinia* sp. and unknown species showed similar antimicrobial activity. The HPLC-DAD analysis showed that extracts may contain phenolics compounds comum in others genus.

5. Acknowledgements

The authors would like to thank the Conselho Nacional de Desenvolvimento Científico e Tecnológico (CNPq), Coordenação de Aperfeiçoamento de Pessoal de Nivel Superior (CAPES) and Fundação de Amparo à Pesquisa do Estado da Bahia (FAPESB). The authors wish to acknowledge the Post Graduation Biotechnology Program of State University of Feira de Santana for financial support.

6. References

Alves, E.G., Vinholis, A. H. C., Casemiro, L. A., Jacometti, N. A., Furtado, C. & Martins, C. H. G. (2008), Estudo comparativo de técnicas de *screening* para avaliação da atividade antibacteriana de extratos brutos de espécies vegetais e de substâncias puras. *Química Nova*, Vol.31, No.5 (April, 2008), pp. 1224-1229, ISSN 0100-4042.

Aly, A.H.; Debbab, A.; Kjer, J. & Proksch, P. (2010). Fungal endophytes from higher plants: a prolific source of phytochemicals and other bioactive natural products. *Fungal diversity*, Vol.41, pp.1-16, ISSN 1560-2745

Arnold, A.E.; Lutzoni, F. (2007). Diversity and host range of foliar fungal endophytes: are tropical leaves biodiversity hotspots? *Ecology*. Vol.88, pp.541-549, ISSN 0012-9658.

Azevedo, J.L. (2001). Endophytic fungi and their roles in tropical plants. *Progress In Microbial Ecology*, pp. 279-287.

Azevedo, J.L.; Azevedo, J.L.; Maccheroni Jr. W.; Pereira, J.O. & Araújo, W.L. (2000). Endophytic microorganisms: a review on insect control and recent advances on tropical plants, *Electronic Journal of Biotechnology*, Vol.3, pp. 40-65.

Baker, D.; Mocek, U. & Garr, C. (2000), Natural Products vs combinatorials: a case study. In: *Biodiversity: new leads for pharmaceutical and agrochemical industries*. Wrigley, S.K.; Hayes, M. A.; Thomas, R. Chrystal, E.J.T. & Nicholson, N. pp. 66-72, The Royal Society of Chemistry, ISBN 0-85404-830-8, Cambridge, UK.

Beuzón C.R. & Holden D.W. (2001), Use of mixed infections with *Salmonella* strains to study virulence genes and their interactions in vivo. *Microbes and Infection*, Vol 3, pp. 1345-1352, ISSN 1286-4579.

Borges, W.S. & Pupo, M.T. Novel (2009). Antraquinone derivates produced by *Phoma sorghina*, an endophytic found in association with the medicinal plant *Tithonia diversifolia* (Asteracea), *Brazilian Journal of Chemistry Society*, Vol.17, pp. 929-934.

Carrol, G.C. (1988), The biology of endophytism in plants with particular reference to woody perennial. In: *Microbiology of phyllosphere*. Fokkema, N.J. & Heuvel, J.V. den., pp. 205-222, Cambridge University, ISBN 0521323444, London.

Chauhan AS, Negi PS, Ramteke RS (2007). Antioxidant and antibacterial activities of aqueous extract of Seabuckthorn (*Hippophae rhamnoides*) seeds. Fitoterapia 78: 590-592.

CLSI. Methods for Dilution Antimicrobial Susceptibility Tests for Bacteria That Grow Aerobically; Approved Standard (2003), 6. ed., ISBN 1- 56238-486-4, Wayne, Pennsylvania USA.

Coates, A.R.M. & Hu, Y. (2007), Novel approaches to developing new antibiotics for bacterial infections. *British Journal of Pharmacology*, Vol.152, pp. 1147-1154, ISSN 0007-1188.

Coelho JCU, Baretta GAP, Okawa L (2007). Seleção e uso de antibioticos em infecções intra-abdominais. Arq. Gastroenterol. 44: 85-90.

Cos, P.; Hermans, N.; de Bruyne, T.; Apers, S.; Sindambiwe, J.B.; Vanden Berghe, D.; Pieters, L. & Vlietinck, A.J. (2002). Further evaluation of Rwandan medicinal plant extracts for their antimicrobial and antiviral activities. *Journal of ethnopharmacology*, Vol.79, pp.155-163.

Cruz MCS, Santos PO, Barbosa Jr AM, Mélo DLFM, Alviano CS,Antoniolli AR, Alviano DS, Trindade RC (2007). Antifungal activity of Brazilian medicinal plants involved in popular treatment of mycoses. J. Ethnopharmacol. 111: 409-412.

Davis, R.A.; Andjic, A.; Kotiw, M. & Shivas, R.G. (2005). Phomoxins B and C: Polyketides from an endophytic fungus of the genus *Eupenicillium*. *Phytochemistry*, Vol.66, p. 2771-2775.

Duarte, M.C.T., Figueira, G. M., Sartoratto, A., Rehder, V. L, & Delarmelina, C. (2005), Anti-Candida activity of Brazilian medicinal plants. *Journal of Ethnopharmacology*, Vol. 97, No 2, (February 2005), pp. 305-311, ISSN 0378-8741.

Duarte MCT, Figueira GM, Pereira B, Magalhães PM, Delarmelina C (2004). Atividade antimicrobiana de extratos hidroalcólicos de espécies da coleção de plantas medicinais CPQBA/UNICAMP. Rev. Bras. Farmacogn. 14: 06-08.

Fennell, C.W., Lindsey, K. L., McGaw, L. J., Sparg,S. G, Stafford, G. I., Elgorashi, E. E., Grace, O. M. & Staden, J. van (2004), Review: Assessing African medicinal plants for efficacy and safety: Pharmacological screening and toxicology. *Journal of. Ethnopharmacology*, Vol. 94, No 2-3 (October 2004), pp. 205-217, ISSN 0378-8741.

Goffin, E.; Ziemons, E.; De Mol, P.; de Madureira, M.; Martins, A.P.C.A.P.; Geneviève; P. Tits, M.; Luc, A. & Frederich, M. (2002). In Vitro Antiplasmodial Activity of Tithonia diversifolia and Identification of its Main Active Constituent: Tagitinin C. *Planta Medica*, Vol.68, pp. 543-545

Guerrant, R.L., Van Gilder, T. & Steiner, T. S. (2001), Practice guidelines for the management of infectious diarrhea. *Clinical Infectious Disease*, Vol. 31, pp. 331–351, ISSN 1058-4838.

Gu, J-Q.; Gills, J.J.; Park, E.J.; Mata-Greenwood, E.; Hawthorne, M.E.; Axelrod, F.; Chavez, P.I.; Fong, H.H.S.; Mehta, R.G.; Pezzuto, J.M. & Kinghorn, A.D. (2002). *Journal of Natural Products*, Vol.65, pp. 532.

Gunatilaka, A.A.L. (2006). Natural products from plant-associated microorganisms: Distribution, structural diversity, bioactivity and implication of their occurrence. *Journal Natural Products*, Vol.69: pp. 509-526.

Gurney, K. A. & Mantle, P. G. (1993), Biosyntesis of 1-N-methylalbonoursin by an endophytic *Streptomyces* sp. Isolated from perennial *ryegrass*. *Journal of Natural Products*. vol.56, No 7 (July 1993), pp. 1194-1198, ISSN 0163-3864.

Hallmann, J.; Quadt-Hallmann, A.; Mahaffee, W.F. & Kloepper, J.W. (1997). Bacterial endophytes in agricultural crops. *Canadian Journal of Microbiology*, Vol.43, pp. 895-914.

Hanson, J.R. (2008). *The Chemistry of Fungi*, RSCPublishing, Brighton..ISBN 978-0-85404-136-7.

Harvey AL (2007). Natural products as a screening resource. Curr. Opin. Chem. Biol. 11: 480-484.

Healy, P. C., Hocking, A., Tran-Dinh, N., Pitt, John, I., Shivas, R. G., Mitchell, J. K., Kotiw, M. & Davis, R.A. (2004), Xanthones from a microfungus of the genus *Xylaria*, *Phytochemistry*, Vol 65, No 16 (August 2004), pp. 2373-2378, ISSN 0031-9422.

Herschler, R.J. United States Patent Office: Enhancing tissue penetration of physiologically active agents with DMSO. Date of access 07/18/11, Available in: < http://www.freepatentsonline.com/3711606.pdf>

Ho, C.K., Huang, Y.L. & Chen, C.C. (2002), Garcinone E, a xanthone derivative, has potent cytotoxic effect against hepatocellular carcinoma cell lines. *Planta Medica*, Vol 68, No 11 (November 2002), pp. 975-979, ISSN 0032-0943.

Hostettmann K, Queiroz EF, Vieira PC (2003). Princípios ativos de plantas superiores, EdUFSCar, São Carlos.

Huang, W. Y.; Cai, Y. Z.; Xing, J.; Corke, H. & Sun, M. (2007), Potential antioxidant resource: endophytic fungi isolated from traditional Chinese medicinal plants. *Economic Botany*, Vol.61, No 4 (December 2007), pp.14-30, ISSN 0013-0001.

Huang, W.Y.; Cai, Y.Z.; Hyede, K.D.; Corke, H. & Sun, M. (2008). Biodiversity of endophytic associated with 29 traditional Chines medicinal plants. *Fungal Diversity*, Vol.33, pp. 61–75, ISSN 1560-2745

Kamisuki, S.; Ishimaru, C.; Onoda, K.; Kuriyama, I.; Ida, N.; Sugawara, F.; Yoshida, H. & Mizushina, Y. (2007). Nodulisporol and Nodulisporone, novel specific inhibitors of human DNA polymerase λ from a fungus, *Nodulisporium* sp. *Bioorganic & Medicinal Chemistry*, Vol.15, No. 9, (May 2007), pp. 3109-3114.

Kaufmann, S.H.E, Raupacha, B. & Finlayb, B. B. (2001), Introduction: microbiology and immunology: lessons learned from *Salmonella*. *Microbes and Infection*, Vol. 3, pp. 1177–1181, ISSN 1286-4579.

Kobayashi, D.Y. & Palumbo, J.D. (2000), Bacterial endophytes and their effects on plants and uses in agriculture. In: Microbial endophytes. Bacon, C.W. & White, J.F., pp. 659-674, Dekker, ISBN 0824788311, New York, N.Y.

Krohn, K. & Riaz, M. (2004), Total synthesis of (+)-xyloketal D, a secondary metabolite from the mangrove fungus Xylaria sp. Tetrahedron Letters, Vol. 45, No. 2 (January 2005), pp. 293-294, ISSN 00404039.

Lee SH, Chang KS, Su MS, Huang YS, Jang HD (2007). Effects of some Chinese medicinal plant extracts on five different fungi. Food Control 18: 1547-1554.

Lin, C.N., Chung, M.I., Liou, S.J., Lee, T.H. & Wang, J.P. (1996), Synthesis and anti-inflammatory effects of xanthone derivatives. *Journal of Pharmacy and Pharmacology*, Vol. 48, No. 5 (May 1996), pp. 532-538, ISSN 0022-3573.

Liu, X., Dong, M., Chen, X., Jiang, M, Lv, X. & Zhou, J. (2008), Antimicrobial activity of an endophytic *Xylaria* sp. YX-28 and identification of its antimicrobial compound 7-amino-4-methylcoumarin. *Applied Microbiology and Biotechnology*, Vol. 78, No 2, pp. 241-247, ISSN 0175-7598.

Lu, H.; Zou, W.X.; Meng, J.C.; Hu, J. & Tan, R.X. (2000). New bioative metabolites produced by *Colletotrichum* sp., an endophytic fungus in *Artemisia annua*. *Plant Science*, Vol.151, pp. 67-73.

Malet-Cascon, L., Romero, F., Espliego-Vazquez, F., Gravalos, D., Fernandez-Puents, J.L. (2003), IB- 00208, a new cytotoxic polycyclic xanthone produced by a marine-derived Actinomadura. I. Isolation of the strain, taxonomy and biological activities. *Journal of Antibiotics*, Vol. 56, No 3, pp. 219-225, ISSN 0021-8820.

Mead, P.S., Slutsker, L., Dietz, V., McCaig, L.F., Bresee, J.S., Shapiro, C., Griffin, P.M. & Tauxe, R.V. (1999), Food-related illness and death in the United States. *Emerging Infection Diseases*, Vol. 5, pp.607–625, ISSN 1080-0059.

Minami, H., Kinoshita, M., Fukuyama, Y., Kodama, M., Yoshizawa, T., Sugiura, M., Nakagawa, K. & Tago, H. (1994), Antioxidant xanthones from *Garcinia subelliptica*. *Phytochemistry*, Vol. 36, No 2, pp. 501-506, ISSN 0031-9422.

More G, Tshikalange TE, Lall N, Botha F, Meyer JJM (2008). Antimicrobial activity of medicinal plants against oral microorganisms. J. Ethnopharmacol. 119: 473-477.

Moricca, S. & Ragazzi, A. (2008), Fungal endophytes in Mediterranean oak forests: a lesson from *Discula quercina*. *Phytopathology,* Vol.98, No 4, pp.380-386, ISSN 0031-949X.

Newman, D.J. & Cragg, G.M. (2007), Natural products as sources of new drugs over the last 25 years. *Jounal of Natural Products*, Vol. 70, No. 3 (March 2007), pp. 461-477, ISSN 0163-3864.

Nohynek, L.J., Alakomi, H., Kähkönen, M.P., Heinonen, M., Helander, I.M., Oksman-Caldentey, K. & Puupponen-Pimiä, R.H.(2006) Berry Phenolics: Antimicrobial Properties and Mechanisms of Action Against Severe Human Pathogens. *Nutrition and cancer*, Vol.54, No 1, pp.18-32, ISSN 1532-7914.

Oliveira RDR, Maffei CML, Martinez R (2001). Infecção urinária hospitalar por leveduras do gênero *Candida*. Rev. Assoc. Med. Bras. 47:231-235.

Orsi, R.O., Sforcin, J.M., Rall V.L.M., Funari S.R.C., Barbosa L., Fernandes, Jr A. (2005) Susceptibility profile of *Salmonella* against the antibacterial activity of propolis produced in two regions of Brazil. *J. Venom. Anim. Toxins incl. Trop. Dis.* Vol.11, No 2, pp109-116, ISSN 16789199.

Ostrosky, E.A., Mizumoto, M. K., Lima, M. E. L., Kaneko,T. M.,Nishikawa, S. O. & Freitas, B. R.(2008) Métodos para avaliação da atividade antimicrobiana e determinação da concentração mínima inibitória (CMI) de plantas medicinais. Revista Brasileira de Farmacognosia, Vol. 18, No. 2 (April-June 2008), pp. 301-307, ISSN 0102-695X.

Peres, V. & Nagem, T.J. (1996), Trioxygenated naturally occurring xanthones. *Phytochemistry*, Vol. 44, No. 2, pp. 191-214, ISSN 0031-9422.

Peres, V. Nagem, T.J. & De Oliveira, F.F. (2000), Tetraoxygenated naturally occurring xanthones. *Phytochemistry,* Vol. 55, No. 7 (December 2000), pp. 683-710, ISSN 0031-9422.

Petrini, O. (1991), Fungal endophyte of tree leaves. In: *Microbial Ecology of Leaves.* Andrews, J. & Hirano, S.S., pp., Springer Verlag, ISBN 0387975799, New York.

Pfaller, M.A. & Barry, A.L. (1994), Evaluation of a novel colorimetric broth microdilution method for antifungal susceptibility testing of yeast isolates. *Journal of Clinical Microbiology*, Vol. 32, No. 8, pp. 1992-1996, ISSN 0095-1137.

Reis, M.O.R. (2006), Avaliação da atividade antimicrobiana *in vitro do extrato hidroalcoólico das folhas de Persea gratissima* Gaertn – Abacateriro (Lauraceae). Dissertação do Mestrado em Promoção da Saúde, Universidade de Franca, Franca-SP.

Ribeiro, F.C.P (2011), Isolamento e identificação de fungos endofíticos de *Mikania laevigata* (Asteraceae) e análise química dos extratos por CLAE-DAD. Biotech Master Thesis; State University of Feira de Santana, Feira de Santana, BA.

Ribeiro, M.G.; Carvalho-Filho, A.S. & Listoni, F.J.P. (2001), Dimetilsulfóxido - DMSO no teste de sensibilidade microbiana *in vitro* em cepas de *Rhodococcus equi* isoladas de afecções pulmonares em potros. *Ciência Rural*, Vol.31, No.5 (January 2001), pp.889-892, ISSN 0103-8478.

Rios, J.L.; Recio, M.C. & Villar A. (1988), Screening methods for natural products with antimicrobial activity: a review of the literature. *Journal of Ethnopharmacology.,* Vol. 23, pp. 127-149, ISSN 0378-8741.

Rocha, L., Marston, A., Kaplan, M.A.C., Stoeckli-Evans, H., Thull, U., Testa, B. & Hostettmann, K. (1994), An antifungal γ-pyrone and xanthones with monoamine oxidase inhibitory activity from Hypericum brasiliense. *Phytochemistry*, Vol. 36, No. 6, pp. 1381-1385, ISSN 0031-9422.

Rocha, A.C.S.; Garcia, D.; Uetanabaro, A.P.T.; Carneiro, R. T. O., Araújo, I.S.; Mattos & C.R.R.; Góes-Neto, A. (2010), Foliar endophytic fungi from *Hevea brasiliensis* and their antagonism on *Microcyclus ulei*. *Fungal Diversity*. Vol.47, pp. 75-84, ISSN 1560-2745.

Rotimi, V. O., Jamal, W., Pal, T., Sonnevend, A., Dimitrov, T.S., Albert, M. J. (2008), Emergence of multidrug-resistant Salmonella spp. and isolates with reduced susceptibility to ciprofloxacin in Kuwait and the United Arab Emirates. *Diagnostic Microbiology and Infectious Disease*, Vol. 60, pp. 71–77, ISSN 0732-8893.

Shulz, B.; Boyle, C.; Draeger, S.; Rommert, A.K. & Krohn, K. (2002). Endophytic fungi: a source of novel biologically active secondary metabolites. *Mycology Research*, Vol.106, pp. 96–1004, ISSN 0953-7562.

Shulz, B. & Boyle, C. (2005) The endophytic continuum. *Mycology Research*, Vol.09, pp.661–686, ISSN 0953-7562.

Smith, C.J., Morin, N.R., Bills, G.F., Dombrowski, A.W., Salituro, G.M., Smith, S.K., Zhao, A. & Macneil,D.J. (2002), Novel sesquiterpenoids from the fermentation of *Xylaria persicaria* are selective ligands for the NPY Y5 receptor. *Journal of Organic Chemistry*, Vol. 67, No. 14, pp. 5001-5004, ISSN 0022-3263.

Souza, A.Q.L.; Souza, A.D.L.; Filho, S.A.; Pinheiro, M.L.B.; Sarquis, M.I.M. & Pereira, J.O. (2004), Atividade antimicrobiana de fungos endofíticos isolados de plantas tóxicas da Amazônia: Palicourea longiflora (aubl.) e Strychnos cogens bentham. *Acta Amazônica*. Vol.43, No. 2, pp.185-195, ISSN 0044-5967.

Souza EAF, Almeida LMM, Guilhermetti E, Mota VA, Rossi RM, Svidzinski T (2007). Freqüência de onicomicoses por leveduras em Maringá, Paraná, Brasil. An. Bras. Dermatol. 82: 151-156.

Stadler, M; Wollweber, H; Muhlbauer, A; Henkel, T; Asakawa,Y.; Hashimoto, T.; Ju, Y.M.; Rogers, J.D., Wetzstein, H.G. & Tichy, H.V. (2001). Secondary metabolite profiles, genetic fingerprints and taxonomy of *Daldinia* and allies, *Mycotaxon*, Vol.77, 379-429.

Stadler, M.; Ju, Y. & Rogers, F.D. Chemotaxonomy of *Entonaema, Rhopalostroma* and other Xylariaceae. *Micological Research*, Vol.3, pp. 239-256, 2004.

Stefanello MEA, Salvador MJ, Ito IY. Macari PAT (2006). Avaliação da atividade antimicrobiana e citotóxica de extratos de *Gochnatia polymorpha* ssp floccosa. Rev. Bras. Farmacogn. 16: 525-530.

Stierle, A.; Strobel, G. & Stierle, D. (1993), Taxol and Taxane Production by *Taxomyces andreane*, an Endophytic Fungus of Pacific Yew. *Science*, Vol. 260, No. 5105 (April 1993), pp. 214-216, ISSN 0036-8075.

Stone, J. K., Becon, C. W. & White Jr, J. F. (2000), An overview of endophytic microbes: endophytism defined. In: *Microbial Endophytes*. Becon, C. W. & White Jr, J. F. pp. 3-29, Marcel Dekker, ISBN 0824788311, New York.

Strege, M.A. (1999), High-performance liquid chromatography-electrospray ionization mass spectrometric analyses for the integration of natural products with modern high-throughput screening. *Journal of Chromatography B: Biomedical Sciences and Applications*, Vol. 725, No. 1, pp. 67-78, ISSN 1570-0232.

Strobel, G. & Daisy, B. (2003), Bioprospecting for microbial endophytes and their natural products. *Microbiology and Molecular Biology Reviews*. Vol. 67, No. 4 (December 2003), pp. 491-502, ISSN 1098-5557.

Strobel, G., Daisy, B. & Castillo, U. (2005), The biological promise of microbial endophytes and their natural products. *Plant Pathology Journal*. Vol. 4, No. 2, pp.161-176, ISSN 1812-5387.

Suryanarayanan, T.S.; Thirunavukkarasu, N.; Govindarajulu, M.B.; Sasse, F.; Jansen, R. & Murali, T.S. (2009), Fungal endophytes and bioprospecting. *Fungal Biology Reviews*. Vol. 23, No. 1-2, pp. 9-19, ISSN 17494613.

Tadeg, H., Mohammed, E., Asres, K., Gebre-Mariam, T. (2005), Antimicrobial activities of some selected traditional Ethiopian medicinal plants used in the treatment of skin disorders. *Journal of Ethnopharmacology*, Vol. 100, No. 1-2, pp. 168-175, ISSN 0378-8741.

Taechowisan, T.; Chuaychot, N.; Chanaphat, S.; Wanbanjob, A. & Shen, Y. (2008). Biological activity of chemical constituents isolated from *Streptomyces* sp. Tc052, an endophytic in *Alpinia galangal*. *International Journal of pharmacology*, Vol.4, No.2, pp. 95-101

Tahergorabi, R., Matak, K. E., Jaczynski, J. (2011), Application of electron beam to inactivate Salmonella in food: Recent developments. *Food Research International*, article in press (February 2011), 10p.

Tan, R.X. & Zou, W.X. (2001), Endophytes: a rich source of functional metabolites. *Natural products Reports*, Vol. 18, No. 4, pp. 448-459, ISSN 0265-0568.

Tejesvi, M.V.; Tamhankar, S.A.; Kini, K.R.; Rao, V.S. & Prakash, H.S. (2009). Phylogenetic analysis of endophytic *Pestalotiopsis* species from ethnopharmaceutically important medicinal trees. *Fungal Diversity*, Vol.38, pp.167–183, ISSN 1560-2745.

Threlfall, E.J., Ward, L.R., Skinner, J.A. & Rowe, B. (1997) Increase in multiple antibiotic resistance in non-typhoidal salmonellas from humans in England and Wales: comparison of data for 1994 and 1996. *Microbial Drug Resistance*, Vol. 3, pp.263–266. ISSN 1076-6294.

Threlfall, E.J. (2002), Antimicrobial drug resistance in Salmonella: problems and perspectives in food- and water-borne infections. *FEMS Microbiology Reviews*, Vol. 26, pp.141-148, ISSN 1574-6976.

Valgas, C., Souza, S. M., Smania, E. F. A. & Smania Jr., A. (2007), Screening methods to determine antibacterial activity of natural products. *Brazilian Journal of Microbiology*, Vol. 38 (February 2007), pp. 369-380, ISSN 1517-8382.

Vanden Bergue, D.A. & Vlietinck, A.J. (1991), Screening methods for antibacterial and antiviral agents from higher plants. In: *Methods in Plant Biochemistry*, Dey, P.M. & Harbone, J.D, pp. 47-69, Academic Press, ISBN 0124610110, London.

Vieira, G.R.T. (2005), Otimização das condições de cultivo de *Polyporus tricholoma* Mont. visando à produção de substâncias antibacterianas. Dissertação do Mestrado - Universidade Federal de Santa Catarina. Florianópolis, SC.

Vuuren SF (2008). Antimicrobial activity of South African medicinal plants. J. Ethnopharmarcol. 119: 462-472.

Wang, X.N., Tan, R.X. & Liu, J.K. (2005), Xylactam, a new nitrogen-containing compound from the fruiting bodies of ascomycete *Xylaria euglossa*. *Journal of Antibiotics*, Vol. 58 (March 2005), pp. 268-270, ISSN 0021-8820.

Weckesser S, Engel K, Simon-Haarhaus B, Wittmer A, Pelz K, Schempp CM (2007). Screening of plant extracts for antimicrobial activity against bacteria and yeasts with dermatological relevance. Phytomedicine, 14: 508-516.

Wedel, S.D., Bender, J.B., Leano, F.T., Boxrud, D.J., Hedberg, C. & Smith, K. (2005) Antimicrobial-drug susceptibility of human and animal Salmonella typhimurium, Minnesota, 1997–2003. *Emerging Infection Diseases*, Vol. 11, pp. 1899–1906, ISSN 1080-0059.

Yu, H.; Zhang, L.; Li, L.; Zheng, C.; Guo, L.; Li, W.; Sun, P. & Qin, L. (2010), Recent developments and future prospects of antimicrobial metabolites produced by endophytes. *Microbiological Research*, Vol. 165, No. 6 (August 2010), pp. 437-449, ISSN 0944-5013.

Zhang, Y.; Mu, J.; Feng, Y.; Kang, Y.; Zhang, J.; Gu, P.; Wang, Y.; Ma, L.; Zhu, Y. (2009), Broad-spectrum antimicrobial epiphytic and endophytic fungi from marine organisms: isolation, bioassay and taxonomy. *Marine drugs*, Vol. 7 (April 2009), pp.97-112, ISSN 1660-3397.

10

Effect of Sanitizers and Oregano Essential Oils to Reduce the Incidence of *Salmonella* and Other Pathogens in Vegetables

Saul Ruiz-Cruz[1*], Gustavo A. González-Aguilar[2],
José de Jesús Ornelas-Paz[3], Luis A. Cira-Chávez[1]
and J. Fernando Ayala-Zavala[2]
*[1]Instituto Tecnológico de Sonora, Departamento de Biotecnología y
Ciencias Alimentarias, Ciudad Obregón, Sonora
[2]Centro de Investigación en Alimentación y Desarrollo, A. C. (CIAD)
Dirección de Tecnología de Alimentos de Origen Vegetal, Hermosillo, Sonora
[3]CIAD Unidad Cuauhtémoc, Chihuahua
México*

1. Introduction

Consumption of fruits and vegetables has increased in the last years because form an important component in human nutrition, being rich sources of phytochemicals compounds that play a significant role in the health to helps prevent many degenerative diseases (González-Aguilar et al., 2008). However, as a consequence of inappropriate manipulation during their growth, development, harvesting, processing, distribution, retail sale, and final preparation, pathogenic microorganisms, may contaminate a product, thus increasing the risk of microbial diseases. In fact, the number of outbreaks and cases of illness caused by consumption of these produce has increased in the last years.

One of the most common disease-causing pathogens in fruits and vegetables is *Salmonella*. This reside in the intestinal tracts of animals, including humans, and are more likely to contaminate raw fruits and vegetables through contact with feces, sewage, untreated irrigation water or surface water (Beuchat, 1996; Wells & Butterfield, 1997). Outbreaks of salmonellosis have been linked to a diversity of fruits and vegetables including tomatoes (Hedberg et al., 1994; Wood et al., 1991), bean sprouts (Mahon et al., 1997; Van Beneden et al., 1999) and melons (Blostein 1991).

Washing is an important step that has been widely adopted by the industry to remove soils and microorganism from the surface (Sapers, 2003). Given that fresh and fresh-cut products are marketed as pre-washed and ready to eat, and not subject to further microbial killing steps, the development and proper application of sanitizing agents or antimicrobial

---

* Corresponding Author

compounds to remove microorganism's pathogen effectively is critical to ensure the safety of these produce (Ruiz-Cruz et al., 20006).

Chlorine has been widely used as a sanitizer during produce washing (Sapers, 2003). However, numerous studies have shown that chlorine used at concentrations permitted by the FDA lacks efficacy in removing human pathogens and spoilage microorganisms (Zhang & Farber, 1996). Additionally, chlorine may react with organic matter in water to form carcinogenic products (Paris et al., 2003). The inadequacies of chlorine as a sanitizer have stimulated interest in finding safer, more effective sanitizers (Ruiz-Cruz et al., 2006). Alternatives to chlorine, such as chlorine dioxide, peroxyacetic acid, acidified sodium chlorite and some essential oils have been already proposed (Beuchat, 1998). Acidified sodium chlorite as shown strong antimicrobial activity against *E. coli* O157:H7, *Salmonella*, *Listeria monocytogenes* and spoilage microorganisms on fresh-cut produce (González et al., 2004; Ruiz-Cruz et al., 2007). Peroxyacetic acid has also been shown to be effective against spoilage and pathogenic microorganisms (Weissinger & Beuchat, 2000). Others sanitizers such as chlorine dioxide and essential oils have been to be very effective in reducing the microbial populations of fresh-cut fruit and vegetable (Burt & Reinders, 2003; Selma et al., 2008).

Moreover, since water wash is often recycled, high organic matter content reduces the activity of sanitizers and increases the likelihood for contamination of fresh vegetables. For this case not all washing methods and washing solutions are equally effective. In order to ensure the safety of fresh produce, it is necessary to evaluate the efficacy of chemical sanitizers and natural antimicrobial compounds in water containing concentrations of organic matter to simulate commercial practices. In view of the importance of *Salmonella* as a cause of food-borne disease and vegetables as a vehicle for its transmission, in our researches are going to look alternatives chemical and naturals to eliminated *Salmonella* and others pathogens in different fruits and vegetables. The objective of this chapter is to evaluate the effectiveness of chemical sanitizers, such as chlorine (Cl, 200 ppm), chlorine dioxide (DC, 5%), peroxyacetic acid (PA, 80 ppm), and acidified sodium chlorite (ASC, 100, 250 and 500 ppm) and oregano essentials oils (OEO, 1 and 2.5 mM) for reducing populations of *Salmonella* and *E. coli* O157:H7 from inoculated cilantro, spinach, lettuce and jalapeño peppers under laboratory and simulated commercial processing wash water conditions.

2. Material and methods

2.1 Bacterial strains and media

Salmonella Typhimurium (ATCC 14028) and *E. coli* O157:H7 (ATCC 43890) were used in this study. Cultures of *Salmonella* and *E. coli* O157:H7 from freezer stocks were grown in tryptic soy broth (Difco Laboratories, Detroit, MI). To suppress growth of microorganisms naturally present on different vegetables, nalidixic acid-resistant strains were obtained and used in this study (Inatsu et al., 2005a; Inatsu et al., 2005b; Ruiz-Cruz et al., 2007). *Salmonella* and *E. coli* O157:H7 strains were adapted to grow on Luria-Bertani broth (LBB; Difco, Becton Dickinson, Sparks, MD) supplemented with 50 µg/mL nalidixic acid (LBB-Nal) and incubated at 37°C. To obtain pure cultures, a loop of *Salmonella* was streaked on Bismuth Sulfite agar (BS; Difco Laboratories, Detroit, MI), *E. coli* O157:H7 on Sorbitol MacConkey agar (SMAC; Difco Becton Dickinson, Sparks, MD). Each agar medium was supplemented with nalidixic acid (50 µg/mL) and plates were incubated at 37°C. After incubation, a single colony from each plate was selected and inoculated into 10 mL of LBB-Nal (*E. coli* O157:H7 and *Salmonella*). Individual strains were grown in each broth at room temperature (30°C with constant agitation at 175 rpm. Cultures were transferred to each broth by loop at two successive 24 h intervals and one overnight (16-18 h) before they were used as inoculants. Cells of each strain were harvested by centrifugation (4,000 X *g*, 15 min) and washed with 2 vol of sterile phosphate-buffered saline and resuspended at a cell density of approximately 10^9 cfu/mL. Volumes of cell suspensions of *Salmonella* or *E. coli* O157:H7 were combined to create a two-strain mixture. The cells were added proportionally to tap water (laboratory conditions) and simulated commercial processing to obtain a dip inoculum solution of approximately 10^7 cfu/mL. The inoculum level was confirmed by replica plating onto selective agar after serial dilution in sterile phosphate-buffered saline, followed by incubation at 37°C for 24 h.

2.2 Vegetable preparation

Fresh cilantro, spinach, lettuce and jalapeño peppers were purchased from a local wholesale market in Cd. Obregón, Sonora, México, transported to the laboratory and used within 24 h following storage at 5°C. Produce were sorted to eliminate damaged, poor quality produce. Vegetable were washed in tap water at ambient temperature to remove residual soil. After washing, cilantro, lettuce and spinach were cut manually in 2 cm pieces and jalapeño peppers into longitudinal strips of 1 cm width and divided into individual 40 g portions contained in nylon mesh bags.

2.3 Inoculation procedure

Since the immersion process is a possible point of contamination in the food industry, dip inoculation is the most suitable method that can be used to simulate such a process (Beuchat et al., 2001). Spinach, lettuce, cilantro and jalapeño peppers were immersed in the inoculum solution (sample:inoculums ratio = 1:7 w/v) and kept under constant agitation for 30 min. After dipping, the samples were drained for 30 s. Samples were then placed into plastic containers and maintained for 1 h at room temperature until washed with the different treatments. Previous studies indicate that bacteria attach to leaf surfaces within 1 h of inoculation (Yang et al., 2003).

2.4 Treatment procedure

The following sanitizer treatments were evaluated for their efficacy in killing or reducing *E. coli* O157:H7 and *Salmonella* on fresh produce cilantro, jalapeño peppers, lettuce and spinach. The treatments were 200 ppm sodium hypochlorite of a commercial bleach preparation, with pH adjusted to 6.5 with HCl (Cloralex®, NL, México, 6% NaOCl), 80 ppm PA; pH 3.5 (Ecolab, St. Paul,Minn., U.S.A.), 100, 250 and 500 ppm ASC; pH 2.8, 2.6 and 2.4, respectively (Sigma-Aldrich, USA), DC 5%; pH 5.4 and 1 and 2 mM OEO (Sigma-Aldrich, USA). Untreated produce samples were used as control. For the simulated commercial water processing, was obtained by repeatedly dipping freshly shredded produce of known mass in a fixed volume of tap water (González et al., 2004; Ruiz-Cruz et al., 2007). Each mesh bag of inoculated produce were dipped into one sanitizer solution (sample to wash water ratio of 1:5 w/v) with contact time of 1 (ASC and OEO) and 2 min (Cl, PA and DC) with constant stirring. After dipping samples were drained for 30 s to remove excess water. Samples of produce weighing 30 g were packaged in ziploc bags and stored at 5°C for 7 days.

2.5 Procedures for microbial enumeration

Samples (10 g) were transferred aseptically into sterile stomacher bags, 90 mL of Dey-Engley (DE) neutralizing broth was added and samples were macerated. Homogenized samples were serially diluted by a factor of ten in sterile phosphate buffer saline. For each dilution, 1 mL was plated on each of sorbitol MacConkey agar and bismuth sulfite agar for *E. coli* O157:H7 and *Salmonella*, respectively and incubated at 37°C for 24 or 36 h. Agars for pathogenic bacterial growth were supplemented with 50 µg/mL nalidixic acid.

3. Results and discussions

Analysis of the fresh produce that had not been inoculated revealed the absence of *Salmonella* and *E. coli* O157:H7. Preliminary studies were conducted to determine the level of inoculum that could be retained on the surface of different produce. The amount of *Salmonella* and *E. coli* O157:H7 attached to the surface of spinach was 6.53 and 6.35 log cfu/g (Fig. 1), jalapeño peppers 6.52 and 6.4 log cfu/g (Fig. 2), cilantro 5.81 and 5.92 log cfu/g (Fig. 3) and lettuce 5 and 5.35 log cfu/g (Fig. 4), respectively.

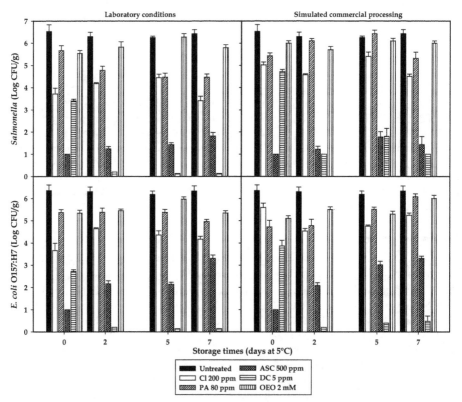

Fig. 1. Efficacy of sanitizers on the reduction of *Salmonella* and *E. coli* O157:H7 populations from artificially inoculated spinach. Bars represent the standard errors of the mean resulting from triplicate experiments. The limit of detection was 0.2 log cfu/g of produce.

All sanitizers reduced significantly (P≤0.05) *Salmonella* and *E. coli* O157:H7 compared with the control (untreated) on day 0, with reduction of 1 – 6 log cfu/g and all produce evaluated. The results of the decontamination for *Salmonella* and *E. coli* O157:H7 on spinach are shown in Fig. 1. The organic matter content in simulated commercial processing reduces the activity of the Cl and DC treatment. However, the effectiveness of Cl was significantly higher that PA and OEO with reductions of 2.5 – 2.7 log cfu/g, but lower that ASC and DC. ASC was the most effective treatment in reducing *Salmonella* and *E. coli* O157:H7. This sanitizer reduced the both pathogen populations to undetectable levels (with 10 cfu/g detection limit), achieving reductions of 5.53 and 5.35 log cfu/g of *Salmonella* and *E. coli* O157:H7, respectively, under both water conditions. The results are similar to those reported by other researchers (González et al., 2004; Ruiz-Cruz et al., 2007). Cells of both pathogens increased by 1.5-3 log cfu/g at the end of storage. On the other hand, interestingly sample treated with DC on tap water conditions lower after 2 days of storage at levels of 0.2-1 log cfu/g. This reduction could be due to damage to the cell by sanitizer and coupled with the cooling temperature, the cells failed to grow.

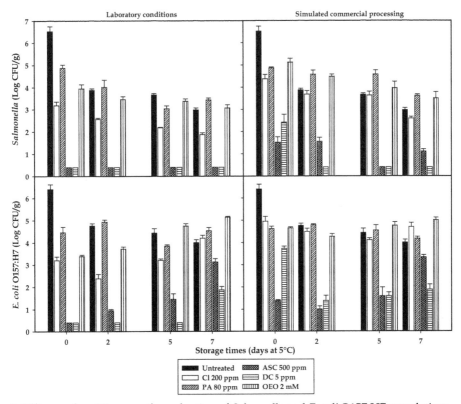

Fig. 2. Efficacy of sanitizers on the reduction of *Salmonella* and *E. coli* O157:H7 populations from artificially inoculated jalapeño peppers. Bars represent the standard errors of the mean resulting from triplicate experiments. The limit of detection was 0.2 log cfu/g of produce.

The use of ASC and DC showed better in inactivating *Salmonella* and *E. coli* O157:H7 on jalapeño peppers that the others sanitizers (Fig. 2). Populations were reduce to undetectable levels (0.4 log cfu/g) at day 0 only under laboratory conditions. Followed by Cl, OEO and PA with reductions of 2-3.2 log cfu/g. The efficacy of DC was affected significantly when was used in simulated commercial processing; however, also caused a higher reduction with 2.7-5 log cfu/g. In general the PA treatment was the least effective in reducing the pathogens; however it effectivity was not affected by the presence of organic matter and this treatment resulted in 1.6 to 2 log cfu/g reduction both pathogens. The cells of *Salmonella* were not recovered during storage time with DC were used under both water conditions. This indicates that the cell could be not retrieved in the agar plate, which caused a reduction of 6 log cfu/g. ASC caused the same effect but only under laboratory conditions. The results are similar to those reports by us previously researched with 250 and 500 ppm of ASC on shredded carrots (Ruiz-Cruz et al., 2007).

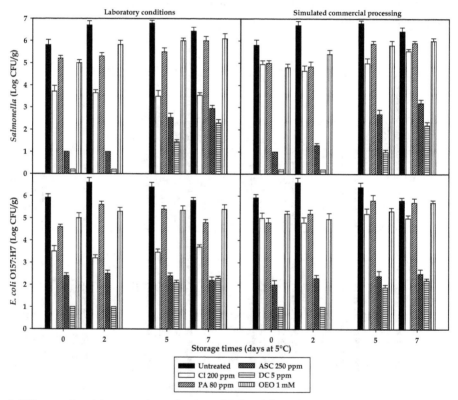

Fig. 3. Efficacy of sanitizers on the reduction of *Salmonella* and *E. coli* O157:H7 populations from artificially inoculated cilantro. Bars represent the standard errors of the mean resulting from triplicate experiments. The limit of detection was 0.2 log cfu/g of produce.

The efficacies of Cl, ASC and DC in reduced *Salmonella* and *E. coli* O157:H7 in cilantro were statistically different ($P \leq 0.05$), when compared with untreated and other treatments (Fig. 3). The effectiveness of Cl treatment was significantly higher than PA and OEO, but lower than ASC and DC in both pathogens. DC and ASC were the most effectives treatments in reducing pathogens on day 0. DC treatments reduced *Salmonella* to undetectable levels (with 0.2 log cfu/g, detection limit), achieving reductions of 5.61 and 4.92 log cfu/g by *Salmonella* and *E. coli* O157:H7, respectively. ASC also caused a strong reduction of both pathogens with values of 4.81 and 3.52 log cfu/g of *Salmonella* and *E. coli* O157:H7, respectively. Allende et al. (2009) reported on reduction of more than 3 log cfu/g were used 100 ppm of ASC on fresh-cut cilantro. Ruiz-Cruz et al. (2007) observed that ASC treatment at 100, 250 and 500 ppm reduced the *Salmonella* and *E. coli* O157:H7 population on shredded carrots in a similar manner to the reduction of *Salmonella* and *E. coli* O157:H7 achieved with 250 ppm, in this study on fresh-cut cilantro. Our study previously, found that 1000 ppm of ASC affected overall quality of shredded carrots (Ruiz-Cruz et al., 2006). Moreover, we found that concentrations up of 250 ppm of ASC affected the quality of fresh-cut cilantro and the concentrations used in this study (250 ppm) maintained the quality of cilantro by 16 days at 5°C (data no show). Moreover, concentrations of ASC above 500 ppm affected the quality of different vegetable (spinach, jalapeño peppers and lettuce). These results confirmed that the effectiveness of ASC to maintain quality and reduce pathogen counts is influenced by the concentration of the sanitizer and the contact time.

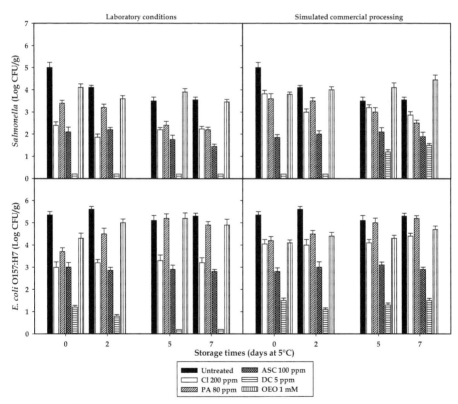

Fig. 4. Efficacy of sanitizers on the reduction of *Salmonella* and *E. coli* O157:H7 populations from artificially inoculated lettuce. Bars represent the standard errors of the mean resulting from triplicate experiments. The limit of detection was 0.2 log cfu/g of produce.

The reduction of *Salmonella* and *E. coli* O157:H7 on lettuce after washing with the different sanitizers were shown in Fig. 4. All treatments caused significant reductions in *Salmonella* and *E. coli* O157:H7 on day 0 under both water conditions compared with the control. Reduction of 1.2 log cfu/g in both pathogens in both water conditions were observed for the OEO treatment, this indicated that its effectivity was not affected by the organic matter. This reduction was maintained throughout the storage. The effectiveness of Cl treatment was significantly higher than PA and OEO, but lower than ASC and DC in reducing *Salmonella* and *E. coli* O157:H7, were used in tap water, causing a reduction of 2.6 and 2.35 log cfu/g, respectively. However, its effectivity was affected in reduced *Salmonella* under simulated commercial processing reducing only 1.18 log cfu/g. PA caused a reduction of 2.4 and 2.6 log cfu/g of *Salmonella* and 1.65 and 1.15 log cfu/g on *E. coli* O157:H7, under tap water and simulated commercial processing, respectively. Similar to the results on the DC treated cilantro, after treated lettuce with DC resulted in a high reduction of *Salmonella* to undetectable levels (0.2 log cfu/g, limit detection), achieving reductions of 4.8 log cfu/g.

4. Conclusions

ASC and DC were the most effective treatments in reducing *Salmonella* and *E. coli* O157:H7 in all produces at all concentrations evaluated. No viable cells of *Salmonella* and *E. coli* O157:H7 were recovered at concentrations of 5% of DC in spinach and lettuce. As well as *Salmonella* in jalapeño peppers treated with DC 5% and ASC 500 ppm, producing a bactericidal effect. However, cells were able to grow during storage, therefore this indicates the ability of the pathogens to adapt to adverse environments present in food is an interesting area that requires more investigation.

The results show that all sanitizers were capable of controlling growth of *Salmonella* and *E. coli* O157:H7 during storage time and can be used by washed these produce. ASC and DC was the most effective sanitizer and have the advantage of being more stable and preserve its efficacy in the presence of organic matter. However, further studies are needed to determine whether these sanitizers might have more lethal effects when lower levels of bacteria are present on produce.

5. Acknowledgments

The authors gratefully acknowledge the financial support from PROMEP, especially to the project ITSON-PTC-056.

6. References

Allende, A.; McEvoy, J.; Tao, Y. & Luo, Y. (2009). Antimicrobial effect of acidified sodium chlorite, sodium chlorite, sodium hypochlorite, and citric acid on *Eschenrichia coli* O157:H7 and natural microflora of fresh-cut cilantro. *Food Control*, 20(3): 230-234. URL: www.sciencedirect.com

Beuchat, L.R. (1996). Pathogenic microorganisms associated with fresh produce. *Journal of Food Protection*, 59(2): 204-216. URL: www.ingentaconnect.com

Beuchat, L.R. (1998). *Surface Decontamination of Fruits and Vegetables Eaten Raw: a Review*. World Health Organization, Food Safety Unit, WHO/FSF/FOS/98.2. Available:

http://who.int/foodsafety/publications/fs management/en/surface decon.pdf [5 July 2011].

Beuchat, L.R.; Farber, J.M.; Garrett, E.; Harris, L.J.; Parish, M.E.; Suslow, T.V. & Busta, F.F. (2001). Standardization of a method to determine the efficacy of sanitizers in inactivating human pathogenic microorganisms on raw fruits and vegetables. *Journal of Food Protection*, 64(7): 1079-1084. URL: www.ingentaconnect.com

Blostein, J. (1991). An outbreak of *Salmonella javiana* associated with consumption of watermelon. *Journal of Environmental Health*, 56(1): 29-31. URL: www.neha.org

Burt, S.A. & Reinders, R.D. (2003). Antibacterial activity of selected plant essential oils against *Escherichia coli* O157:H7. *Letters in Applied Microbiology*, 36(3): 162-167. URL: www.wiley.com

González, R.J.; Luo, Y.; Ruiz-Cruz, S. & McEvoy, J.L. (2004). Efficacy of sanitizers to inactivate *Escherichia coli* O157:H7 on fresh-cut carrot shreds under simulated process water conditions. *Journal of Food Protection*, 67(11): 2375-2380. URL: www.ingentaconnect.com

González-Aguilar, G.A.; Robles-Sánchez, R.M.; Martínez-Tellez, M.A.; Olivas, I.G.; Alvarez-Parrilla, E. & de la Rosa, L.A. (2008). Bioactive compounds in fruits: health benefits and effect of storage conditions. *Steward Postharvest Review*, 4(3): 1-10. URL: www.ingentaconnect.com

Hedberg, C.W.; MacDonald, K.L. & Osterholm, M.T. (1994). Changing epidemiology of food-borne disease: a Minnesota perspective. *Clinical Infectious Diseases*, 18(5): 671-682. URL: www.idsociety.org

Inatsu, Y.; Bari, M.L.; Kawasaki, S.; Isshiki, K. & Kawamoto, S. (2005a). Efficacy of acidified sodium chlorite treatments in reducing *Escherichia coli* O157:H7 on chinese cabbage. *Journal of Food Protection*, 68(2): 251-255. URL: www.ingentaconnect.com

Inatsu, Y.; Maeda, Y.; Bari, M.L.; Kawasaki, S. & Kawamoto, S. (2005b). Prewashing with acidified sodium chlorite reduces pathogenic bacteria in lightly fermented Chinese cabbage. *Journal of Food Protection*, 68(5): 999-1004. URL: www.ingentaconnect.com

Mahon, B.E.; Ponka, A.; Hall, W.N.; Komatsu, K.; Dietrich, S.E.; Siitonen, A.; Cage, G.; Hayes, P.S.; Lambert-Fair, M.A.; Bean, N.H.; Griffin, P.M. & Slutsker, L. (1997). An international outbreak of *Salmonella* infections caused by alfalfa sprouts grown from contaminated seeds. *Journal Infectious Diseases*, 175(4): 876-882. URL: www.idsociety.org

Parish, M.E.; Beuchat, L.R.; Suslow, T.V.; Harris, L.J.; Garrett, E.H. & Farber, J.N. (2003). Methods to reduce/eliminate pathogens from fresh and fresh-cut produce. *Comprehensive Reviews in Food Science and Food Safety*, 2(supplement): 161-173. URL: www.wiley.com

Ruiz-Cruz, S.; Luo, Y.; Gonzalez, R.J.; Tao, Y. & González-Aguilar, G.A. (2006). Acidified sodium chlorite as an alternative to chlorine to control microbial growth on shredded carrots while maintaining quality. *Journal of the Science of Food and Agriculture*, 86(12): 1887-1893. URL: www.wiley.com

Ruiz-Cruz, S.; Acedo-Félix, E.; Díaz-Cinco, M.; Islas-Osuna, M. & González-Aguilar, G.A. (2007). Efficacy of sanitizers in reducing *E. coli* O157:H7, *Salmonella* spp. and *Listeria monoytogenes* populations on fresh-cut carrots. *Food Control*, 18(11): 1383-1390. URL: www.sciencedirect.com

Sapers, G.M. (2003). Washing and sanitizing raw materials for minimally processed fruit and vegetable products, *in* Novak, J.S.; Sapers, G.M. & Juneja, V.K. (Ed.), *Microbial Safety of Minimally Processed Foods*, CRC Press, Boca Raton, FL, pp. 221-253.

Selma, M.V.; Ibañez, A.M.; Allende, A.; Cantwell, M. & Suslow, T. (2008). Effect of gaseous ozone and hot water on microbial and sensory quality of cantaloupe and potential transference of *Escherichia coli* O157:H7 during cutting. *Food Microbiology*, 25(1): 162-168. URL: www.sciencedirect.com

Van Beneden, C.A.; Keene, W.E.; Strang, R.A.; Werker, D.H.; King, A.S.; Mahon, B.; Hedberg, K.; Bell, A.; Kelly, M.T.; Balan, V.K.; Mac Kenzie, W.R. & Fleming, D. (1999). Multinational outbreak of *Salmonella enterica* serotype Newport infections due to contaminated alfalfa sprouts. *The Journal of the American Medical Association*, 281(2): 158-162. URL: http://jama.ama-assn.org

Wells, J.M. & Butterfield, J.E. (1997). *Salmonella* contamination associated with bacterial soft rot of fresh fruits and vegetables in the marketplace. *Plant Disease*, 81(8): 867-872. URL: http://apsjournals.apsnet.org

Weissinger, W.R. & Beuchat, L.R. (2000). Comparison od aqueous chemical treatments to eliminate *Salmonella* on alfalfa seeds. *Journal of Food Protection*, 63(11): 1475-1482. URL: www.ingentaconnect.com

Wood, R.C.; Hedberg, C. & White, K. (1991). A multistate outbreak of *Salmonella javiana* infections associated with raw tomatoes. In: CDC Epidemic Intelligence Service, 40th Ann. Conf., p. 69 (abstracts). Atlanta, GA, U.S. Dept. of Health and Human Services, Public Health Service.

Yang, H.; Swem, B.L. & Li, Y. (2003). The effect of pH on inactivation of pathogenic bacteria on fresh-cut lettuce by dipping treatment with electrolyzed water. *Journal of Food Science*, 68(3): 1013–1017. URL: www.wiley.com

Zhang, H. & Farber, J.M. (1996). The effects of various disinfectants against *Listeria monocytogenes* on fresh-cut vegetables. *Food Microbiology*, 13(4): 311-321. URL: www.sciencedirect.com

11

Antibiotic Resistance in *Salmonella*: A Risk for Tropical Aquaculture

Renata Albuquerque Costa[1], Fátima Cristiane Teles de Carvalho[2]
and Regine Helena Silva dos Fernandes Vieira[3]
[1]Doctoral Student in Fisheries Engineering. C.A.P.E.S. Federal University of Ceará
[2]Doctoral Student in Tropical Marine Sciences. Federal University of Ceará
[3]Prof. Dr. Sea Sciences Institute. Federal University of Ceará
Brazil

1. Introduction

Salmonelas are rod-shaped, non-spore-forming Gram-negative facultative anaerobes measuring 0.7-1.5 by 2-5 μm. With the exception of the serovars Gallinarum and Pullorum, salmonelas are motile organisms. They are classified according to morphology and staining pattern and are divided into serotypes and serovars based on their reaction to somatic (O) and flagellar (H) antigens (Bremer et al., 2003). According to Kumar et al.(2003), the genus *Salmonella* has over 2,000 serovars. Two of these—Saintpaul and Newport—have been isolated from seafood (Ponce et al., 2008).

The prevalence of specific *Salmonella* serovars is related to food type. Thus, the serovars Weltevreden and Rissen are predominant in seafoods, as shown by Kumar et al. (2009) in a study on the distribution and phenotypical characterization of *Salmonella* serovars isolated from samples of fish, crustaceans and mollusks from India.

High incidences of *Salmonella* in seafoods have been reported worldwide (Kumar et al., 2010; Asai et al., 2008) in association with outbreaks of fever, nausea, vomiting and diarrhea (Ling et al., 2002). Since *Salmonella* inhabits the intestinal tract of warm-blooded animals, its presence in aquaculture livestock is most likely due to the introduction of fecal bacteria into culture ponds (Koonse et al., 2005). In fact, in a study on *Salmonella* in shrimp, Shabarinath et al. (2007) concluded this pathogen is generally found in rivers and marine/estuarine sediments exposed to fecal contamination.

The quality of aquaculture products may be compromised by exposure to pathogens and biological or chemical contaminants. The latter include chemical agents commonly used in aquaculture, such as veterinary antibiotics, antiseptics and anesthetics. Few antibiotics have been adapted to or developed specifically for use in aquatic organisms. Thus, in Europe several classes of antibiotics may be used in aquaculture, including sulfonamides, quinolones, macrolides, tetracyclines and emamectin. This, however, poses a considerable risk of release of antimicrobials into the environment and eventually of the development of resistance in pathogenic bacteria (Fauconneau, 2002).

The second half of the 20th century saw two major events in the epidemiology of salmonellosis: the appearance of human infections caused by food-borne *S. enteritidis* and by *Salmonella* strains with multiple resistance (Velge et al., 2005). In fact, Angulo et al. (2000) suggested that the factors determining resistance to multiple antibiotics in strains of *S.* Typhimurium DT104 may first have developed in bacteria in the aquaculture environment, possibly as the result of the regular use of antibiotics in fodder.

The present study is a review of the literature on resistant *Salmonella* strains in aquaculture and an assessment of the risk this represents for human health. In addition, information was collected on the incidence of resistant *Salmonella* strains isolated from shrimp farm environments in Northeastern Brazil.

2. Methods of isolation, identification and evaluation of antibacterial susceptibility in *Salmonella*

2.1 Isolation and identification of *Salmonella*

Salmonella may be detected in samples from aquaculture environments using the traditional method described by Andrews and Hammack (2011). The method includes pre-enrichment of 25-g aliquots in lactose broth, selective enrichment in broth (*eg*, tetrathionate and Rappaport-Vassiliadis or tetrathionate and selenite cystine) and selective plating on MacConkey and Hektoen enteric agar. Typical *Salmonella* colonies grown during the selective enrichment stage are screened biochemically with triple sugar iron agar (TSI), lysine iron agar (LIA) or sulfide indole motility agar (SIM). Serotyping is done with commercially available antisera (Koonse et al., 2005), O:H polyvalent antiserum (Carvalho et al., 2009) or somatic (O), flagellar (H) and capsular (Vi) antisera (Kumar et al, 2009).

In addition, molecular biology techniques may be used for rapid detection of *Salmonella* in foods: TaqMan PCR (Kimura et al., 1999), PCR amplification of a 152-bp segment of the gene *hns* (Kumar et al., 2003), real-time PCR (Malorny et al., 2004), PCR, dot blot hybridization, RAPD and ERIC-PCR (Shabarinath et al., 2007), PCR amplification of the gene *invA* (Upadhyay et al., 2010) and uniplex and multiplex PCR (Raj et al., 2011).

2.2 Antibiogram, MIC and plasmid curing

The phenotypical susceptibility of *Salmonella* to antibiotics may be determined by the method of disk diffusion on Muller-Hinton agar (Kha et al., 2006). When testing salmonellas from aquaculture environments, the selection of antibiotics depends on the origin of the isolates, but usually covers a range of families, including the tetracyclines, sulfonamides, quinolones, macrolides and aminoglycosides (Ponce et al., 2008; Carvalho et al., 2009). The classification of bacteria according to susceptibility or resistance to antibiotics is based on the criteria of the Clinical and Laboratory Standards Institute (CLSI, 2009). The antibacterial resistance index (ARI) may be calculated following the recommendations of Jones et al. (1986). Multiple antibacterial resistance (MAR) may be calculated using the methodology described in Krumperman (1983).

Antibacterial susceptibility may also be estimated by determining the minimum inhibitory concentration (MIC) based on macrodilution of Mueller-Hinton broth (CLSI, 2009).

Salmonella strains with phenotypical profile of antibacterial resistance may be submitted to plasmid curing in Luria-Bertani broth supplemented with acridine orange dye at 100 µg mL⁻¹. The method makes it possible to determine whether resistance stems from chromosomal or plasmidial elements (Molina-Aja et al., 2002).

2.3 Determination of resistance genes and plasmid profile

Polymerase chain reaction (PCR) has been used to detect genes encoding resistance to tetracycline in *Salmonella* strains from fish farms. Restriction enzymes used in PCR include *Sma*I (for detecting the gene *tetA*), *Sal*I (for *tetC*), *Sph*I (for *tetB*, *tetD* and *tetY*), *Eco*RI (for G) and *Nde*II (for *tetE*, *tetH* and *tetI*) (Furushita et al., 2003).

The extraction of plasmidial DNA from salmonelas is usually done by alkaline lysis, as proposed by Birnboim and Doly (1979), with or without modification, or with acidic phenol, as described by Wang and Rossman (1994). For small plasmids, the extraction product may be submitted to electrophoresis in 1% agarose gel following the protocol of Akiyama et al. (2011). The protocol for electrophoresis of mega-plasmid DNA molecules in 1% agarose gel is described in Ponce et al. (2008).

3. Results

3.1 *Salmonella* in tropical aquaculture

Salmonelas are recognized worldwide as one of the main etiological agents of gastroenteritis in humans. Despite variations in the regulation of microbiological quality of foods around the world, the largest importers of seafoods only buy products completely free from *Salmonella*, based on the claim that salmonelas are not part of the indigenous microbiota of aquatic environments and that, therefore, the presence of salmonelas in aquatic organisms is associated with poor sanitation and inadequate hygiene practices (Dalsgaard, 1998).

Several studies published in the 1990s reported *Salmonella* in shrimp farming environments in tropical countries. Reilly and Twiddy (1992) found *Salmonella* in 16% of their shrimp samples and 22.1% of their pond water and sediment samples collected on farms in Southeast Asia. Weltevreden was the most abundant *Salmonella* serovar identified, followed by Anatum, Wandsworth and Potsdam. According to the authors, the incidence of *Salmonella* was higher in ponds located near urban areas and, not surprisingly, the bacterial load increased during the rainy season. Bhaskar et al. (1995) detected *Salmonella* in 37.5%, 28.6% and 37.4% of shrimp, sediment and water samples, respectively, collected from semi-intensive grow-out ponds in India.

In contrast, despite detecting high indices of thermotolerant and total coliforms, Dalsgaard et al. (1995) found no *Salmonella* in water, sediment and shrimp samples from sixteen different penaeid shrimp farms in Thailand.

Hatha and Rao (1998) reported only one *Salmonella*-positive sample out of 1,264 raw shrimp. They believed the presence of the bacteria was due to pond contamination from different sources, including the use of untreated fertilizer of animal origin. Likewise, Hatha et al. (2003) found the incidence of *Salmonella* to be low in shrimp farm products exported by India.

Koonse et al. (2005) investigated the prevalence of *Salmonella* in six major shrimp-producing countries in Southeast Asia (n=2), Central Asia (n=1), Central America (n=1), North America (n=1) and the Pacific (n=1). In four of these countries, *Salmonella* was detected in 1.6% of shrimp samples, and two serovars were identified (Paratyphi B var. Java and Weltevreden Z6). The authors highlighted the need to control or eliminate potential sources of fecal matter polluting the water bodies adjacent to the grow-out ponds.

In Brazil, the microbiological quality of shrimp (*Litopenaeus vannamei*) farmed in Ceará was evaluated by Parente et al. (2011) and Carvalho et al. (2009), both of whom detected *Salmonella* in shrimp and water samples (Table 1). The authors associated the presence of salmonelas with discharge of fecal matter into the respective estuaries where the farms are located. The detection of *Salmonella* in estuaries in Ceará is not an isolated finding. Farias et al. (2010) found salmonelas in samples of the bivalve *Tagelus plebeius* collected in the estuary of the Ceará river and identified the serovars Bredeny, London and Muechen. Similar findings were reported by Silva et al. (2003) in a study on *Salmonella* in the oyster *Crassostrea rhizophorae* obtained from natural oyster grounds in the estuary of the Cocó river, on the outskirts of Fortaleza, Ceará.

Country	Sample	N°	Sorovars	Source
Brazil	Water and Shrimp	3	*S*. ser. Saintpaul e *S*. ser. Newport	Parente et al. (2011)
Brazil	Fish	30	*S*. ser. Agona, *S*. ser. Albany, *S*. ser. Anatum, *S*. ser. Brandenburg, *S*. ser. Bredeney, *S*. ser. Cerro, *S*. ser. Enteretidis, *S*. ser. Havana, *S*. ser. Infantis, *S*. ser. Livingstone, *S*. ser.London, *S*. ser. Mbandaka, *S*. ser. Muenchen, *S*. ser. Newport, *S*. ser. Saintpaul, *S*. ser. Thompson, *S*. ser. O4,5:i:-, *S*. ser. O4,5:-:1,7, *S*. O:17	Ribeiro et al., 2010
Brazil	Water, Sediment and Shrimp	23	*S*. ser. Anatum, *S*. ser. Newport, *S*. ser. Soahanina e *S*. ser. Albany	Carvalho et al. (2009)
Vietnam	Shrimp	29	*S*. ser. Bovismorbificans, *S*. ser. Derby, *S*. ser. Dessau, *S*. ser. Lexington, *S*. ser. Schleissheim, *S*. ser. Tennessee, *S*. ser. Thompson, *S*. ser. Virchow, *S*. ser. Weltevreden, *S*. ser. II heilbron	Ogasawara et al. (2008)
India	Shrimp	54	*S*. ser. Bareilly, *S*. ser. Braenderup, *S*. ser. Brancaster, *S*. ser. Derby, *S*. ser. Kottbus, *S*. ser. Lindenburg, *S*. ser. Mbandaka, *S*. ser. Oslo, *S*. ser. Rissen, *S*. ser. Takoradi, *S*. ser. Typhi, *S*. ser. Typhimurium, *S*. ser. Weltevreden, *Salmonella* VI	Kumar et al. (2009)

*N°: number of positive samples.

Table 1. *Salmonella* in tropical seafood.

Thus, Shabarinath et al. (2007), who also detected *Salmonella* in shrimp, concluded that since salmonelas inhabit the intestinal tract of warm-blooded animals, their presence in rivers and in marine/estuarine sediments exposed to fecal contamination is not surprising.

Tropical fish species may also be infected with salmonelas (Ponce et al., 2008; Heinitz et al., 2000; Ogbondeminu, 1993); in fact, microorganisms of this genus have recently been associated with farmed catfish (McCoy et al., 2011).

3.2 Antimicrobial susceptibility profile of *Salmonella*

The use of antibiotics for prophylaxis in aquaculture not only favors the selection of resistant bacteria in the pond environment, thereby changing the natural microbiota of pond water and sediments, but also increases the risk of transferring resistance genes to pathogens infecting humans and terrestrial animals (Cabello, 2006). Thus, Le and Munekage (2005) reported high levels of drug residues (sulfamethoxazole, trimetoprim, norfloxacin and oxolinic acid) in pond water and sediments from tiger prawn farms in Northern and Southern Vietnam due to indiscriminate use of antibiotics.

According to Seyfried et al. (2010), autochthonous communities in aquatic environments may serve as a reservoir for elements of antibacterial resistance. However, the contribution of anthropic activities to the development of such reserves has not been fully clarified.

Holmström et al. (2003) reported the use, often indiscriminate, of large amounts of antibiotics on shrimp farms in Thailand, and concluded that at a regional scale human health and the environmental balance may be influenced by such practices. Adding to the impact, many of the antibiotics used for prophylaxis in shrimp farming are very persistent and toxic.

Heuer et al. (2009) presented a list of the major antibacterials used in aquaculture and their respective routes of administration: amoxicillin (oral), ampicillin (oral), chloramphenicol (oral, bath, injection), florfenicol (oral), erythromycin (oral, bath, injection), streptomycin (bath), neomycin (bath), furazolidone (oral, bath), nitrofurantoin (oral), oxolinic acid (oral), enrofloxacin (oral, bath), flumequine (oral), oxytetracycline (oral, bath, injection), chlortetracycline (oral, bath, injection), tetracycline (oral, bath, injection) and sulfonamides (oral).

Current aquaculture practices can potentially impact human health in variable, far-reaching and geographically specific ways. On the other hand, the increasing flow of aquaculture products traded on the global market exposes consumers to contaminants, some of which from production areas (Sapkota et al., 2008).

Antibacterial susceptibility in microorganisms associated with aquaculture livestock is an increasingly frequent topic in the specialized literature (Molina-Aja et al., 2002; Peirano et al., 2006; Akinbowale et al., 2006; Costa et al., 2008; Newaja-Fyzul et al., 2008; Dang et al., 2009; Del Cerro et al., 2010; Fernández-Alarcón et al., 2010; Patra et al., 2010; Vieira et al., 2010; Tamminem et al., 2011; Laganà et al., 2011; Millanao et al., 2011; Rebouças et al., 2011; Dang et al., 2011).

In this respect, salmonelas are one of the most extensively investigated groups of intestinal bacteria. Thus, in China salmonelas isolated from fish ponds were resistant to ampicillin

(20%), erythromycin (100%), cotrimoxazole (20%), gentamicin (20%), nalidixic acid (40%), penicillin (100%), streptomycin (20%), sulfanomides (40%), tetracycline (40%) and trimethoprim (20%) (Broughton and Walker, 2009).

Ubeyratne et al. (2008) detected *Salmonella* resistant to erythromycin, amoxicillin and sulfonamides in shrimp (*Penaeus monodon*) farmed in Sri Lanka. Likewise, Ogasawara et al. (2008) found salmonelas resistant to oxytetracycline and chloramphenicol in Vietnamese shrimp samples but concluded ARI values were not as high as in neighboring or developing countries.

Low ARI values were also reported by Boinapally and Jiang (2007) who in a single sample of shrimp imported to the US detected *Salmonella* resistant to ampicillin, ceftriaxone, gentamicin, streptomycin and trimethoprim. This is in accordance with published findings for shrimp in tropical regions, where the major exporters of farmed shrimp are located.

Zhao et al. (2003) evaluated the profile of antibacterial resistance in salmonelas isolated from seafood from different countries and found that most of the resistant bacteria came from Southeast Asia. The authors believe the use of antibiotics in aquaculture, especially in Southeast Asia, favors the selection of resistant *Salmonella* strains which may find their way into the US market of imported foods.

In Brazil, Ribeiro et al. (2010) reported an antibacterial resistance index of 15.1% among salmonelas isolated from an aquaculture system. The *Salmonella* serovars Mbandaka (n=1) and Agona (n=2) were resistant to tetracycline, Albany (n=1) was resistant to sulfamethoxazole-trimethoprim, and London (n=2) was resistant to chloramphenicol. In addition, Carvalho et al. (2009) collected samples from three penaeid shrimp farms in Ceará (Northeastern Brazil) and found *Salmonella* serovars Newport and Anatum to be resistant to tetracycline and nalidixic acid. Water and sediment samples collected in the vicinity of the three farms contained the *Salmonella* serovars Newport, Soahanina, Albany and Anatum, which were likewise resistant to tetracycline and nalidixic acid, suggesting the ponds were contaminated by water drawn from the estuaries.

Bacterial resistance in *Salmonella* may be of either chromosomal or plasmidial nature (Frech e Schwarz, 1999; Mirza et al., 2000; Govender et al., 2009; Tamang et al., 2011; Glenn et al., 2011). In bacteria, the acquisition and diffusion of resistance genes may be influenced by exchanges of DNA mediated by conjugative plasmids and by the integration of resistance genes into specialized genetic elements (Carattoli et al., 2003).

Evidence of plasmidial mediation of antibacterial resistance in *Salmonella* has been available since the 1970s and 1980s (Anderson e Threlfall, 1974; Frost et al., 1982). Thus, Anderson et al. (1977) detected three types of resistance plasmids in *Salmonella* strains from different countries. According to the authors, plasmids of the F_{Ime} type confer resistance to penicillin, ampicillin and streptomycin, whereas, for example, resistance to furazolidone in all *Salmonella* isolates from Israel was considered to be chromosomal. Mohan et al. (1995) determined the plasmid profile of *Salmonella* strains isolated from different regions in India and found a large diversity of small plasmids (2.7 to 8.3 kb) in strains resistant to ampicillin, chloramphenicol, kanamycin, streptomycin, sulphamethoxazole, tetracycline and trimethoprim.

In one study, salmonelas isolated from food animals were found to carry CMY-2, a plasmid-mediated AmpC-like β-lactamase (Winokur et al., 2001). Doublet et al. (2004) found *florR* (a florfenicol resistance gene) and *bla*$_{CMY-2}$ plasmids to be responsible for resistance to wide-spectrum cephalosporines in salmonelas isolated from clinical samples, animals and foods in the US. The authors added that the use of phenicols in animal farming environments may place a selective pressure on organisms and favor the dissemination of *bla*$_{CMY-2}$ plasmids. In addition, *florR* is known to confer cross-resistance to chloramphenicol.

Kumar et al. (2010) found evidence that tropical seafood can serve as vehicle for resistant salmonela strains, some of which resistant to as many as four antibiotics (sulfamethizole, carbenicillin, oxytetracycline and nalidixic acid). The authors also identified low-molecular-weight plasmids in the *Salmonella* serovars Braenderup, Lindenburg and Mbandaka.

Six isolates of *Salmonella* serovar Saintpaul from samples of shrimp and fish from India, Vietnam and Saudi Arabia presented one or more resistance plasmids of varying size (2.9 to 86 kb). One of these carried a *Incl1* plasmid (Akiyama et al., 2011).

As discussed above, the indiscriminate use of antibiotics in aquaculture is one of the major causes of the emergence of resistant bacteria in the environment. Several of the mechanisms of resistance in *Salmonella* have been investigated, especially with regard to beta-lactams (Alcaine et al., 2007) and quinolones (Piddock et al., 1998; Piddock, 2002)—two families of antibiotics widely used in aquaculture.

4. Conclusion

The growing incidence of *Salmonella* in tropical aquaculture environments is a worldwide concern which may have local impacts (in the culture area) or global impacts (considering the dynamics of the international seafood market). Human health and environmental balance are further threatened by the emergence of salmonelas resistant to antibiotics employed in farming, in some cases mediated by mobile genetic elements. The elimination of sources of fecal pollution from tropical areas used for aquaculture seems to be the main strategy for minimizing the risk of transference of salmonelas to foods destined for human consumption. As a final consideration, studies should be encouraged on the presence, antibacterial susceptibility and mechanisms of resistance in salmonelas occurring in tropical areas destined for culture of fish, crustaceans and mollusks.

5. References

Akinbowale OL, Peng H, Barton MD. Antimicrobial resistance in bacteria isolated from aquaculture sources in Australia. Journal of Applied Microbiology, v. 100, p. 1103-1113, 2006.

Akiyama T, Khan AA, Cheng CM, Stefanova R. Molecular characterization of *Salmonella* enteric serovar Saintpaul isolated from imported seafood, pepper, environmental and clinical samples. Food Microbiology, v. 28, p. 1124-1128, 2011.

Alcaine SD, Warnick LD, Martin W. Antimicrobial Resistance in Nontyphoidal *Salmonella*. Journal of Food Protection, v. 70, n. 3, p. 780-790, 2007.

Anderson ES, Threlfall EJ, Carr JM, Mcconnell MM, Smith HR. Clonal distribution of resistance plasmid-carrying Salmonella typhimurium, mainly in the Middle East. Journal of Hygiene, v. 79, p. 425-448, 1977.

Anderson ES, Threlfall EJ. The characterization of plasmids in the enterobacteria. Journal of Hygiene, v. 72, p. 471-87, 1974.

Andrews WH, Hammack T. *Salmonella*. In: Bacteriological Analytical Manual. U.S. Food and Drug Administration. 2011. Available in: http://www.fda.gov/Food/Science Research/LaboratoryMethods/BacteriologicalAnalyticalManualBAM/default.htm.

Angulo FJ, Griffin PM. Changes in antimicrobial resistance in *Salmonella enterica* serovar Typhimurium. Emerging Infectious Diseases, v. 6, n. 4, p. 436-437, 2000.

Asai Y, Kaneko M, Ohtsuka K, Morita Y, Kaneko S, Noda H, Furukawa I, Takatori K, Hara-kudo Y. *Samonella* prevalence in seafood imported into Japan. Journal of Food Protection, v. 71, n. 7, p. 1460-1464, 2008.

Bhaskar N, Setty TMR, Reddy GVS, Manoj YB, Anantha CS, Raghunath BS, Antony JM. Incidence of *Salmonella* in cultured shrimp *Penaeus monodon*. Aquaculture, v. 138, p. 257-266, 1995.

Bhaskar N, Setty TMR, Mondal S, Joseph MA, Raju CV, Raghunath BS, Anantha CS. Prevalence of bacteria of public health significance in the cultured shrimp Penaeus monodon. Food Microbiology, v. 15, p. 511-519, 1998.

Birnboim HC, Doly J. A rapid alkaline extraction procedure for screening recombinant plasmid DNA. Nucleic Acids Research, v. 7, n. 6, p. 1513-1523, 1979.

Boinapally K, Jiang X. Comparing antibiotic resistance in commensal and pathogenic bacteria isolated from wild-caught South Carolina shrimps vs. farm-raised imported shrimps. Canadian Journal of Microbiology, v. 53, n. 7, p. 919-924, 2007.

Bremer PJ, Fletcher GC, Osborne C. *Salmonella* in seafood. New Zealand Institute for Crop & Food Research Limited, 2003.

Broughton EI, Walker DG. Prevalence of antibiotic-resistant *Salmonella* in fish in Guangdong, China. Foodborne pathogens and disease, v. 6, n. 4, p. 519-521, 2009.

Cabello FC. Heavy use of prophylactic antibiotics in aquaculture: a growing problem for human and animal health and for the environment. Environmental Microbiology, v. 8, p. 1137-1144, 2006.

Carattoli A. Plasmid-mediated antimicrobial resistance in *Salmonella enterica*. Current Issues in Molecular Biology, v. 5, p. 113-122, 2003.

Carvalho FCT, Barreto NSE, Reis CMF, Hofer E, Vieira RHSF. Susceptibilidade antimicrobiana de *Salmonella* spp. Isoladas de fazendas de carciniculturas no Estado do Ceará. Revista Ciência Agronômica, v. 40, n. 4, p. 549-556, 2009.

CLSI. Clinical and Laboratory Standards Institute. Performance Standards for Antimicrobial Susceptibility Testing; Twentieth Informational Supplement: Supplement M100-S19, Wayne, PA, USA, 2009.

Costa RA, Vieira GHF, Silva GC, Vieira RHSF, Sampaio SS. Susceptibilidade "in vitro" a antimicrobianos de estirpes de *Vibrio* spp isoladas de camarões (*Litopenaeus vannamei*) e de água de criação destes animais provenientes de uma fazenda de camarões no Ceará - Nota prévia. Brazilian Journal of Veterinary Research and Animal Science, v. 45, n. 6, p. 458-462, 2008.

Dang H, Zhao J, Song L, Chen M, Chang Y. Molecular characterizations of chloramphenicol- and oxytetracycline-resistant bacteria and resistance genes in mariculture waters of China. Marine Pollution Bulletin, v. 58, n. 7, p. 987-994, 2009.

Dang ST, Petersen A, Van Truong D, Chu HT, Dalsgaard A. Impact of medicated feed on the development of antimicrobial resistance in bacteria at integrated pig-fish farms in Vietnam. Applied and Environmental Microbiology, v. 77, n. 13, p. 4494-4498, 2011.

Dalsgaard A. The occurrence of human pathogenic *Vibrio* spp. and *Salmonella* in aquaculture. International Journal of Food Science and Technology, v. 33, p. 127-138, 1998.

Dalsgaard A, Huss HH, H-Kittikun A, Larsen JL. Prevalence of *Vibrio cholerae* and *Salmonella* in a major shrimp production area in Thailand. International Journal of Food Protection, v. 28, p. 101-113, 1995.

Del Cerro A, Márquez I, Prieto JM. Genetic diversity and antimicrobial resistance of *Flavobacterium psychrophilum* isolated from cultured rainbow trout, *Onchorynchus mykiss* (Walbaum), in Spain. Journal of Fish Diseases, v. 33, n. 4, p. 285-291, 2010.

Doublet B, Carattoli A, Whichard JM, White DG, Baucheron S, Chaslus Dancla E, Cloeckaert A. Plasmid-mediated florfenicol and ceftriaxone resistance encoded by the floR and bla$_{CMY-2}$ genes in Salmonella enterica serovars Typhimurium and Newport isolated in the United States. FEMS Microbiology Letters, v. 233, n. 2, p. 301-305, 2004.

Fauconneau B. Health value and safety quality of aquaculture products. Revue de Médecine Vétérinaire, v. 153, n. 5, p. 331-336, 2002.

Farias MF, Rocha-Barreira CA, Carvalho FCT, Silva CM, Reis EMF, Costa RA, Vieira RHSF. Condições microbiológicas de *Tagelus plebeius* (Lightfoot 1786) (Mollusca: Bivalvia: Solecurtidae) e da água no estuário do rio Ceará, em Fortaleza-CE. Boletim do Instituto de Pesca, v. 36, n. 2, p. 135-142, 2010.

Fernández-Alarcón C, Miranda CD, Singer RS, López Y, Rojas R, Bello H, Domínguez M, González-Rocha G. Detection of the *flo*R gene in a diversity of florfenicol resistant Gram-negative bacilli from freshwater salmon farms in Chile. Zoonoses and Public Health, v. 57, n. 3, p.181-188, 2010.

Frech G, Schwarz S. Plasmid-encoded tetracycline resistance in *Salmonella enteric* subsp. *enterica* serovars *choleraesuis* and *typhimurium*: identification of complete and truncated Tn1721 elements. FEMS Microbiology Letters, v. 176, p. 97-103, 1999.

Frost JA, Rowe B, Ward LR, Threlfall EJ. Characterization of resistance plasmids and carried phages in an epidemic clone of multi-resistant *Salmonella typhimurium* in India. Journal of Hygiene, v. 88, p. 193-204, 1982.

Furushita M, Shiba T, Maeda T, Yahata M, Kaneoka A, Takahashi Y, Torii K, Hasegawa T, Ohta M. Similarity of tetracycline resistance genes isolated from fish farm bacteria to those from clinical isolates. Applied and Environmental Microbiology, v. 69, n. 9, p. 5336–5342, 2003.

Glenn LM, Lindsey RL, Frank JF, Meinersmann RJ, Englen MD, Fedorka-Cray PJ, Frye JG. Analysis of antimicrobial resistance genes detected in multidrug-resistant *Salmonella enterica* serovar Typhimurium isolated from food animals. Microbial Drug Resistance, 2011. DOI:10.1089/mdr.2010.0189.

Govender N, Smith AM, Karstaedt AS, Keddy KH. Plasmid-mediated quinolone resistance in *Salmonella* from South Africa. Journal of Medical Microbiology, v. 58, p. 1393-1394, 2009.

Hatha AAM, Rao NPB. Bacteriological quality of individually quick-frozen (IQF) raw and cooked ready-to-eat shrimp produced from farm raised black tiger shrimp (*Penaeus monodon*). Food Microbiology, v. 15, p. 177-183, 1998.

Hatha AAM, Maqbool TK, Kumar SS. Microbial quality of shrimp products of export trade produced from aquacultured shrimp. International Journal of Food Microbiology, v. 82, p. 213-221, 2003.

Heinitz ML, Ruble RD, Wagner DE, Tatini SR. Incidence of *Salmonella* in fish and seafood. Journal of Food Protection, v. 63, n. 5, p. 579-592, 2000.

Heuer OE, Kruse H, Grave K, Collignon P, Karunasagar I, Angulo FJ. Human health consequences of use of antimicrobial agents in aquaculture. Clinical Infectious Diseases, v. 49, p. 1248-1253, 2009.

Holmström K, Gräslund S, Wahlström A, Poungshompoo S, Bengtsson BE, Kautsky N. Antibiotic use in shrimp farming and implications for environmental impacts and human health. International Journal of Food Science and Technology, v. 38, p. 255-266, 2003.

Jones JG, Gardener S, Simon BM, Pickup RW. Factors affecting the measurement of antibiotic resistance in bacteria isolated from lake water. Journal of Applied Microbiology, v. 60, n. 5, p. 455-462, 1986.

Khan AA, Cheng CM, Khanh TV, Summage-West C, Nawaz MS, Khan SA. Characterization of class 1 integron resistance gene cassettes in Salmonella enteric serovars Oslo and Bareily from imported seafood. Journal of Antimicrobial and Chemotherapy, v. 58, p. 1308-1310, 2006.

Kimura B, Kawasaki S, Fujii T, Kusunoki J, Ithoh T, Flood SJ. Evaluation of TaqMan PCR assay for detecting Salmonella in raw meat and shrimp. Journal of food protection, v. 62, n. 4, p. 329-335, 1999.

Koonse B, Burkhardt III W, Chirtel S, Hoskin GP. Salmonella and the sanitary quality of aquacultured shrimp. Journal of Food Protection, v. 68, n. 12, p. 2527-2532, 2005.

Krumperman PH. Multiple antibiotic resistance indexing of Escherichia coli to indentify high-risk sources of fecal contamination of foods. Applied and Environmental Microbiology, v. 46, p. 165-170, 1983.

Kumar HS, Sunil R, Venugopal MN, Karunasagar I, Karunasagar I. Detection of Salmonella spp. in tropical seafood by polymerase chain reaction. International Journal of Food Microbiology, v. 88, p. 91-95, 2003.

Kumar R, Surendran PK, Thampuran N. Analysis of antimicrobial resistance and plasmid profiles in Salmonella serovars associated with tropical seafood of India. Foodborne Pathogens and Disease, v. 6, n. 5, p. 621-625, 2009.

Kumar R, Surendran PK, Thampuran N. Distribuition and genotypic characterization of Salmonella serovars isolated from tropical seafood of Cochin, India. Journal of Applied Microbiology, v. 106, p. 515-524, 2009.

Kumar R, Surendran PK, Thampuran N. Rapid quantification of Salmonella in seafood using real-time PCR assay. Journal of Microbiology and Biotechnology, v. 20, n. 3, p. 569-573, 2010.

Le TX, Munekage Y, Kato S. Antibiotic resistance in bacteria from shrimp farming in mangrove areas. The Science of the Total Environment, v. 349, p. 96-105, 2005.

Laganà P, Caruso G, Minutoli E, Zaccone R, Santi D. Susceptibility to antibiotics of Vibrio spp. and Photobacterium damsela ssp. piscicida strains isolated from Italian aquaculture farms. The New Microbiologica, v. 34, n. 1, p. 53-63, 2011.

Ling ML, Goh KT, Wang GCY, Neo KS, Chua T. An outbreak of multidrug-resistant Salmonella enterica subsp. enterica serotype Typhimurium, DT104L linked to dried anchovy in Singapore. Epidemiology and Infection, v. 128, p. 1-5, 2002.

Malorny B, Paccassoni E, Fach P, Bunge C, Martin A, Helmuth R. Diagnostic Real-Time PCR for Detection of Salmonella in Food. Applied and Environmental Microbiology, v. 70, n. 12, p. 7046-7052, 2004.

McCoy E, Morrison J, Cook V, Johnston J, Eblen D, Guo C. Foodborne agents associated with the consumption of aquaculture catfish. Journal of Food Protection, v. 74, n. 3, p. 500-516, 2011.

Millanao AB, Barrientos MH, Gómez GC, Tomova A, Buschmann A, Dölz H, Cabello FC. Uso inadecuado y excesivo de antibióticos: Salud pública y salmonicultura em Chile. Revista Médica de Chile, v. 139, p. 107-118, 2011.

Mirza S, Kariuki S, Mamun KZ, Beeching NJ, Hart CA. Analysis of plasmid and chromosomal DNA of multidrug-resistant Salmonella enterica Serovar Typhi from Asia. Journal of Clinical Microbiology, v. 38, n. 4, p. 1449-1452 2000.

Molina-Aja A, García-Gasca A, Abreu-Grobois A, Bolán-Mejía C, Roque A, Gomez-Gil B. Plasmid profiling and antibiotic resistance of Vibrio strains isolated from cultured penaeid shrimp. FEMS Microbiology Letters, v. 213, p. 7-12, 2002.

Mohan VP, Sharma KB, Agarwal DS, Purnima G, Pillai PR. Plasmid profile and phage type of Salmonella typhimurium strains encountered in different regions of India. Comparative Immunology, Microbiology and Infectious Diseases, v. 18, n. 4, p. 283-290, 1995.

Newaj-Fyzul A, Mutani A, Ramsubhag A, Adesiyun A. Prevalence of bacterial pathogens and their anti-microbial resistance in Tilapia and their pond water in Trinidad. Zoonoses and Public Health, v. 55, n. 4, p. 206-213.

Ogasawara N, Tran TP, Ly TLK, Nguyen TT, Iwata T, Okatani AT, Watanabi M, Taniguchi T, Hirota Y, Hayashidani H. Antimicrobial susceptibilities of Salmonella from domestic animals, food and human in the Mekong delta, Vietnam. The Journal of veterinary medical science, v. 70, n. 11, p. 1159-1164, 2008.

Ogbondeminu FS. The occurrence and distribution of enteric bacteria in fish and water of tropical aquaculture ponds in Nigeria. Journal of Aquaculture in the Tropics, v. 8, n. 1, p. 61-66. 1993.

Parente LS, Costa RA, Vieira GHF, Reis EMF, Hofer E, Fonteles AA, Vieira RHSF. Bactérias entéricas presentes em amostras de água e camarão marinho Litopenaeus vannamei oriundos de fazendas de cultivo no Estado do Ceará, Brasil. Brazilian Journal of Veterinary Research and Animal Science, v. 48, n. 1, p. 46-53, 2011.

Patra S, Das TK, Gosh SCh, Sarkar D, Jana BB. Cadmium tolerance and antibiotic resistance of Pseudomonas sp. isolated from water, sludge and fish raised in wastewater-fed tropical ponds. Indian Journal of Experimental Biology, v. 48, n. 4, p. 383-393, 2010.

Peirano G, Agerso Y, Aarestrup FM, Reis EMF, Rodrigues DP. Occurrence of integrons and antimicrobial resistance genes among Salmonella enterica from Brazil. Journal of Antimicrobial Chemotherapy, v. 58, p. 305-309, 2006.

Piddock LJV, Ricci V, McLaren I, Griggs DJ. Role of mutations in the gyrA and parC genes of nalidixic-acid-resistant Salmonella serotypes isolated from animals in the United Kingdom. The Journal of antimicrobial chemotherapy, v. 41, n. 6, p. 635-642, 1998.

Piddock LJV. Fluoroquinolone resistance in Salmonella serovars isolated from humans and food animals. FEMS Microbiology Reviews, v. 26, n. 1, p. 3-16, 2002.

Ponce E, Khan AA, Cheng C-M, Summage-West C, Cerniglia CE. Prevalence and characterization of Salmonella enteric serovar Weltevreden from imported seafood. Food Microbiology, v. 25, p. 29-35, 2008.

Raj KT, Jeyasekaran G, Shakila RJ, Thangarani AJ, Sukumar D. Multiplex polymerase chain reaction assay for the detection of Salmonella enteric serovars in shrimp in 4 h. Journal of Bacteriology Research, v. 3, n. 3, p. 56-62, 2011.

Rebouças RH, Sousa OV, Lima AS, Vasconcelos FR, Carvalho PB, Vieira RHSF. Antimicrobial resistance profile of Vibrio species isolated from marine shrimp farming environments (Litopenaeus vannamei) at Ceará, Brazil. Environmental Research, v. 111, p. 21-24, 2011.

Reilly PJ, Twiddy DR. Salmonella and Vibrio cholerae in brackishwater cultured tropical prawns. International Journal of Food Microbiology, v. 16, n. 4, p. 293-301, 1992.

Ribeiro RV, Reis EMF, Reis CMF, Freitas-Almeida AC, Rodrigues DP. Incidence and antimicrobial resistance of enteropathogens isolated from an integrated aquaculture system. Letters in Applied Microbiology, v. 51, p. 611-618, 2010.

Sapkota A, Sapkota AR, Kucharski M, Burke J, McKenzie S, Walker P, Lawrence R. Aquaculture practices and potential human health risks: Current knowledge and future priorities, v. 34, p. 1215-1226, 2008.

Seyfried EE, Newton RJ, Rubert IV KF, Pedersen JA, McMahon KD. Occurrence of tetracycline resistance genes in aquaculture facilities with varying use of oxytetracycline. Microbial Ecology, v. 59, p. 799-807, 2010.

Shabarinath S, Sanath Kumar H, Khushiramani R, Karunasagar I, Karunasagar I. Detection and characterization of *Salmonella* associated with tropical seafood. International Journal of Food Microbiology, v. 114, n. 2, p. 227-33, 2007.

Silva A.I.M., Vieira R.H.S.F., Menezes F.G.R., Lima L.N.G.C., Nascimento S.M.M., Carvalho F.C.T. Bactérias fecais em ostras, *Crassostrea rhizophorae*. Arquivos de Ciências do Mar, v. 36, p. 63-66, 2003.

Tamang M.D., Nam H.M., Kim T.S., Jang G.C., Jung S.C., Lim S.K. Emergence of extended-spectrum {beta}-lactamase (CTX-M-15 and CTX-M-14) - producing nontyphoid *Salmonella* with reduced susceptibility to ciprofloxacin among food animals and humans in Korea. Journal of Clinical Microbiology, v. 49, n. 7, p. 2671-2675, 2011.

Tamminen M., Karkman A., Lõhmus A., Muziasari W.I., Takasu H., Wada S., Suzuki S., Virta M. Tetracycline resistance genes persist at aquaculture farms in the absence of selection pressure. Environmental Science & Technology, v. 45, n. 2, p. 386-391, 2010.

Upadhyay B.P., Utrarachkij F., Thongshoob J., Mahakunkijcharoen Y., Wongchinda N., Suthienkul O., Khusmith S. Detection of *Salmonella* invA gene in shrimp enrichment culture by polymerase chain reaction. The Southeast Asian Journal of Tropical Medicine and Public Health, v. 41, n. 2, p. 426-435, 2010.

Ubeyratne K.H., Hildebrandt G., Kleer J., Khattiya R., Padungtod P. Microbiological quality of marketed *Penaeus Monodon* shrimps in north western province, Sri Lanka. Proceedings, The 15th Congress of FAVA. OIE Joint Symposium on Emerging Diseases Bangkok, Thailand, p. P63-P65, 2008.

Vieira R.H.S.F., Carvalho E.M.R., Carvalho F.C.T, Silva C.M., Sousa O.V., Rodrigues D.P. Antimicrobial susceptibility of *Escherichia coli* isolated from shrimp (*Litopenaeus vannamei*) and pond environment in northeastern Brazil. Journal of Environmental Science and Health. Part. B, Pesticides, Food Contaminants, and Agricultural Wastes, v. 45, n. 3, p. 198-203, 2010.

Velge P., Cloeckaert A., Barrow P. Emergence of *Salmonella enterica* serotype Enteritidis and multiple antibiotic resistance in other major serotypes. Veterinary Research, v. 36, p. 267-288, 2005.

Wang Z., Rossman T.G.. Large-scale supercoiled plasmid preparation by acidic phenol extraction. Biotechniques, v. 16, n. 3, p. 460-463, 1994.

Winokur P.L., Vonstein DL, Hoffman L.J., Uhlenhopp E.K., Doern G.V. Evidence for Transfer of CMY-2 AmpC β-Lactamase Plasmids between *Escherichia coli* and *Salmonella* isolates from food animals and humans. Antimicrobial Agents and Chemotherapy, v. 45, n. 10, p. 2716-2722, 2001.

Zhao S., Datta A.R., Ayers S., Friedman S., Walker R.D., White D.G. Antimicrobial-resistant Salmonella serovars isolated from imported foods. The International Journal of Food Microbiology, v. 84, p. 87–92, 2003.

Part 3

Genetics

12

Salmonella as a Unique Tool for Genetic Toxicology

Mehmet Karadayı, Özlem Barış and Medine Güllüce
Biology Department of Atatürk University, Erzurum
Turkey

1. Introduction

In molecular biology and genetics, mutations are described as sudden and spontaneous or induced changes in a genomic sequence (Brown, 2007). They have wide effects on all living organisms from bacteria with a single prokaryotic cell construction to multicellular and eukaryotic organisms including human being with high-level cellular differentiation. Mutations occur also in the genomic materials (DNA or RNA) of viruses and affect their functionality (Hartl & Jones, 1998; Lewin, 2004). When a mutation happens, it can basically result in several different types of change in DNA (or RNA for some viruses) sequences; these can have no effect, alter the product of gene, and prevent the gene from functioning properly or completely. Alterations in the product of gene and partial or total loss of gene function generally result in a disadvantageous situation for the organism, which cause various symptoms and ailments affect the maintenance of life (Brown, 2007). Previous studies made to understand the relations between mutations and their negative effects on human being clearly showed that some diseases, such as most forms of cancer, heart disease and mental disorders, have a partly or completely genetic basis closely related to mutagenesis (Bertram, 2000; Alberts et al., 2002; Lodish et al., 2007). Therefore, recent investigations have mainly focused on mutation classification, understanding mutagenesis mechanisms, determination of mutagenic agents and prevention strategies (Cox, 1976; Albertini et al., 1990; Davidson et al., 2002; Akiyama, 2010; Evans et al., 2010; Gulluce et al., 2010; Lynch, 2010; Waters et al., 2010; Lange et al., 2011; Loeb, 2011; Pao & Girard, 2011). Thus, the identification of substances capable of inducing mutations has become an important procedure in safety assessment. In the research studies, mutations can be divided in two main groups to get more comprehensive results according to their size. First group is described as gene mutations, where only single base is modified, or one or a relatively few bases are inserted or deleted (Brown, 2007). Other one consists of chromosome mutations, which are including chromosome breaks, large deletions, rearrangements, or gain or loss of whole chromosome (Hartl & Jones, 1998; Lewin, 2004).

Mutation test systems also divide in long-term and short-term systems according to obtaining of the results (Wickramasnghe, 1979; Mortelmans & Zeiger, 2000; Zeiger et al., 2005). The long-term mutagenicity tests, which use *in vivo* researches with various experimental animals, give more reliable results than short-term mutagenicity test systems. However, they are not preferred as beginning test systems due to their high cost and long time requirements, where mutagenic potential of many synthetic and natural chemicals are checked (Wickramasnghe, 1979; Gulluce et al., 2010). In these studies, the short-term test systems, which eliminate disadvantages of the long-term test systems, are more suitable and acceptable. Many short-term studies result in gaining reliable and alternative data under controlled *in vitro* conditions. Another important advantage is that short-term mutation test systems are not only correlated with other short-term test systems, but also long-term systems (Mortelmans & Zeiger, 2000). Thus, the combinations of the mutagenicity test systems play a key role to get more meaningful results.

The bacterial mutation assays are known as most important short-term systems in order to determine mutagenic and antimutagenic potential of natural or synthetic chemicals related to gene mutations (Ames et al., 1973a, 1973b; Maron & Ames, 1983; Mortelmans & Riccio, 2000; Mortelmans & Zeiger, 2000). The *Salmonella* bacterial reverse mutation assay is one of the simplest, the most meaningful and acceptable short-term mutagenicity and antimutagenicity test systems. The test was initially developed by Ames in 1971. Therefore, it is also called as the Ames mutagenicity assay or the Ames/*Salmonella* mutagenicity assay (Ames et al., 1973a, 1973b; Maron & Ames, 1983; Zeiger, 2004, 2010). The main advantages of the assay, which employs mutant *Salmonella typhimurium* tester strains as model prokaryotic organisms, can be ordered as inexpensive applications enable studying a large number of test materials, quickly resulting (approximately 48 hours) allows making replicates in a short time, divers tester strains with several gene mutations allow to research the molecular effect mechanism of test materials, additional mutations in each strain result in more sensitivity such as *rfa* or *uvrB* and mesophile character of *Salmonella* allows to study several test materials affective at human body temperature. Furthermore, combination of the cytochrome-based P450 metabolic oxidation system, which usually consists of a 9000×g supernatant fraction of a rat liver homogenate (S-9 microsomal fraction), with the *Salmonella* mutagenicity test allows to determine some mutagenic agents, which are biologically inactive unless they are metabolized to active forms (Ames et al., 1973b; Mortelmans & Zeiger, 2000; Zeiger et al., 2005). Thus, the assay is used world-wide in genetic toxicology laboratories as a beginning mutation test to determine mutagenic and antimutagenic potentials of various chemicals.

The present study includes an introduction to use of *Salmonella* strains in genetic toxicology, principles of *Salmonella* bacterial reverse mutation assay, the most popular assay procedures with explanatory figures and clues for experimental design.

2. Scientific background

2.1 Mutations and their effects on living organisms

Genetic materials of all living organisms are dynamic structures that change and rearrange many times as a result of cumulative effects of mutations. Mutations, described as sudden and spontaneous or induced changes in a genomic sequence, are classified in two main groups depend on their physical effect sizes on the genome (Hartl & Jones, 1998; Lewin, 2004; Klug et al., 2005; Brown, 2007). First group is large-scale mutations in chromosomal level, including numerical and structural anomalies. Numerical anomalies are also called as aneuploidy, means an abnormal number of chromosomes. The most known examples for aneuploidy are monosomy (missing a chromosome from a pair), nullisomy (missing a pair of chromosomes), trisomy and polysomy (having one or more than two chromosomes of a pair). Down and Turner syndromes are important examples related to chromosomal anomalies in humans. An individual with Down syndrome has a developmental disorder caused by having three copies of chromosome 21. Therefore, it is also called as Trisomy 21.

Another disorder is Turner syndrome, an example of a monosomy where the individual is born with only one X chromosome (Klug et al., 2005).

Structural anomalies of the large-scale mutations include deletions, duplications, inversions and translocations. A deletion is a loss of one or more pieces from a chromosome after DNA breaks induced by various physical or chemical agents (Klug et al., 2005). Genetic information loss together with deletions causes serious disorders in humans, for example, Wolf-Hirschhorn syndrome, also known as deletion 4p syndrome, and Jacobsen syndrome, also known as deletion 11q syndrome (Hirschhorn et al., 1965; Jacobsen et al., 1973). Duplication is described as a phenomenon that a chromosome has extra copies of a chromosomal region, which may affect phenotype by altering gene function and transcriptional dosage (Zhang, 2003; Mao & Pevsner, 2005). Because most embryonic processes requires sensitively balanced protein levels, many duplications lead to developmental defects such as *Bar* eye mutation in *Drosophila* and Charcot-Marie-Tooth disease in humans (Sutton, 1943; Latour et al., 1997). An inversion type chromosomal mutation occurs when a portion of the chromosome breaks off, 180° rotates and reattaches, resulting in an inverted genetic material. There is little knowledge about the linkage between inversions and disease formation, and it is believed that many affect mechanisms of inversions directly associated with deletions. Juvenile Polyposis of Infancy, a rare genetic disorder, is a good example for a disease evolved by the cumulative effects of inversions (a paracentric inversion in 10q) and deletions (a deletion in 10p) (Gimelli et al., 2003; Antonacci et al., 2009; Vargas-Gonzales et al., 2010). The last group of the structural anomalies is translocations, defined as an exchange of segments among the non-homologues chromosomes. Several forms of cancer, leukemia and lymphoma are the best known disorders related to translocations (Li et al., 1999; Kurzrock et al., 2003; Anton et al., 2004). Figure 1 illustrates structural chromosome mutations.

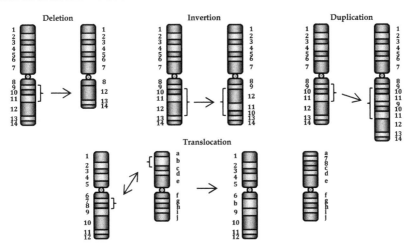

Fig. 1. Structural chromosome mutations

Small-scale mutations, also known as gene mutations, include three main groups: point mutations, which are the most common type of the gene mutations and replace one nucleotide with another, insertions, which add one or a few extra nucleotides into the DNA, and deletions, which remove one or a few nucleotides from the DNA (Brown, 2007).

Point mutations are also divided into two categories as transitions and transversions. Transitions, which are purine-to-purine or pyrimidine-to-pyrimidine changes (A↔G or C↔T), are more common type of the point mutations than transversions, which are purine-to-pyrimidine or pyrimidine-to-purine changes (A↔C, A↔T, G↔C or G↔T) (Brown, 2007).

Contrary to "small-scale" word in their names, these mutations can cause wide-range significant changes in genomes and phenotypes of living organisms with mutated genetic materials. For example, a point mutation may result in a synonymous change that causes forming a new codon specifying the same amino acid as the unmutated codon, a non-synonymous change that causes a missense mutation where a new codon specifies a different amino acid from the unmutated codon, a nonsense mutation where the change converts an amino acid specifying codon into a termination codon, or a readthrough mutation where the change converts a termination codon into an amino acid specifying codon. Except synonymous changes, also called as silent mutations because the mutated gene codes for exactly the same protein as the unmutated gene, the other three types of point mutations have significant impacts on the genome and related phenotypes by effecting amino acid sequence of the coding protein (Hartl & Jones, 1998; Alberts et al., 2002; Lewin, 2004; Brown, 2007; Lodish et al., 2007). The effects of point mutations on the coding region of a gene are shown in Figure 2.

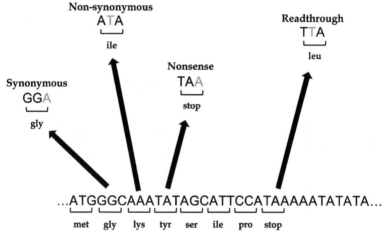

Fig. 2. The effects of point mutations on the coding region of a gene (Brown, 2007).

Insertion and deletion types of small-scale mutations affect the coding capabilities of the gene in a different way. It is defined as a frameshift mutation, caused by addition or deletion of a number of nucleotides that is not evenly divisible by three from a DNA sequence. Because codons consist of three nucleotides, an insertion or deletion type mutation can disrupt the reading frame, resulting in a completely different translation from the unmutated gene. Thus, insertion or deletion mutations generally have more significant effects on the protein function than the point mutations because the translated protein have completely different sequence from the mutated point to the end. An exception occurs that the number of inserted or deleted nucleotides is three or a multiple of three, which results in addition or deletion of one or more codons (Alberts et al., 2002; Lewin, 2004; Brown, 2007; Lodish et al., 2007). Figure 3 illustrates two possible effect mechanisms of the insertion or deletion type mutations on the coding region of a gene.

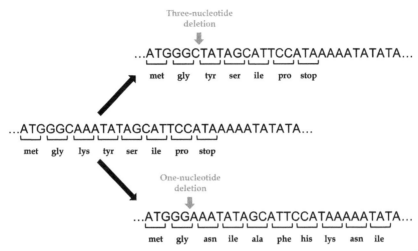

Fig. 3. Two possible effect mechanisms of the insertion or deletion type mutations on the coding region of a gene (Brown, 2007).

Phenotypic results of mutations can be deleterious or advantageous for the affected organism. Many hereditable disorders are either directly caused by the mutations or indirectly associated with the mutagenesis. Cancer formation can be given as a satisfactory example to demonstrate deleterious effect of mutations. Recent studies clearly showed that there is a strict connection between mutagenesis and the formation of the several cancer types (Davidson et al., 2002). In this manner, mutations provide a unique resource for all tumors that show genomic instability with few exceptions. On the other hand, mutated organisms can gain various advantages as a result of mutagenesis. Gain of antibiotic resistance in microorganisms and HIV/AIDS protective mutations on the SDF1, CCR5 and CCR2 genes in the human genome are well examples for the advantageous mutations (Stephan et al., 1998; Galvani & Slatkin, 2003; Apostolakis et al., 2005). These are also very important for evaluation of the organisms. Therefore, mutagenesis can be considered as one of the most important evolutionary sources. For example; simple sequence repeats (SSRs, also called microsatellites and minisatellites) are defined as advantageous mutators in adaptive evolution. Recent studies showed that temperature compensation of circadian rhythm in *Drosophila*, adaptive divergence among barley and wheat populations, social behavior in voles, skeletal morphology in domestic dogs and sporulation efficiency and cell adhesion in yeast are closely related to SSRs, which are mutation-prone DNA tracts composed of tandem repetitions of relatively short motifs (Kashi & King, 2006).

2.2 The causes of mutations

Mutations are divided into spontaneous and induced alterations according to their formation sources. Spontaneous mutations arise from replication errors due to defective replication enzymes and alternative tautomer forms of nucleotide bases. These are rare types of mutations. On the other hand, induced mutations, common types of mutations, are caused by various mutagens. In the molecular mechanism of induced mutations, a physical or chemical mutagen reacts with the DNA strand, causing a structural change that affects the base-pairing capability of the altered nucleotide. The most important types of physical mutagens are ultraviolet radiation of wavelength 260 nm, ionizing radiation and heat shock. However, base analogs such as 5-bromouracil (5-bU), deaminating agents such as nitrous acid, alkylating agents such as ethylmethane sulfonate (EMS) and intercalating agents such as (ethidium bromide) are the most well-known chemical mutagen classes (Brown, 2007).

Chemical mutagens are more frequent agents because thousands of natural or synthetic chemicals have been introduced for daily use in many areas including medicine, pharmacy, food and cosmetics. The count of new chemicals is increasing day-by-day, and each chemical can be considered as a potential mutagen before tested. Therefore, many test systems for detecting of chemical mutagens have been developed and frequently used in the laboratories around the world (Zieger, 2000; World Health Organization [WHO], 2007).

2.3 The mutagenicity and antimutagenicity test systems

The deleterious effects of mutations enforce the determination of mutagenic chemicals. There are many assay systems for this purpose, and a new chemical is tested for mutagenic potential before introduced to use. The main groups of the assay systems are long-term and short-term assay systems (Wickramasnghe, 1979; Zeiger et al., 2005; Mortelmans & Zeiger, 2000).

The long-term assay systems mainly include *in vivo* applications performed with experimental animals. These are the most comprehensive and reliable test systems. However, the long-term assay systems are not preferred as the beginning mutation test systems due to their high — cost and time consuming properties (Wickramasnghe, 1979; Zeiger et al., 2005).

On the other hand, the short-term assay systems mainly include *in vitro* assays performed with bacterial strains, cytological cell-line cultures and biotechnology based applications. Relatively inexpensive and time-saver nature of the short-term assay systems makes them good candidates for preliminary mutagen determination studies performed with fairly huge numbers of synthetic or natural chemicals. Furthermore, these assays can identify substances inhibiting mutagens and mutations (called as antimutagens) with some modifications. Therefore, a mutagenicity test system can be also considered as antimutagenicity test system (Ames et al., 1973a, 1973b; Wickramasnghe, 1979; Fenech, 2000; Maron & Ames, 1983; Mortelmans & Riccio, 2000; Mortelmans & Zeiger, 2000; Zeiger et al., 2005; Rossi et al., 2007; Ozbek et al., 2008a, 2008b; Gulluce et al., 2010).

2.4 Ames/*Salmonella* test system

The Ames/*Salmonella* test system, also called as Ames test, was developed by Ames *et al.* in the beginning of 1970s. The test, which employs histidine auxotroph *Salmonella* strains originated from *Salmonella typhimurium* LT-2 by chemical and radiation induced mutations, was initially designed as a spot test for determination of mutagenic chemicals, then as a more sensitive method: plate incorporation test (Ames et al., 1973a, 1973b; Maron & Ames, 1983; Gee et al., 1994; Mortelmans & Zeiger, 2000; Tijs, 2008).

In the molecular mechanism of the test system, a tester strain carries a unique gene mutation at the histidine operon that makes the strain histidine-dependent to grow, and a mutagenic chemical reacts with the mutated site resulting in a reverse mutation. Thus, the strain regains histidine production ability and the bacterial cells can grow in the absence of histidine. Therefore, the test is often referred as a reversion assay (Ames et al., 1973a, 1973b; Maron & Ames, 1983; Gee et al., 1994; Mortelmans & Zeiger, 2000; Tijs, 2008).

After its introduction to the scientific world, Ames test has been widely accepted as a short-term bacterial test system for determining chemicals that can cause gene mutations. The test has many advantages for identification of the chemicals that cause gene mutations. These advantages can be listed in:

- Short-term resulting: allows making replicates and obtaining more reliable results in a short duration. It takes only about 48 hours.
- Low-cost: allows studying a large number of test materials inexpensively.
- Various tester strains with several gene mutations: enable to research the molecular effect mechanism of test materials
- Additional mutations and genetic alterations: allow gaining more sensitivity for various chemicals.
- Mesophile character of *Salmonella* strains: enables to study mutagenic potential of the chemicals at human body temperature.

Apart from all the maintained advantages, Ames/*Salmonella* test system is very versatile, and many modifications has been developed to determine mutagenic potencies of various materials such as environmental chemicals, environmental mixtures, body fluids, foods, drugs and physical agents. The most common assay procedures are the spot test: a primal method for determination of chemical mutagens, the standard plate incorporation method: an easily resulting and more comprehensive method than the spot test, the pre-incubation method: developed for performing more effective studies with lower volumes of test materials, the desiccator assay modifications: developed to study volatile materials and gases, and the modified *Salmonella* microsuspension assay (Kado): a highly sensitive method for testing the materials that are available only in small amounts (Kado et al., 1983; Hughes et al., 1987; Zeiger et al., 1992; Araki et al., 1994; Mortelmans & Zeiger, 2000; Tijs, 2008).

Although *Salmonella* has prokaryotic cell structure, combination of the cytochrome-based P450 metabolic oxidation system with the Ames/*Salmonella* test system allows determining some mutagenic agents, which are biologically inactive unless they are metabolized to active forms (Ames et al., 1973b; Mortelmans & Zeiger, 2000). Moreover, all the procedures of the test system can be altered to identify antimutagenic agents, inhibit mutagenesis and protect the organisms against deleterious effects of the mutagens, with some modifications (Nagabhushan et al., 1987; Bala & Grover, 1989; Edenharder et al., 1999; Edenharder & Grünhage, 2003; Ozbek et al., 2008a, 2008b; Gulluce et al., 2010) (see 3.8).

3. Material and methods

3.1 Supplies and equipment

The following items are required for performing the Ames/*Salmonella* mutagenicity and antimutagenicity protocols.

3.1.1 Supplies

- Magnetic stir bars
- Sterile glass test tubes (100×16 mm) and racks
- Sterile microbiological loops
- Sterilizing membrane filters (0.2 μm)

- Sterile syringes (5, 10 and 50 ml)
- Sterile Petri dishes (100×15 mm)
- Disposable spectrophotometer cuvettes
- Solvents, reagents, media and positive control chemicals
- General laboratory glassware (bottles, flasks and graduated cylinders)
- Dispensers for delivering top agar, buffer and S-9 mix to the test tubes
- Sterile cryogenic storage vials for freezing down permanent and working cultures
- General laboratory safety items (biohazard waste bags, goggles or protective eye wear, gloves, lab coats)
- Glass pipettes (1, 2, 5 and 10 ml), automatic micropipettes (adjustable volumes up to 200 and 500 µl) and pipette tips

3.1.2 Equipment

- Autoclave
- Manual or electronic colony counter
- Spectrophotometer for monitoring cell density
- Centrifuge (up to 8000 rpm)
- Liquid and solid waste disposal
- Magnetic stirrers
- Desiccator and vacuum pump
- Balances
- Biological/chemical safety cabinet equipped with gas line for keeping aseptic techniques while inoculating cultures
- Ultra-low temperature freezer set at -86 °C or liquid nitrogen tank for long term storage of frozen permanent cultures
- Refrigerator (4 °C) and freezer (-20 °C)
- Water purification system to generate distilled water
- Water bath set at 43 °C to 48 °C to maintain temperature of top agar
- Incubator for incubating the agar plates
- Shaking incubator for incubating the liquid cultures and growing the overnight cultures
- Boiling water bath or microwave oven for melting top agar

3.2 Reagents and media

Glucose solution (10% w/v): The solution is used as carbon source for the GM agar plates. Dissolve 100 g dextrose (D-glucose) in 700 ml of distilled water by stirring on a magnetic stirrer. Add additional water to bring the final volume to 1000 ml and distribute in 50 ml aliquots. Autoclave 121 °C for 20 min and store at 4 °C.

Vogel-Bonner medium E (VB salts 50×): The solution is used as salt source for the GM agar plates. Add 10 g magnesium sulfate ($MgSO_4 \cdot H_2O$), 100 g citric acid monohydrate ($C_6H_8O_7 \cdot H_2O$), 500 g potassium phosphate dibasic (K_2HPO_4) and 175 g sodium ammonium phosphate ($Na_2NH_2PO_4 \cdot 4H_2O$) in the order indicated to 650 ml of warm water making sure that each salt is dissolved thoroughly by stirring before adding the next salt. Add additional water to bring the final volume to 1000 ml and distribute in 20 ml aliquots. Autoclave 121 °C for 30 min and store at room temperature in the dark.

GM agar plates: The medium is used as bottom agar for mutagenicity and antimutagenicity assays. Add 15 g agar to 900 ml of distilled water and autoclave for 30 min at 121 °C. When cooled to approximately 65 °C, add 20 ml of sterile Vogel-Bonner medium E and mix well; then add 50 ml of sterile glucose solution and mix thoroughly. Pour nearly 25 ml of the medium into sterile 100×15 mm petri dishes and store at 4 °C for several weeks by packing with sealed plastic bags after solidified. Note that the plates should be warmed up to room temperature and examined for excess moisture before use. Put the plates with too much moisture overnight in an incubator set at 37 °C prior to use.

Histidine/biotin solution (0.5 mM): The solution is used to supplement top agar with adequate biotin and a trace amount of histidine. Dissolve 124 mg D-biotin and 96 mg L-histidine·HCl in 1000 ml of boiling water. Sterilize the solution by filtration through a membrane filter with 0.2 µm pore size or autoclaving for 20 min at 121 °C. Store at 4 °C in a glass bottle.

Top agar supplemented with histidine/biotin: The solution is used to apply the bacteria, chemicals and buffer or S9 mix to the bottom agar. Dissolve 6 g agar and 6 g sodium chloride (NaCl) in 900 ml of distilled water by heating. Add 100 ml of histidine/biotin solution (0.5 mM) and dispense 200 ml aliquots in screw-cap bottles. Autoclave for 20 min at 121 °C and store at room temperature in the dark. Melt the top agar in a microwave oven or boiling water bath before use.

Nutrient broth: Oxoid nutrient broth no. 2 or Difco nutrient broth can be used to grow the tester strains overnight. Follow the manufacturer's instructions for preparing the medium. Dispense 50 ml in Erlen Meyer flasks with 125 ml capacity or 5 ml in 100×16 mm test tubes, autoclave for 20 min at 121 °C and store in the dark at room temperature.

Nutrient agar plates: The medium is used for streaking newly received cultures for single colonies, checking crystal violet sensitivity due to presence of *rfa* mutation and testing viability of bacteria. Add 15 g agar to 1000 ml of nutrient broth medium and dissolve by heating. Autoclave for 20 min at 121 °C. After cooled to 65 °C, pour nearly 25 ml of the medium into sterile 100×15 mm petri dishes and store at 4 °C by packing with sealed plastic bags.

Sodium phosphate buffer (0.1 mM – pH 7.4): The solution is used to perform mutagenicity and antimutagenicity assays in the absence of metabolic activation. In the first step, prepare Reagent A (0.1 M sodium phosphate monobasic solution: 13.8 g $NaH_2PO_4·H_2O$ in 1000 ml of distilled water) and Reagent B (0.1 M sodium phosphate dibasic reagent: 14.2 g $Na_2HPO_4·H_2O$ in 1000 ml of distilled water). After that, mix 120 ml of Reagent A and 880 ml Reagent B and swirl well. Adjust pH to 7.4 using Reagent A/B and dispense 100 ml aliquots in screw-cap bottles. Autoclave for 30 min at 121 °C and store at room temperature in the dark.

Metabolic activation system (S-9 mix): The solution is used to perform mutagenicity and antimutagenicity assays in the presence of metabolic activation. Moltox metabolic activation system products can be used. Follow the manufacturer's instructions for preparation and storage of the related solutions.

Biotin solution (0.01%, w/v): The solution is used to prepare enriched GM agar plates for biotin auxotrophy check. Dissolve 10 mg D-biotin in 100 ml of boiling distilled water. Sterilize using a membrane filter with 0.2 µm pore size and store at 4 °C.

Histidine solution (0.5%, w/v): The solution is used to prepare enriched GM agar plates for histidine auxotrophy check. Dissolve 500 mg L-histidine in 100 ml of distilled water. Autoclave for 15 min at 121 °C and store at 4 °C.

Ampicillin solution (0.8%, w/v): The solution is used to prepare enriched GM agar plates for examining presence of plasmid pKM101 in several tester strains such as TA97, TA98, TA100 and TA102. Dissolve 8 mg ampicillin in 100 ml of warm (65 °C) distilled water and sterilize using membrane filter with 0.2 μm pore size. Store at 4 °C.

Tetracycline solution (0.8%, w/v): The solution is used to prepare enriched GM agar plates for examining presence of plasmid pAQ1 in TA102. Dissolve 8 mg tetracycline in 100 ml of 0.02 N hydrochloric acid (HCl) and sterilize using membrane filter with 0.2 μm pore size. Store at 4 °C in the dark due to the light sensitivity of tetracycline.

Enriched GM agar plates: Each medium contains essential nutrients and antibiotics for the strain check and preparation of stock cultures' master plates.

- Biotin plates (B): Prepare GM agar medium. After autoclaving, add 8 ml of sterile biotin solution (0.01%, w/v), mix well and pour nearly 25 ml of the medium into sterile 100×15 mm petri dishes.
- Histidine plates (H): Prepare GM agar medium. After autoclaving, add 8 ml of sterile histidine solution (0.5%, w/v), mix well and pour nearly 25 ml of the medium into sterile 100×15 mm petri dishes.
- Biotin/histidine plates (BH): Prepare GM agar medium. After autoclaving, add 8 ml of sterile biotin solution (0.01%, w/v) and 8 ml of sterile histidine solution (0.5%, w/v), mix well and pour nearly 25 ml of the medium into sterile 100×15 mm petri dishes.
- Biotin/histidine/ampicillin plates (BHA): Prepare GM agar medium. After autoclaving, add 8 ml of sterile biotin solution (0.01%, w/v), 8 ml of sterile histidine solution (0.5%, w/v) and 3 ml of ampicillin solution, mix well and pour nearly 25 ml of the medium into sterile 100×15 mm petri dishes.
- Biotin/histidine/tetracycline plates (BHT): Prepare GM agar medium. After autoclaving, add 8 ml of sterile biotin solution (0.01%, w/v), 8 ml of sterile histidine solution (0.5%, w/v) and 0.25 ml of tetracycline solution, mix well and pour nearly 25 ml of the medium into sterile 100×15 mm petri dishes.
- Biotin/histidine/ampicillin/tetracycline plates (BHAT): Prepare GM agar medium. After autoclaving, add 8 ml of sterile biotin solution (0.01%, w/v), 8 ml of sterile histidine solution (0.5%, w/v), 3 ml of ampicillin solution and 0.25 ml of tetracycline solution, mix well and pour nearly 25 ml of the medium into sterile 100×15 mm petri dishes.

Crystal violet solution (0.1%, w/v): The solution is used to confirm the presence of the *rfa* mutation in all of the tester strains. Dissolve 100 mg crystal violet in 100 ml of distilled water. Mix well and store at 4°C in an amber glass bottle to protect against light.

All reagents and solutions, reported here, have been previously described by Mortelmans and Zeiger (2000).

3.3 Bacterial strains

Salmonella typhimurium TA1535, TA1537, TA1538, TA97, TA98, TA100, TA102 and TA104 are the most common tester strains used in the Ames/*Salmonella* test system. All strains are histidine auxotroph because of a mutation in the histidine operon. The tester strains also have additional mutations and genetic alterations that provide more sensitivity for chemical mutagens. These are *uvrB*, *rfa* mutations and introduction of pKM101 and pAQ1 plasmids.

The *uvrB* mutation, which is present in all strains except TA102, arise from a deletion type mutation through the *uvrB-bio* genes that eliminates the accurate DNA repair and makes the cells biotin dependent. All strains have the *rfa* mutation that affects the bacterial cell wall, resulting in a defective lipopolysacharide layer that provides more permeability to bulky chemicals. Existence of pKM101 plasmid in TA97, TA98, TA100, and TA102 provides ampicillin resistance and sensitivity for chemical and induced mutagenesis associated with error-prone recombinational DNA repair pathway. TA102 strain also has multicopies of pAQ1 plasmid carrying *hisG428* mutation, which provides tetracycline resistance and sensitivity for detection of DNA cross-linking agents (Mortelmans & Zeiger, 2000). Table 1 presents genotypes of the tester strains.

Strain	DNA Target	*uvrB*	*rfa*	Plasmid	Reversion Event
TA1535	*hisG46* -G-G-G-	+	+	-	Base-pair substitutions
TA1537	*hisC3067* -C-C-C- +1 frameshift	+	+	-	Frameshifts
TA1538	*hisD3052* -C-G-C-G-C-G-C-G- -1 frameshift	+	+	-	Frameshifts
TA97	*hisD6610* -C-C-C-C-C-C- +1 frameshift	+	+	pKM101	Frameshifts
TA98	*hisD3052* -C-G-C-G-C-G-C-G- -1 frameshift	+	+	pKM101	Frameshifts
TA100	*hisG46* -G-G-G-	+	+	pKM101	Base-pair substitutions
TA102	*hisG428* TAA (ochre)	-	+	pKM101 pAQ1	Base-pair substitutions
TA104	*hisG428* TAA (ochre)	+	+	-	Base-pair substitutions

Table 1. Genotypic properties of the most common *Salmonella* tester strains.

There are also additional tester strains (TA7001-7006 and TA7041-7046 series), which developed by Gee et al. (1994), to identify specific transitional and transversional base-pair substitutions induced by various mutagenic agents. Table 2 presents genotypic properties of these strains.

Strain	DNA Target	uvrB	rfa	Plasmid	Reversion Event	Amino acid change
TA7001	hisG1775	+	+	pKM101	AT → GC	Asp-153 → Gly-153 (GAT → GGT)
TA7002	hisC9138	+	+	pKM101	TA → AT	Ile-217 → Lys-217 (ATA → AAA)
TA7003	hisG9074	+	+	pKM101	TA → GC	Val-153 → Gly-153 (GTT → GGT)
TA7004	hisG9133	+	+	pKM101	GC → AT	Gly-169 → Asp-169 (GGG → GAT)
TA7005	hisG9130	+	+	pKM101	CG → AT	Ala-169 → Asp-169 (GCG → GAT)
TA7006	hisC9070	+	+	pKM101	CG → GC	Arg-163 → Gly 163 (CGA → GGA)
TA7041	hisG1775	+	-	pKM101	AT → GC	Asp-153 → Gly-153 (GAT → GGT)
TA7042	hisC9138	+	-	pKM101	TA → AT	Ile-217 → Lys-217 (ATA → AAA)
TA7043	hisG9074	+	-	pKM101	TA → GC	Val-153 → Gly-153 (GTT → GGT)
TA7044	hisG9133	+	-	pKM101	GC → AT	Gly-169 → Asp-169 (GGG → GAT)
TA7045	hisG9130	+	-	pKM101	CG → AT	Ala-169 → Asp-169 (GCG → GAT)
TA7046	hisC9070	+	-	pKM101	CG → GC	Arg-163 → Gly 163 (CGA → GGA)

Table 2. Genotypic properties of *Salmonella* tester strains developed by Gee et al. (1994).

The test system performed with TA700x tester strains is called as AMES II (Kamber et al., 2009). The set of TA7041-7046 strains is not suitable to test mutagenic and antimutagenic potential of chemicals due to lack of *rfa* mutation and their instable genotypes.

3.4 Positive control chemicals

Chemicals divide in two groups according to their affect mechanisms. These groups are direct and indirect acting positive controls.

Many direct acting agents has been introduced as positive controls because of their high specificity for the tester strains. The most common direct-acting positive control chemicals for Ames/*Salmonella* test system are listed in Table 3.

2-Aminoanthracene (2-AA; CAS# 613-13-8), 2-Aminofluorene (2-AF; CAS# 153-78-6) and Aflatoxin B1 (AFB1; CAS# 1162-65-8) are frequently used indirect-acting positive controls that requires metabolic activation before react with the *Salmonella* tester strains (Mortelmans & Zeiger, 2000; Ozbek et al., 2008b; Limem et al., 2010).

Chemical	Mechanism of Genotoxicity	Tester Strain	Reference	CAS#
4-Nitro-*o*-phenilenediamine (4-NPD)	Teratogenic and intercalating agent	TA1538 TA98	Ben Sghaier et al., (2010) Kaur et al., (2010) Mortelmans & Zeiger (2000)	99-56-9
4-Nitroquinoline 1-oxide (4-NQO)	Causing DNA lesions	TA1538 TA100 TA98	Oh et al., (2008) Ozbek et al., (2008a) Brennan & Schiestl (1998)	56-57-5
9-Aminoacridine (9-AA)	DNA intercalating agent	TA1537 TA97	Gulluce et al., (2010) Miadokova et al., (2009) Mortelmans & Zeiger (2000)	90-45-9
Methyl methane sulfonate (MMS)	Alkylating agent	TA102 TA104	Mortelmans & Zeiger (2000) Dellai et al., (2009) Zahin et al., (2010)	66-27-3
Mitomycin C (MTC)	DNA cross-linker and alkylating agent	TA102	Biso et al., (2010) Mortelmans & Zeiger (2000) Zhang et al., (2011)	50-07-7
N-Methyl-*N'*-nitro-*N*-nitrosoguanidine (MNNG)	Alkylating agent	TA100	Caldini et al., (2005) Duh et al., (2009) Oh et al., (2008)	70-25-7
Sodium azide (NaN$_3$)	L-azidoalanine mediated base substitution	TA1535 TA100	Bulmer et al., (2007) Gulluce et al., (2010) Mortelmans & Zeiger (2000)	26628-22-8

Table 3. The frequently used direct-acting chemicals for the *Salmonella* tester strains.

3.5 Genetic analysis of the *Salmonella* tester strains

When a new strain received, its genotypic characteristics (*his*, *rfa* and *uvrB-bio*), spontaneous mutation rate and the presence of pKM101 and pAQ1 plasmids should be checked before preparation of frozen cultures for long term storage. For this purpose, follow these steps:

3.5.1 Inchoative stages for genetic analysis

- Add 1 mL of sterile nutrient broth to rehydrate the lyophilized culture.
- Transfer 10 μL of the rehydrated culture to nutrient agar plate and strake the inoculum to get individual colonies that serve as main sources for the genetic analysis of the tester strains.
- Transfer the rest portion of the rehydrated culture to 4 mL of nutrient broth. This broth culture serve as a back-up point in case of there is no growth on the nutrient agar plates.
- Incubate the cultures overnight at 37 °C. Then, check the agar plates and broth cultures for bacterial growth.

- At least two purification steps should be made to get more reliable results. Pick one healthy looking colony and streak it again on nutrient agar plates or GM agar plates supplemented with excess of biotin and histidine. If the tester strain carries pKM101 and/or pAQ1 plasmids, GM agar plates should be also supplemented with ampicillin and/or tetracycline, respectively. However, growth of the tester strains on the supplemented GM agar plates takes more time (approximately 48 h) than nutrient agar plates; it is recommended because using of them reduces contamination risks.

3.5.2 Genetic analysis

Five mandatory steps for all strains and additional one or two steps for plasmid carrying strains should be made to perform the best reliable genetic analysis. For this purpose, inoculate 5 mL of nutrient broth with a single colony after purification steps, and incubate the culture overnight 37 °C. Then, follow these steps for a complete strain check:

- In the 1st step, streak a loop of the overnight culture on the surface of a GM agar plate supplemented with excess of biotin, which demonstrates the histidine dependence (*his* mutation) of all the *Salmonella* tester strains. After an incubation period at 37 °C for 24-48 h, there should be no growth on the plate (see Figure 4a).
- In the 2nd step, streak a loop of the overnight culture on the surface of a GM agar plate supplemented with excess of histidine, which demonstrates the biotin dependence (*bio* mutation) of all the *Salmonella* tester strains except TA102 strain. After an incubation period at 37 °C for 24-48 h, there should be no growth on the plate. Due to lack of the *bio* mutation, TA102 strain can be growth on a GM agar plate supplemented with excess of histidine (see Figure 4b).
- In the 3rd step, streak a loop of the overnight culture on the surface of a GM agar plate supplemented with excess of biotin and histidine, which demonstrates the biotin and histidine dependence (*bio* and *his* mutations) of all the *Salmonella* tester strains. After an incubation period at 37 °C for 24-48 h, there should be growth on the plate (see Figure 4c).
- In the 4th step, streak a loop of the overnight culture on the surface of a GM agar plate supplemented with excess of biotin and histidine. Place a sterile filter paper disk in the middle of the plate and apply 10 μL crystal violet solution (0.1%, w/v) onto the disk. After an incubation period at 37 °C for 24-48 h, all strains show a zone of growth inhibition surrounding the disk, which demonstrates the presence of *rfa* mutation (see Figure 4d).
- In the 5th step, streak a loop of the overnight culture on the surface of a GM agar plate supplemented with excess of biotin and histidine. Unseal the top and cover the half of the plate with sterile aluminum foil. Expose the plate to a low level of UV irradiation for a short time (approx. 8-10 seconds) that kills the *uvrB* strain but not its isogenic DNA repair proficient strain. After an incubation period at 37 °C for 24-48 h, there should be normal growth on the non-exposed part of the plate but not on the exposed part. It demonstrates the presence of *uvrB* mutation. It is known that the source of *uvrB* mutation, a deletion mutation, also covers the biotin gene region. Therefore, if a strain shows a positive *bio* mutation result in the 2nd step, there is no need to check the presence of the *uvrB* mutation for this strain (see Figure 4e).
- In the 6th step, streak a loop of the overnight culture on the surface of a GM agar plate supplemented with excess of biotin, histidine and ampicillin. After an incubation period at 37 °C for 24-48 h, there should be growth on the plate, which demonstrates the

presence of pKM101 plasmid in the tester strains TA97, TA98, TA100, TA102 and TA104 (see Figure 4f).

- In the 7th step, streak a loop of the overnight culture on the surface of a GM agar plate supplemented with excess of histidine and tetracycline. After an incubation period at 37°C for 24-48 h, there should be growth on the plate, which demonstrates the presence of pAQ1 plasmid in the tester strain TA102 (see Figure 4f).

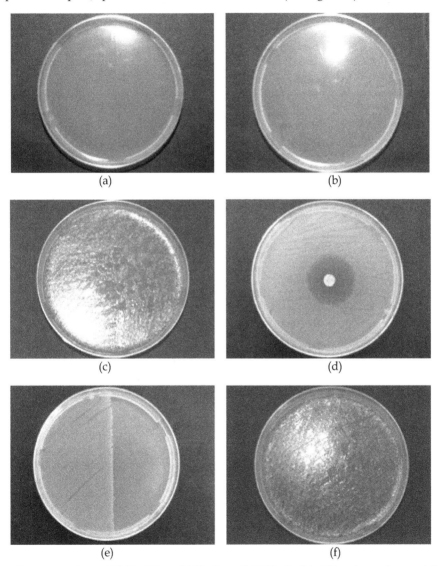

(a)

(b)

(c)

(d)

(e)

(f)

Fig. 4. Demonstration of (a) histidine, (b) biotin and (c) biotin/histidine dependence of the *Salmonella* tester strains, and presence of (d) the *rfa* mutation, (e) the *uvrB* mutation, (f) pKM101/pAQ1 plasmids.

3.5.3 Spontaneous mutation rates

Each laboratory has its characteristic spontaneous mutation rates for the all tester strains, and these values show a wide-range variation among the laboratories. Therefore, the spontaneous mutant frequency should be determined for all strains and recorded. It serves as historical control values provide choosing suitable strains for mutagenicity and antimutagenicity assays. Table 4 shows a sample of acceptable control values for the most common *Salmonella* tester strains.

Strain	Number of revertants	
	Without metabolic activation	With metabolic activation
TA97	75-200	100-200
TA98	20-50	20-50
TA100	75-200	75-200
TA102	100-300	200-400
TA104	200-300	300-400
TA1535	5-20	5-20
TA1537	5-20	5-20
TA1538	5-20	5-20

Table 4. Spontaneous revertant control values for the most common *Salmonella* tester strains (Mortelmans & Zeiger, 2000).

3.6 Long term storage of the tester strains

The *Salmonella* tester strains should be stored in a freezer at -80 °C or liquid nitrogen. Healthy looking single colonies should be chosen to prepare the frozen stock cultures. Dimethylsulfoxide (DMSO) or glycerol is suggested as cryoprotective agent. The final concentration of the cryoprotective should be at least 10% (v/v) (Mortelmans & Zeiger, 2000).

3.7 Viability assay and determination of test concentrations

Cytotoxic properties of the test materials toward the *Salmonella* tester strains should be determined before performing mutagenicity and antimutagenicity assays. The viability assay includes observations for *Salmonella* colonies on plates after 48 h incubation at 37 °C. Following three main characteristics for the tester strains should be taken into account.

- Thinning of the background lawn
- Absence of background lawn
- Presence of pinpoint non-revertant colonies

These characteristics indicate toxic levels of the test chemicals, and applicable dose ranges should be determined by repeating of the viability assay with lower concentrations of the test chemicals (Mortelmans & Zeiger, 2000).

3.8 Mutagenicity and antimutagenicity assays

Various test procedures for Ames/ *Salmonella* test system have been developed to determine mutagenic and antimutagenic potency of synthetic and natural chemicals. These procedures mainly are based on the physical properties or quantity of the test chemical. For example; the desiccator assay has been developed for gases and volatile substances, and Kado assay allows studying the chemicals in small amounts. However, the standard plate incorporation method is the most common application procedure of the Ames/ *Salmonella* test system (Kado et al., 1983; Hughes et al., 1987; Zeiger et al., 1992; Araki et al., 1994; Mortelmans & Zeiger, 2000; Tijs, 2008).

3.8.1 The standard plate incorporation method

The method consists of exposing the tester strains to the test chemical directly on a glucose agar plate. The main advantages of the method can be listed in giving easy, reproducible, reliable and comprehensive results.

Follow these steps for performing mutagenicity assay (Mortelmans & Zeiger, 2000):

1. Steps taken prior to performing the experiment
   - Inoculate the tester strain from frozen culture into 5 mL of nutrient broth and incubate the new culture overnight at 37 °C.
   - Prepare an appropriate number of labeled GM agar plates and sterile tubes for each test chemical.
   - Prepare metabolic activation system and keep on ice until use.
   - Prepare chemical dilutions.
   - Melt top agar supplemented with 0.05 mM histidine and biotin and maintain at 43 °C to 48 °C.
2. Add following items respectively into sterile glass tubes maintained at 43 °C and mix well each addition*.
   - 2 mL of molten top agar
   - 0.5 mL of S-9 mix (for the test performed with metabolic activation system) or buffer (without activation)
   - 0.05 mL of the test chemical dilution
   - 0.05-0.10 mL overnight culture of the tester strain (approx. $1\text{-}2 \times 10^8$ bacteria per tube – A_{540} 0.1-0.2)
3. Mix well the tubes and pour onto the surface of GM agar plates
4. When the top agar is solidify, invert and incubate the cultures at 37 °C for 48 h
5. Count the colonies after incubation and express the results as the number of revertant colonies per plate.

*Notes: This step includes two additional groups which are negative controls and positive controls. The negative controls do not include 0.05 mL of the test chemicals, but include the solvent at equal quantity. Positive controls also do not include 0.05 mL of the test chemicals, but include the suitable positive mutagen solution for the tester strain at equal quantity.

The procedures of mutagenicity assay are all applicable to the antimutagenicity assay. The only procedural difference is the addition of the suitable positive mutagen solution to the all test chemical groups (Nagabhushan et al., 1987; Bala & Grover, 1989; Edenharder et al., 1999; Edenharder & Grünhage, 2003; Ozbek et al., 2008a, 2008b; Gulluce et al., 2010).

4. Conclusion

In conclusion, the mutant *Salmonella* strains are beneficial for humanity contrary to their pathogenic wild-type strains. The histidine auxotrophic *Salmonella typhimurium* strains, object of the present study, provide a possibility to determine natural and synthetic chemicals with mutagenic properties. Similarly, these are also valuable for identification of antimutagenic chemicals after minor technical modifications. When a chemical, precious for industrial or health applications, is found or synthesized, determination of its genotoxic properties has a great importance. In this perspective, the Ames test allows making relatively cheap and reliable applications resulting in a short time.

5. Acknowledgement

The authors express their thanks to Dr. Tülin Arasoğlu and the Microbiology & Molecular Biology Research Team of Biology Department, Atatürk University.

6. References

Akiyama, M. (2010). ABCA12 Mutations and Autosomal Recessive Congenital Ichthyosis: A Review of Genotype/Phenotype Correlations and of Pathogenetic Concepts. *Human Mutation*, 31, (10), 1090–1096, ISSN 1059-7794.

Albertini, R. J.; Nicklas, J. A.; Oneill, J. P. & Robison, S. H. (1990). In vivo Somatic Mutations in Humans – Measurement and Analysis. *Annual Review of Genetics*, 24, 305-326, ISSN 0066-4197.

Alberts, B.; Johnson, A.; Lewis, J.; Raff, M.; Roberts, K. & Walter, P. (2002). *Molecular Biology of The Cell* (4th edition), Garland Science, ISBN 978-0815332183, USA.

Ames, B. N.; Lee, F. D. & Durston, W. E. (1973a). An Improved Bacterial Test System for the Detection and Classification of Mutagens and Carcinogens. *Proceedings of the National Academy of Sciences of the United States of America*, 70, (3), 782-786, ISSN-0027-8424.

Ames, B. N.; Durston, W. E.; Yamasaki, E. & Lee, F. D. (1973b). Carcinogens are Mutagens: A Simple Test System Combining Liver Homogenates for Activation and Bacteria for Detection. *Proceedings of the National Academy of Sciences of the United States of America*, 70, (8), 2281-2285, ISSN-0027-8424.

Anton, E.; Blanco, J.; Egozcue, J. & Vidal F. (2004). Sperm FISH studies in seven male carriers of Robertsonian translocation t(13;14)(q10;q10). *Human Reproduction*, 19, (6), 1345–1351, ISSN 1460-2350.

Antonacci, F.; Kidd, J. M.; Marques-Bonet, T.; Ventura, M.; Siswara, P.; Jiang, Z. & Eichler, E. E. (2009). Characterization of six human disease-associated inversion polymorphisms. *Human Molecular Genetics*, 18, (14), 2555–2566, ISSN: 1460-2083.

Araki, A.; Noguchi, T.; Kato, F. & Matsushima, T. (1994). Improved method for mutagenicity testing of gaseous compounds by using a gas sampling bag. *Mutation Research/Fundamental and Molecular Mechanisms of Mutagenesis*, 307, (1), 335–344, ISSN 0027-5107.

Bala, S. & Grover, I. S. (1989). Antimutagenicity of some citrus fruits in *Salmonella typhimurium*. *Mutation Research/Genetic Toxicology*, 222, (3), 141-148, ISSN: 1383-5718.

Ben Sghaier, M.; Boubaker, J.; Skandrani, I.; Bouhlel, I.; Limem, I.; Ghedira, K. & Chekir-Ghedira, L. (2010). Antimutagenic, antigenotoxic and antioxidant activities of phenolic-enriched extracts from *Teucrium ramosissimum*: Combination with their phytochemical composition. *Environmental Toxicology and Pharmacology*, 31, (1), 220–232, ISSN: 1382-6689.

Bertram, J. (2000). The molecular biology of cancer, *Molecular Aspects of Medicine* 21, (6), 167–223, ISSN 0098-2997.

Biso, F. I.; Rodrigues, C. M.; Rinaldo, D.; dos Reis, M. B.; Bernardi, C. C.; de Mattos, J. C. P.; Caldeira-de-Araujo, A.; Vilegas, W.; Colus, I. M. de S. & Varanda, E. A. (2010). Assessment of DNA damage induced by extracts, fractions and isolated compounds of *Davilla nitida* and *Davilla elliptica* (Dilleniaceae). *Mutation Research/Genetic Toxicology and Environmental Mutagenesis*, 702, (1), 92–99, ISSN: 1383-5718.

Brennan, R. J. & Schiestl, R. H. (1998). Free radicals generated in yeast by the *Salmonella* test-negative carcinogens benzene, urethane, thiourea and auramine O. *Mutation Research/Fundamental and Molecular Mechanisms of Mutagenesis*, 403, (1-2), 65–73, ISSN 0027-5107.

Brown, T. A. (2007). *Genomes 3* (3rd edition), Garland Science, ISBN 978-0815341383, USA.

Bulmer, A. C.; Ried, K.; Coombes, J. S.; Blanchfield, J. T.; Toth, I. & Wagneri K.-H. (2007). The anti-mutagenic and antioxidant effects of bile pigments in the Ames *Salmonella* test. *Mutation Research/Genetic Toxicology and Environmental Mutagenesis*, 629, (2), 122–132, ISSN: 1383-5718.

Caldini, G.; Trotta, F.; Villarini, M.; Moretti, M.; Pasquini, R.; Scassellati-Sforzolini, G. & Cenci, G. (2005). Screening of potential Lactobacilli antigenotoxicity by microbial and mammalian cell-based tests. *International Journal of Food Microbiology*, 102, (1), 37– 47, ISSN 0168-1605.

Cox, E. C. (1976). Bacterial Mutator Genes and Control of Spontaneous Mutation. *Annual Review of Genetics*, 10, 135-156, ISSN 0066-4197.

Davidson, J. F.; Guo, H. H. & Loeb, L. A. (2002). Endogenous mutagenesis and cancer. *Mutation Research/Fundamental and Molecular Mechanisms of Mutagenesis*, 509, (1-2), 17–21, ISSN 0027-5107.

Dellai, A.; Ben Mansur, H.; Limem, I.; Bouhlel, I.; Ben Sghaier, M.; Boubaker, J.; Ghedira, K. & Chekir-Ghedira, L. (2009). Screening of antimutagenicity via antioxidant activity in different extracts from the flowers of *Phlomis crinita* Cav. ssp *mauritanica* munby from the center of Tunisia. *Drug and Chemical Toxicology*, 32, (3), 283-292, ISSN 1525-6014.

Duh, P.-D.; Wu, S.-C.; Chang, L.-W.; Chu, H.-L.; Yen, W.-J. & Wang, B.-S. (2009). Effects of three biological thiols on antimutagenic and antioxidant enzyme activities. *Food Chemistry*, 114, (1), 87–92, ISSN: 0308-8146.

Edenharder, R.; Worf-Wandelburg, A.; Decker, M. & Platt, K. L. (1999). Antimutagenic effects and possible mechanisms of action of vitamins and related compounds against genotoxic heterocyclic amines from cooked food. *Mutation Research/Genetic Toxicology and Environmental Mutagenesis*, 444, (1), 235–248, ISSN: 1383-5718.

Edenharder, R. & Grünhage, D. (2003). Free radical scavenging abilities of flavonoids as mechanism of protection against mutagenicity induced by *tert*-butyl hydroperoxide or cumene hydroperoxide in *Salmonella typhimurium* TA102. *Mutation Research/Genetic Toxicology and Environmental Mutagenesis*, 540, (1), 1–18, ISSN: 1383-5718.

Evans, T. J.; Yamamoto, K. N.; Hirota, K. & Takeda, S. (2010). Mutant cells defective in DNA repair pathways provide a sensitive high-throughput assay for genotoxicity. *DNA Repair*, 9, (12), 1292–1298, ISSN 1568-7864.

Fenech, M. (2000). The in vitro micronucleus technique. *Mutation Research/Fundamental and Molecular Mechanisms of Mutagenesis*, 455, (1-2), 81–95, ISSN 0027-5107.

Galvani, A. P. & Slatkin, M. (2003). Evaluating plague and smallpox as historical selective pressures for the CCR5-Delta 32 HIV-resistance allele. *Proceedings of the National Academy of Sciences of the United States of America*, 100, (25), 15276-15279, ISSN-0027-8424.

Gee, P.; Maron, D. M. & Ames, B. N. (1994). Detection and Classification of Mutagens: A Set of Base-Specific Salmonella Tester Strains. *Proceedings of the National Academy of Sciences of the United States of America*, 91, (24), 11606-11610, ISSN-0027-8424.

Gimelli, G.; Pujana, M. A.; Patricelli, M. G.; Russo, S.; Giardino, D.; Larizza, L.; Cheung, J.; Armengol, L.; Schinzel, A.; Estivill, X. & Zuffardi, O. (2003). Genomic inversions of human chromosome 15q11–q13 in mothers of Angelman syndrome patients with class II (BP2/3) deletions. *Human Molecular Genetics*, 12, (8), 849–858, ISSN: 1460-2083.

Gulluce, M.; Agar, G.; Baris, O,; Karadayi, M.; Orhan, F. & Sahin, F. (2010). Mutagenic and Antimutagenic Effects of Hexane Extract of some *Astragalus* Species Grown in the Eastern Anatolia Region of Turkey. *Phytotherapy Research* 24, (7), 1014-1018, ISSN 1099-1573.

Hartl, D. L. & Jones, W. J. (1998). *Genetics: Principles and Analysis* (4th edition), Jones & Bartlett Publishers, ISBN 978-0763704896, UK.

Hirschhorn, K.; Cooper, H. L. & Firschein, I. L. (1965). Deletion of short arms of chromosome 4-5 in a child with defects of midline fusion. *Humangenetik*, 1, (5), 479–82, ISSN: 0018-7348.

Hughes, T. J.; Simmons, D. M.; Monteith, L. G. & Claxton, L. D. (1987). Vaporization technique to measure mutagenic activity of volatile organic chemicals in the Ames/*Salmonella* assay, *Environmental Mutagenesis*, 9, (4), 421–441, ISSN: 1383-5718.

Jacobsen, P.; Hauge, M.; Henningsen, K.; Hobolth, N.; Mikkelsen, M. & Philip, J. (1973). An (11; 21) translocation in four generations with chromosome 11 abnormalities in the offspring. A clinical, cytogenetical, and gene marker study. *Human Heredity*, 23, (6), 568–85, ISSN 1423-0062.

Kado, N. Y.; Langley, D. & Eisenstadt, E. (1983). A simple modification of the *Salmonella* liquid incubation assay: increased sensitivity for detecting mutagens in human urine, *Mutation Research Letters*, 122, (1), 25–32, ISSN: 0165-7992.

Kamber, M.; Fluckiger-Isler, S.; Engelhardt, G.; Jaeckh, R. & Zeiger, E. (2009). Comparison of the Ames II and traditional Ames test responses with respect to mutagenicity, strain specificities, need for metabolism and correlation with rodent carcinogenicity. *Mutagenesis*, 24, (4), 359–366, ISSN: 1464-3804.

Kashi, Y. & King, D. G. (2006). Simple Sequence Repeats as Advantageous Mutators in Evolution. *Trends in Genetics*, 22, (5), 253–259, ISSN 0168-9525.

Kaur, P.; Chandel, M.; Kumar, S.; Kumar, N.; Singh, B. & Kaur, S. (2010). Modulatory role of alizarin from *Rubia cordifolia* L. against genotoxicity of mutagens. *Food and Chemical Toxicology*, 48, (1), 320–325, ISSN: 0278-6915.

Klug, S. W.; Cummings, M. R. & Spencer, C. (2005) *Concepts of Genetics* (8th Edition), Benjamin Cummings, ISBN 978-0131699441, USA.

Kurzrock, R.; Kantarjian, H. M.; Druker, B. J. & Talpaz, M. (2003). Philadelphia chromosome-positive leukemias: from basic mechanisms to molecular therapeutics. *Annals of Internal Medicine*, 138, (10), 819–830, ISSN: 1539-3704.

Lange, S. S.; Takata, K. & Wood, R. D. (2011). DNA polymerases and cancer. *Nature Reviews Cancer*, 11,(2), 96-110, ISSN 1474-175X.

Latour, P.; Fabreguette, A.; Ressot, C.; Blanquet-Grossard, F.; Antonia, J. C.; Calvas, P.; Chapon, F.; Corbillon, E.; Ollagnon, E.; Sturtz, F.; Boucherat, M.; Chazot, G.; Dautigny, A.; Pham-Dinh, D. & Vandenberghe, A. (1997). New mutations in the X-linked form of Charcot-Marie-Tooth disease. *European Neurology*, 37, (1), 38–42, ISSN 1421-9913.

Lewin, B. (2004). *Gene VIII* (8th edition), Pearson Education, ISBN 978-0131439818, USA.

Li, J. Y.; Gaillard, F.; Moreau, A.; Harousseau, J. L.; Laboisse, C.; Milpied, N.; Bataille, R. & Avet-Loiseau, H. (1999). Detection of translocation t(11;14)(q13;q32) in mantle cell lymphoma by fluorescence in situ hybridization. *The American Journal of Pathology*, 154 (5): 1449–1452, ISSN 1525-2191.

Limem, I.; Bouhlel, I.; Bouchemi, M.; Kilani, S.; Boubaker, J.; Ben-Sghaier, M.; Skandrani, I.; Behouri, W.; Neffati, A.; Ghedira, K. & Chekir-Ghedira, L. (2010). *Phlomis mauritanica* Extracts Reduce the Xanthine Oxidase Activity, Scavenge the Superoxide Anions, and Inhibit the Aflatoxin B1-, Sodium Azide-, and 4-Nitrophenyldiamine-Induced Mutagenicity in Bacteria. *Journal of Medicinal Food*, 13, (3), 717-724,

Lodish, H.; Berk, A.; Kaiser, C. A.; Krieger, M.; Scott, M. P.; Bretscher, A.; Ploegh, H. & Matsudaira, P. (2007). *Molecular Cell Biology* (6th edition), W. H. Freeman, ISBN 978-0716776017, USA.

Loeb, L. A. (2011). Human cancers express mutator phenotypes: origin, consequences and targeting. *Nature Reviews Cancer*, 11, (6), 450-457, ISSN: 1474-175X.

Lynch, M. (2010). Evolution of the mutation rate. *Trends in Genetics*, 26, (8), 345–352, ISSN 0168-9525.

Maron, D. & Ames, B. N. (1983). Revised methods for the *Salmonella* mutagenicity test. *Mutation Research/Environmental Mutagenesis and Related Subjects*, 113, (3-4), 173–215, ISSN 0165-1161.

Mao, R. & Pevsner, J. (2005). The use of genomic microarrays to study chromosomal abnormalities in mental retardation. Developmental Disabilities Research Reviews, 11, (4), 279–285, ISSN 1940-5529.

Miadokova, E.; Chalupa, I.; Vlckova, V.; Sevcovicova, A.; Nodova, S.; Kopaskova, M.; Hercegova, A.; Gasperova, P.; Alfoldiova, L.; Komjatiova, M.; Csanyiova, Z.; Galova, E.; Cellarova, E. & Vlcek, D. (2009). Genotoxicity and Antigenotoxicity Evaluation of Non-photoactivated Hypericin. *Phytotherapy Research*, 24, (1), 90–95, ISSN 1099-1573.

Mortelmans, K. & Riccio, E. S. (2000). The bacterial tryptophan reverse mutation assay with *Escherichia coli* WP2. *Mutation Research/Fundamental and Molecular Mechanisms of Mutagenesis*, 455, (1-2), 61–69, ISSN 0027-5107.

Mortelmans, K. & Zeiger, E. (2000). The Ames *Salmonella*/microsome mutagenicity assay. *Mutation Research/Fundamental and Molecular Mechanisms of Mutagenesis*, 455, (1-2), 29–60, ISSN 0027-5107.

Nagabhushan, M.; Amonkar, A. J. & Bhide, S. V. (1987). Mutagenicity of gingerol and shogaol and antimutagenicity of zingerone in *Salmonella*/microsome assay. *Cancer Letters*, 36, (2), 221-233, ISSN: 0304-3835.

Oh, H.-Y.; Kim, S.-H.; Choi, H-J.; Chung, M. J. & Ham, S.-S. (2008). Antioxidative and antimutagenic activities of 70% ethanol extract from masou salmon (*Oncorhynchus masou*). *Toxicology in Vitro*, 22, (6), 1484–1488, ISSN: 0887-2333.

Ozbek, T.; Gulluce, M.; Agar, G.; Adiguzel, A.; Ozkan, H.; Sahin, F. & Baris, O. (2008a). Investigation of The Antimutagenic Effects of Methanol Extract of Astragalus L. Species Growing in Eastern Anatolia Region of Turkey. *Fresenius Environmental Bulletin*, 17, (12A), 2052-2058, ISSN 1018-4619.

Ozbek, T.; Gulluce, M.; Sahin, F.; Ozkan, H.; Sevsay, S. & Baris, O. (2008b). Investigation of the Antimutagenic Potentials of the Methanol Extract of *Origanum vulgare* L. subsp *vulgare* in the Eastern Anatolia Region of Turkey. *Tukish Journal of Biology*, 32, (4), 271-276, ISSN 1303-6092.

Pao, W. & Girard, N. (2011). New driver mutations in non-small-cell lung cancer. *Lancet Oncology*, 12, (2), 175-180, ISSN 1470-2045.

Rossi, D.; Aiello, V.; Mazzoni, L.; Sensi, A. & Calzolari, E. (2007). *In vitro* short-term test evaluation of catecholestrogens genotoxicity. *Journal of Steroid Biochemistry and Molecular Biology*, 105 (2007) 98–105, ISSN: 0960-0760.

Stephens, J. C.; Reich, D. E. & Goldstein, D. B. (1998). Dating the origin of the CCR5-Delta32 AIDS-resistance allele by the coalescence of haplotypes. *The American Journal Human Genetics*, 62, (6), 1507–1515, ISSN 0002-9297.

Sutton, E. (1943). Bar Eye in Drosophila melanogaster: a Cytological Analysis of Some Mutations and Reverse Mutations. *Genetics*, 28, (2), 97–107, ISSN 1943-2631.

Tejs, S. (2008). The Ames test: a methodological short review. *Environmental Biotechnology*, 4, (1), 7-14, ISSN 1432-0614.

Vargas-Gonzales, R.; Torre-Mondragon, L.; Aparicio-Rodriguez, J. M.; Paniagua-Morgan, F.; Lopez-Hernandez, G.; Garrido-Hernandez, M. A. & Nunez-Barrera, S. (2010). Juvenile Polyposis of Infancy Associated with Paracentric Inversion and Deletion of Chromosome 10 in a Hispanic Patient: A Case Report. *Pediatric and Developmental Pathology*, 13, (6), 486-491, ISSN: 1615-5742.

Wickramasnghe, R. H. (1979). Short-term Mutagenicity Test Systems for Detecting Carcinogens Report of an International Symposium in Dortmund. *International Archives of Occupational Environmental Health*, 44, (1), 61-64, ISSN 0340-0131.

World Health Organization [WHO], (2007). Mutagenicity Testing for Chemical Risk Assessment. *Harmonization Project DRAFT Document for Public and Peer Review*, September 2007. World Health Organization Press.

Zahin, M.; Aqil, F. & Ahmad, I. (2010). Broad spectrum antimutagenic activity of antioxidant active fraction of *Punica granatum* L. peel extracts. *Mutation Research/Genetic Toxicology and Environmental Mutagenesis*, 703, (2), 99–107, ISSN: 1383-5718.

Zeiger, E.; Anderson, B.; Haworth, S.; Lawlor, T. & Mortelmans, K. (1992). *Salmonella* mutagenicity tests: V. Results from the testing of 311 chemicals. *Environmental and Molecular Mutagenesis*, 19, (21), 1–14, ISSN 1098-2280.

Zieger, E. (2000). Current strategies for detecting carcinogens by using genetic toxicity tests, *Developments in Animal and Veterinary Sciences*, 31, 113-122, ISSN: 0167-5168.

Zeiger, E. (2004). History and Rationale of Genetic Toxicity Testing: An Impersonal, and Sometimes Personal, View. *Environmental and Molecular Mutagenesis*, 44, (5), 363–371, ISSN 1098-2280.

Zeiger, E.; Gollopudi, B. & Spencer, B. (2005). Genetic toxicity and carcinogenicity studies of glutaraldehyde—a review. *Mutation Research/Reviews in Mutation Research*, 589, (2), 136–151, ISSN 1383-5742.

Zeiger, E. (2010). Historical Perspective on the Development of the Genetic Toxicity Test Battery in the United States. *Environmental and Molecular Mutagenesis*, 51, (8-9), 781–791, ISSN 1098-2280.

Zhang, J. (2003). Evolution by gene duplication: an update. *Trends in Ecology and Evolution*, 18, (6), 292–298, ISSN: 0169-5347.

Zhang, Z.; Fu, J.; You, B.; Zhang, X, Zhou, P. & Zhou, Z. (2011). In vitro genotoxicity of danthron and its potential mechanism. *Mutation Research/Genetic Toxicology and Environmental Mutagenesis*, 722, (1), 39–43, ISSN: 1383-5718.

13

Reticulate Evolution Among the Group I *Salmonellae*: An Ongoing Role for Horizontal Gene Transfer

Eric W. Brown, Rebecca L. Bell, Marc W. Allard, Narjol Gonzalez-Escalona,
Andrei Perlloni, Joseph E. LeClerc and Thomas A. Cebula
Center for Food Safety and Applied Nutrition
Food and Drug Administration, College Park, MD
USA

1. Introduction

Salmonella enterica is responsible for 1.4 million cases of foodborne salmonellosis in the United States annually making it the number one causative agent of bacterial foodborne illnesses (CDC, 2007). Infection can occur after eating undercooked meat, poultry and eggs that have been contaminated with *Salmonella* (CDC, 2007). In recent years several outbreaks have occurred in the United States that were associated with *Salmonella* contamination of produce, the most recent being a *S. enterica* Saintpaul outbreak associated with tomatoes, jalapeño and serrano peppers that sickened over 1400 individuals (CDC, 2008). The movement of several serovars of *Salmonella* into previously naïve niches (*i.e.,* produce-growing environs) suggests that the pathogen is readily adapting to new environments. An understanding of the reticulate evolutionary mechanisms that underpin the acquisition and composition of the requisite genetic and phenotypic features of *Salmonella* is essential to more accurate risk assessment of this pathogen (Hohmann, 2001).

It is now widely accepted that horizontal gene transfer (HGT) has driven the emergence of more aggressive and virulent strains of *Salmonella* in the environment, on the farm, and in the food supply. Such assault by various salmonellae has fueled the in-depth examination of specific genotypes and conditions that permit reticulate evolutionary change and the rise of deleterious phenotypes (LeClerc et el., 1996; 1998; 1999; Cebula and LeClerc, 1997). The hypermutable phenotype represents one scheme by which reticulate evolution of the bacterial chromosome may occur (Trobner and Piechoki, 1984; Haber et al., 1988; Haber and Walker, 1991; LeClerc et al., 1996; Matic et al., 1997; Radman et al., 1999; Cebula and LeClerc; 2000; Funchain et al., 2000). Methyl-directed mismatch repair (MMR) defects, leading to a mutator or hypermutable phenotype, are found in more than 1% of the isolates within naturally-occurring populations of *Salmonella enterica* (LeClerc et al., 1996) and at even greater frequencies in the food supply where oxidative and other anti-microbial stressors are applied (Cebula et al., 2001). Up to 73% of the MMR defects found in feral settings are due to lesions within the *mutS* gene, resulting in increased nucleotide substitution rates, enhanced DNA transposition, and, perhaps most importantly, a relaxation of the internal barriers that normally restrict homeologous recombination following HGT of foreign DNA (Cebula and LeClerc, 1997; Radman et al., 1999).

This latter role, as a major sentinel for recombination, led to a substantial focus on the genetics and evolution of the *mutS* gene and its adjacent sequences located at 63 min on the *Salmonella* chromosome (Brown et al, 2002; 2003; Kotewicz et al., 2003; 2003). Phylogenetic analyses of *mutS* alleles from strains of the SAR (*Salmonella* reference) collections (i.e., SARA, SARB, and SARC)—largely taken to represent the extent of genetic variability within the species (Boyd et al., 1993; 1996; Beltran et al., 1991)—have revealed striking levels of phylogenetic discordance between trees derived from *mutS* alleles and whole-chromosome trees of the same strains based on MLEE (multilocus enzyme electrophoresis) analysis (Brown et al., 2002, 2003). These differences were interpreted as numerous examples of HGT among *mutS* alleles in *Salmonella*. Similar observations have been made among sequences abutting the *mutS* gene in *Salmonella, E. coli,* and *Shigella* spp (Kotewicz et al., 2002; 2003; Brown et al., 2001b). Our laboratory showed previously that the 61.5 min *mutS-rpoS* region retains a novel and highly polymorphic 2.9 kb sequence in the genome of all *E. coli* O157:H7 strains, *Shigella dysenteriae* type 1, and several other *E. coli* strains (LeClerc et al., 1999) but not in *Salmonella enterica* (Kotewicz et al., 2003). This highly polymorphic stretch of DNA (previously coined the *mutS-rpoS* "unusual region") is varied in its distribution among enteric bacterial lineages and is absent in others entirely (Kotewicz et al., 2003). Sequence analysis of the region revealed an IS1 insertion element in place of the *prpB* gene in *S. dysenteriae* type 1 suggesting the existence of a recombinational crossover in the *mutS-rpoS* region for this strain (LeClerc et al., 1999). Evidence for additional crossovers in the same region were also obtained for other *E. coli* strains (Brown et al., 2001b). These findings support the notion that HGT helped forge current relationships among *Salmonella* and other enteric pathogens in this region and throughout numerous other locales in the *Salmonella* chromosome.

Indeed, as evidenced from global efforts involving whole-genome sequencing, microarray, and multi-locus sequence typing, the substantial impact that HGT has played in structuring the chromosome of *Salmonella enterica* is now indisputable (Porwollik and McClelland, 2003; Fricke et al., 2011; Kelly et al., 2009; Hall, 2010). Previous estimates indicate that at least one-quarter of the *Salmonella* genome may have been forged through HGT and reticulate evolutionary events (Porwollik and McClelland, 2003), although this number seems conservative from current views. In addition to the 61.5 min region surrounding *mutS*, HGT has played a key role in structuring many other regions of the *Salmonella* chromosome. Notably, SPI elements (*Salmonella* pathogenicity islands) have likely been acquired through HGT (Groisman and Ochman, 2000; Ochman et al., 2000; Hacker and Kaper, 2000; Baumler et al., 1997). For example, the SPI-1 pathogenicity island, comprising the genes encoding a type III secretion system, was probably acquired early in *Salmonella* evolution (Kingsley and Baumler, 2000; Li et al., 1995), yet several *inv–spa* alleles seem to have converged horizontally more recently between *S. enterica* groups IV and VII (Boyd et al., 1997; Brown et al., 2002). Additionally, type 1 pilin genes that encode fimbrial adhesins retain unusually low GC contents and aberrant DNA sequence phylogenies relative to other *fim* genes (Boyd and Hartl, 1999). Other studies focusing on numerous housekeeping gene loci have reported evolutionary histories for these genes that are strikingly decoupled from *S. enterica* strain history (Nelson and Selander, 1994; Thampapillae et al., 1994; Brown et al., 2002; Boyd et al.,

Christensen and Olsen, 1998; Groisman et al., 1992; Li et al., 1994; Liu and Sanderson, 1996; Nelson and Selander, 1994; Nelson et al., 1992; 1997).

The now incontrovertible connection between horizontal transfer and MMR gene evolution has led to the thesis that genetic exchange of *mutS* alleles could simultaneously quiet the mutator phenotype while rescuing adaptive changes from the population (LeClerc et al., 1996; Denamur et al., 2000). Consistent with this hypothesis, the *mutS* gene is evolutionarily scrambled by HGT in subspecies I *Salmonella enterica*. Our laboratories documented the prevalence of horizontal gene transfer (HGT) among strains of *Salmonella enterica* (Brown et al., 2002; 2003). In comparing across and within subspecies of *Salmonella*, a recombination gradient was noted wherein the incidence of HGT was inversely correlated with the genetic diversity separating individual strains. It appears that a genetic threshold exists that tolerates free exchange of sequences within a framework delimited by sequence variation and niche diversity of individual strains. We demonstrated this through identification of intragenic (patch-like) recombination as the primary outcome across disparate *Salmonella* subspecies and assortative (whole-allele) recombination which caused extensive reassortment of alleles among more genetically homogeneous populations of group I *Salmonella* pathogens, all sharing a common niche restricted to warm-blooded mammals.

A torrent of scientific information has accrued over the past decade to support the important role of HGT in the genetic and evolutionary diversification of *S. enterica* subspecies, serovars, and individual pathogenic clones (McQuiston et al., 2008; Octavia and Lan, 2006; Lan et al., 2009; Fricke et al., 2011). Our understanding in reconstructing the horizontal acquisitions of important features including those involved in virulence, drug resistance, and other adaptations that foster an enhanced fitness for *Salmonella* persistence in foods, animals, and people is expanding at a pace which we could not have foreseen even a decade ago (Sukhnanand et al., 2005). It is important to recall however that reticulate evolutionary pressures do not subside once selectively advantageous traits are gained. Rather, horizontal exchange likely continues to dapple the evolutionary landscape between even the most closely related salmonellae (Brown et al., 2003). Here, we provide results of several previously unreported phylogenetic studies that evidence (i) the continued role of HGT in the intra-operon shuffling of SPI-1 alleles among subspecies I *S. enterica* strains; (ii) the often under-appreciated role for HGT and recombination in the homogenization of allele structure in a closely related population of *S. enterica*; and (iii) the panmictic and reticulate nature of restriction-modification (R-M) genes among group I salmonellae. This last finding, noting free exchange of R-M (*i.e., hsd*) alleles, provides phylogenetic evidence of the compatibility of *S. enterica* subspecies I R-M complexes, likely accounting for the documented successful HGT of entire gene sequences among closely (*e.g.*, intra-subspecies) related strains as DNA exchange between strains that shared or recently shared common R-M alleles would not be subject to substantial restriction (Sharp et al., 1992).

2. Reticulate evolution in SPI-1 of *Salmonella enterica* subspecies I

Salmonella pathogenicity island 1 (SPI-1) specifies a type III secretion system essential for host cell invasion and macrophage apoptosis (Galan and Curtiss, 1989; Galan and Collmer, 1999). SPI-1 comprises a cluster of virulence genes (*e.g.*, the *inv/spa* gene cluster) that encode, in part, the "needle complex", a key delivery component for transporting virulence associated effector molecules into the host cell (Galan and Collmer, 1999). The disparate phylogenetic distribution, lack of chromosomal synteny, and diverse base compositions of SPI-1 and its homologues indicate that these sequences were obtained independently across enteric species of bacteria. It is presumed that SPI-1 was present in the last common ancestor of all *Salmonella* lineages. Horizontal acquisition of the *inv/spa* gene cluster, however, is thought to have been a pivotal event for the emergence of *Salmonella* as a pathogenic species (Boyd et al., 1997; Groisman and Ochman, 2000). The gene complex lies adjacent to the polymorphic *mutS–rpoS* region of the chromosome. We and others previously presented phylogenetic evidence for intragenic recombination of sequences within several SPI-1 invasion loci (Boyd et al., 1997; Brown et al., 2002), primarily among *S. enterica* subspecies IV and VII. However, in order to determine the extent to which HGT may have disrupted SPI-1 evolution across the more ecologically and genetically homologous group I salmonellae, we examined nine SPI-1 invasion loci from nearly half of the SARB reference collection of strains (Boyd et al., 1993), composed exclusively of subspecies I *Salmonella* serovars.

2.1 SPI-1 gene evolution is decoupled from *Salmonella* chromosome evolution

Using a cladistic approach (Forey et al., 1992; Allard et al., 1999; Bell et al., 2011), the nucleotide sequences from nine invasion gene sequences were subjected to phylogenetic analysis. The resultant invasion gene phylogenies were then compared to phylogenetic groupings from the *mdh* gene, a chromosomal anchor locus that is taken largely to reiterate chromosome evolution within subspecies I (Boyd et al., 1994) and MLEE (multi-locus enzyme electrophoresis), also applied here as a metric of strain/chromosome evolution for the group I salmonellae (Boyd et al., 1993). As shown in Fig. 1, strains composing single SARB *mdh* and MLEE lineages were, for the most part, distributed across disparate *inv/spa* gene clades for all nine invasion genes tested indicating that many of these strains, although linked tightly in chromosome evolution, retain invasion gene alleles with unrelated evolutionary histories, presumably as a result of HGT.

Evolutionary incongruence between *inv/spa* genes and the *Salmonella* chromosome was affirmed using the ILD (incongruence length difference) test, which evaluates the likelihood of a common evolutionary history between genes (Farris et al., 1995; LeCointre et al., 1998; Brown et al., 2001a). Seven of the nine invasion genes yielded significant ILD scores ($p < 0.05$), indicating that a hypothesis of congruence could be rejected for these strains and further reinforcing the discordance evident in the clade comparisons. The only exceptions were *invB* ($p = 0.08$) and *spaP* ($p = 0.59$), albeit both still retained cladistic signatures of HGT from broken clade structures in the tree analysis.

2.2 SPI-1 gene evolution is decoupled from *mutS* gene evolution

The *mutS* gene, downstream and adjacent to SPI-1 in *S. enterica*, has been shuffled extensively by HGT (Brown et al., 2003). In order to determine whether *mutS* may have been linked in the recombination now evident among SPI-1 genes, cladistic comparisons were made between *mutS* phylogeny and *inv/spa* gene phylogeny revealing substantial incongruence between *inv/spa* trees and *mutS* trees. Six of these comparisons are shown in the form of tanglegrams (Fig. 2). Again, strains composing SARB *mutS* clades were distributed across disparate *inv/spa* gene clades for all nine invasion genes tested, and seven of nine *inv/spa* genes were further confirmed as discordant with *mutS* based on ILD testing. Taken together, these findings indicate that *inv/spa* gene sequences and *mutS* sequences from the same strains are decoupled in their evolution. These data suggest that reticulate evolution has repeatedly forged this contiguous region of the *Salmonella* chromosome such that different strains appear to have been affected by assortative (allelic) HGT between the two loci.

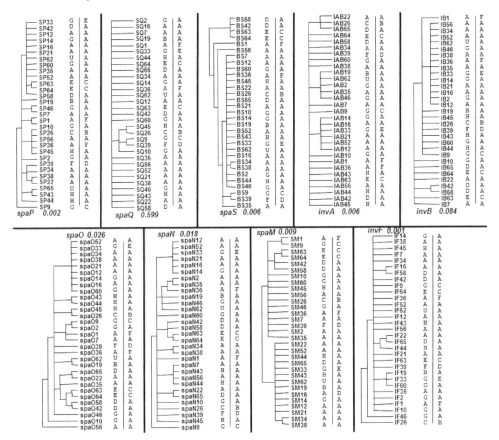

Fig. 1. Phylogenetic discordance between SPI-1invasion genes and the *Salmonella* chromosome. *mdh* and MLEE comparisons are shown to each of nine different *inv/spa* genes indicated. Identical letters denote strains from the same *mdh* or MLEE lineage. It is important to note that letters are only relevant to their respective data column and do not cross-over columns. The column to the left of the dividing line designates *mdh* clade assignments for the respective *S. enterica* strain while the column of letters to the right of the divider corresponds to MLEE clade assignments. The number at the base of each tree denotes the ILD score (p-value) relative to a comparison for congruence between the respective *inv/spa* gene and the *mdh* gene sequence alignment for the same strains. Trees shown were rooted using *S. bongori* as an outgroup. Nucleotide sequence alignments were performed using CLUSTAL X (Thompson et al., 1998). Most parsimonious trees were generated in PAUP* v.10 (Swofford et al., 2002).

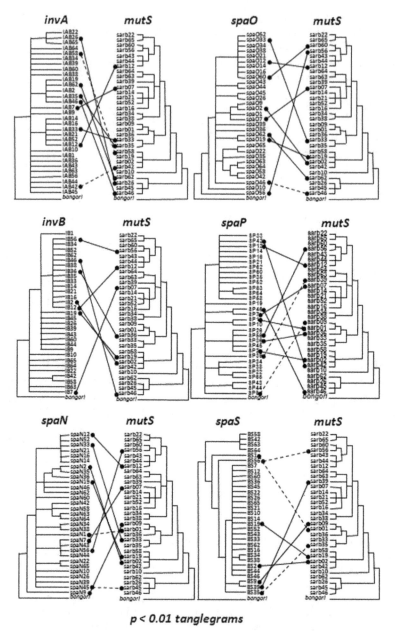

p < 0.01 tanglegrams

Fig. 2. Tanglegrams of several invasion gene and *mutS* revealing the phylogenetic incongruence between *inv/spa* genes and the *mutS*, which lies adjacent to SPI-1 on the *Salmonella* chromosome. Lines connect the discordant, potentially recombinagenic (incongruent) strains. *inv/spa* to *mutS* comparisons with an ILD score of p < 0.01 were displayed. Trees shown were again rooted using *S. bongori* as an outgroup taxa.

2.3 Intra-island HGT within the SPI-1 region of subspecies I *Salmonella* strains

In order to determine the presence and extent to which HGT has shuffled individual alleles within SPI-1 among more closely related subspecies I strains, a pairwise ILD approach was adopted wherein congruence was scored for individual comparisons of all nine of the *inv/spa* genes included in this study (Fig. 3). Several findings were noteworthy. Although no individual invasion gene showed unanimous evolutionary discordance with its neighbors, three *inv/spa* loci (*invA*, *invB*, and *spaP*) were incongruent (p < 0.10) with a significant majority of other genes. *invA* and *invB* showed discordance with all other loci except *spaN* and *spaQ*, while *spaP* showed discordance to all but *spaM* and *spaQ*. Conversely, with the exception of *spaQ*, no *inv/spa* gene was congruent with every other. Thus, a hypothesis of extensive intra-island shuffling begins to emerge with an evolutionary decoupling of individual invasion loci one from another. Additional tree comparisons buttressed this conclusion. Akin to the selfish operon theory (Lawrence and Roth, 1996), these data suggest that the SPI-1 region is a chromosomal mosaic, composed of *inv/spa* gene sequences that have converged within this island but with each retaining unique evolutionary paths.

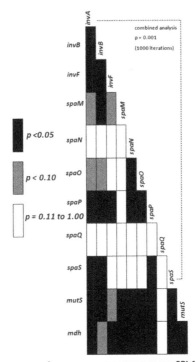

Incongruence measures among SPI-1 loci

Fig. 3. ILD test results for intragenic comparisons among *inv/spa* invasion genes. ILD tests (Farris et al., 1995) were performed with 1000 partitions using the Partition Heterogeneity command in PAUP* v.10 (Swofford et al., 2002). A p-value of 0.05 or less allows for a rejection of the null hypothesis of congruence (vertical evolution) and accepts the alternative hypothesis of incongruence which is interpreted among bacterial phylogeny as evidence for HGT (LeCointre et al., 1998).

2.4 Key observations

i.   The *inv/spa* complex of *S. enterica* subspecies I appears to have undergone extensive intra-island allelic shuffling due to HGT. This suggests that the SPI-1 region is a mosaic composed of SPI-1 gene sequences with distinct evolutionary origins.

ii.   Invasion genes within this *Salmonella* population are not only decoupled phylogenetically from *mutS* and other flanking sequences but also from the chromosomes of group I *S. enterica* strains, suggesting that these genes have been re-assorted by HGT.

iii. Much of the recombination observed here appears to be assortative transfer, a finding that contrasts to the *inv* genes in *S. enterica* as a whole, where tree structure was largely intact with HGT limited mostly to subspecies IV and VII (Boyd et al., 1997; Brown et al., 2002).

iv.   Allele shuffling appears to be most prominent within the subspecies I taxonomic boundary and not across other subspecies of *S. enterica*. This finding is consistent with a relaxed and compatible restriction-modification system among more closely related *Salmonella* strains (Brown et al., 2003).

3. HGT homogenizes the *mutS* gene among 'Typhimurium' complex strains

Here, we present phylogenetic and genetic analyses of *Salmonella* reference collection A (SARA), also known as the Typhimurium strain complex—the most homogeneous *S. enteric* reference collection, consisting solely of five closely related subspecies I serovars (Typhimurium, Paratyphi B, Muenchen, Saintpaul, and Heidelberg) (Beltran et al., 1991). Given the evolutionary similarity shared among these pathogens and trend noted previously that highlight the inverse relationship between *Salmonella* diversity and recombination, one would expect to observe an even greater role for HGT in the population structure of the *S. enterica* SARA collection of pathogens.

3.1 Cladistic evidence for horizontal exchange of *mutS* alleles among 'Typhimurium' complex strains

As was done for SPI-1 gene sequences, a phylogenetic tree was derived from 72 SARA *mutS* sequences and was compared to phylogenetic trees derived from multi-locus enzyme electrophoresis (MLEE) and *mdh* (malate dehydrogenase) gene sequences for the same strains. Phylogenies derived from horizontally exchanged sequences display evolutionary discordance (incongruence) when compared to *mdh* and MLEE trees. In the tree shown, six clades of *mutS* alleles were observed and compared to the distribution of four *mdh* and six MLEE multi-strain containing clades (Fig. 4). Two of the four SARA *mdh* clades were found to be displaced into multiple clades on the *mutS* tree. Two additional *mdh* clades were found to have converged into a single *mutS* clade, suggesting that HGT may have homogenized *mutS* diversity of these particular *mutS* lineages. Similarly, strains from five of the six MLEE lineages were displaced into separate clades on the *mutS* tree. The only exception was a single clade of MLEE SARA strains (A57, A58, A59, and A60), which was also found intact in the *mutS* tree except for the inclusion of SARA strain A56. Nonetheless, numerous examples of evolutionary discordance between the 1.1 kb *mutS* segment and the chromosome of the 'Typhimurium' complex strains indicate that horizontal exchanges of *mutS* alleles have accumulated during the rather shallow radiation of even these highly homogeneous group I pathogens. As an aside, it was noteworthy that full-length *mutS* alleles were horizontally transferred among SARA *S. enterica* strains, lending further credence to a model for R-M compatibility among closely related *S. enterica* serovars and strains.

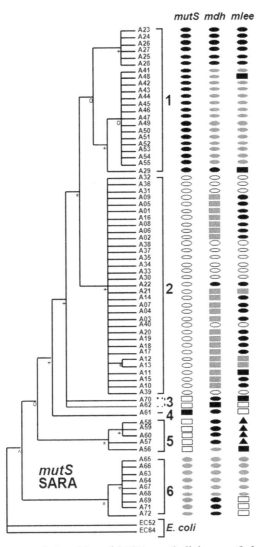

Fig. 4. Most-parsimonious relationships of SARA *mutS* alleles. *mutS* clades are bracketed and numbered to the right of the tree. Distributions of *mutS*, *mdh*, and MLEE clades are presented in column form. Note that strains originating from the same clade retain a common shape and common internal shading. Bootstrap nodal support values (Felsenstein et al., 1985) are presented on the *mutS* tree as follows: ^, 76-100%; *, 51-75%; +, 26-50%; o, 1-25%. In this case, *mdh* and MLEE are taken to represent the evolution of the strain in general (Boyd et al., 1994; Beltran et al., 1991). The tree shown is rooted with two *E. coli* outgroups.

3.2 Homogenization of *mutS* sequence diversity among *S*. Typhimurium and *S*. Heidelberg strains

Curiously, a single clade in the SARA *mutS* tree was found to comprise three distinct *Salmonella* serovars. In this clade, every strain representing *S*. Typhimurium (n=21) and *S*. Heidelberg (n=11), along with a single strain of *S*. Saintpaul, converged into a single evolutionary lineage of *mutS* alleles. In the SARA *mdh* tree (Fig. 5), *mdh* alleles for these same SARA serovars formed three disparate clades in the tree such that *S*. Typhimurium strains clustered only with other *S*. Typhimurium and *S*. Heidelberg strains only with other *S*. Heidelberg. *S*. Saintpaul strains formed a single lineage at the tip of the tree with strains of *S*. Muenchen and a single *S*. Paratyphi B. It should be noted that these distinct clades retained substantial statistical support with bootstrap values around 90% (Felsenstein, 1985). Thus, phylogenetic comparison of *mutS* and *mdh* sequences supported the notion that these serovars have converged into a single *mutS* clade, possibly as a result of the repeated HGT of only one or a few preferred *mutS* alleles.

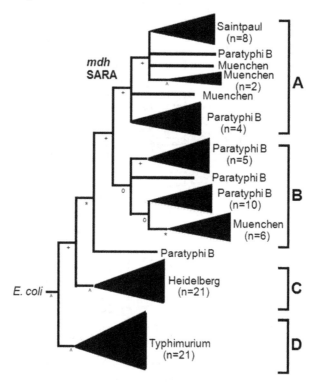

Fig. 5. Phylogenetic tree revealing the most-parsimonious relationships of SARA *mdh* alleles. *mdh* clades are bracketed and lettered while SARA serovars are labeled to the right of the tree. For sample sizes greater than one, multiple strains of the same serovar are depicted as a cone on the tree terminal nodes. Note that strains originating from the same clade are designated by a common bracket and letter. Bootstrap nodal support values are presented on the *mdh* tree as follows: ^, 76-100%; *, 51-75%; +, 26-50%; o, 1-25%. Note the bifurcations between specific clusters in the tree, signaling sequence diversity among distinct serovars using the *mdh* gene.

In order to further investigate the genetic structure of this converged clade, we examined *mutS* sequence homogeneity across the strains composing this lineage as well as the remaining *mutS* alleles of the SARA collection (Fig. 6). Evaluation of polymorphic positions in the *mutS* alignment revealed several findings consistent with homogeneous clade structure surrounding these serovars. First, five substitutions were observed across the entire 1,115 bp sequence for all 33 strains that define this *mutS* clade (#2). Second, with the exception of the polymorphism at position 913 in SARA strains 12 and 13, no clade #2 substitution was retained by more than one strain. Thus, none of the substitutions present within this clade partitioned any member serovar from another. The near structural uniformity of this clade at the nucleotide level further suggests that HGT has homogenized *mutS* alleles among these particular serovars. This is consistent with the thesis of Dykhuizen and Green (1991) who reminded that recombination can not only diversify the genome but can also homogenize it as well.

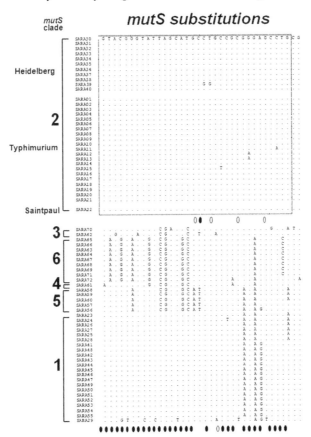

Fig. 6. *mutS* nucleotide sequence homogeneity among *S. enterica* serovars Typhimurium, Heidelberg, and a strain of Saintpaul. Periods indicate exact nucleotide identity to the reference sequence at the top of the alignment while listed nucleotides represent actual substitutions. The synonymous/nonsynonymous status (blackened ovals indicate synonymous change) of each substitution is noted below the alignment. Nucleotide sequences were generated using a PCR-based approach and automated CE-sequencing technology.

3.3 Distinct roles for HGT across various taxonomic tiers of *S. enterica*

With the inclusion of the SARA analysis reported here, we have been able to define varying roles for HGT across three taxonomically distinct populations of *S. enterica* (SARA, B, and C) (Fig. 7). Within *S. enterica* as a whole, a model for HGT begins to emerge that tolerates near-free HGT among closely-related subspecies I strains. As genetic divergence increases across serovars, however, the extent of HGT appears to decrease. The analysis reported here suggested two unique findings for SARA, the most genetically monomorphic population.

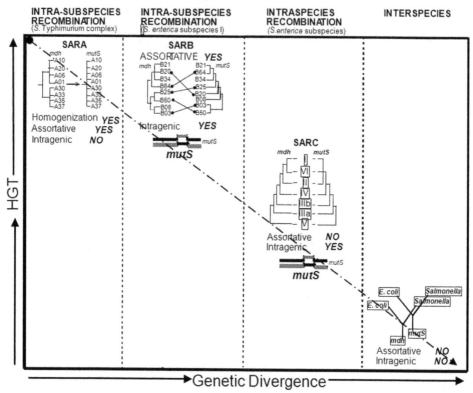

Fig. 7. Model for the frequency and effects of HGT among various taxonomic tiers of *Salmonella enterica*. Graphic representation of the various effects of HGT on the taxonomically distinct SARA, SARB, and SARC strain collections as well as an interspecies comparison. The *S. enterica* collections are plotted relative to genetic divergence versus the extent of HGT observed. Specific effects and trends associated with the HGT occurring at each taxonomic level are noted below each of the *Salmonella* populations shown.

First, SARA revealed evidence for a substantial convergence of *mutS* alleles between distinct serovars suggesting, that, recombination can have a homogenizing effect on sequence diversity in this population. Second, despite yielding numerous examples of assortative (allelic) exchange, SARA appears to be—at least from a phylogenetic perspective—refractory to intragenic (mosaic) HGT within the *mutS* gene. Thus, the SARA and SARB groups seem to have been influenced more extensively by HGT than SARC possibly because they are not so diverged that exchange is inhibited due to extreme niche or R-M (restriction-modification) system variability. Moreover, it is also possible that much of the HGT among SARA strains have gone undetected here since identical alleles would leave no phylogenetic footprint following an exchange event.

3.4 Key observations

i.   Horizontal gene transfer of *mutS* alleles in *Salmonella* appears to play a prominent role in the evolutionary structure of the five closely-related serovars representing the SARA ('Typhimurium' complex) collection, a finding consistent with extensive HGT that has been documented among subspecies I serovars in general (Brown et al., 2003).

ii.   Cladistic analysis of SARA strains revealed the first example of a substantial convergence of *mutS* alleles from disparate serovars into a single clade. This suggests that HGT is homogenizing allele diversity among certain *Salmonella* strains and serovars—an observation reminiscent of allele homogenization observed for the *E. coli polA* gene (Patel and Loeb, 2000).

iii. Among closely related 'Typhimurium' complex strains, *mutS* alleles appear to have shuffled largely as single units rather than in intragenic segments. One explanation for this might be a more recent evolutionary divergence of the five serovars composing the highly homogeneous 'Typhimurium' strain complex. Alternatively, recombination of highly homologous mosaic segments of the *mutS* gene would do little to obscure phylogeny and likely go undetected in these analyses.

iv.   Retrospective comparison of SARA HGT patterns with that of SARB and SARC strains yields a gradated model for HGT whereby different taxonomic tiers of *Salmonella* are subject to different HGT effects. The differences appear coupled to the extent of genetic diversity that defines these three different "tiers" of *Salmonella* population structure.

4. HGT among restriction-modification (R-M) genes of subspecies I salmonellae

The restriction and modification (R-M) system is a defense mechanism developed by bacteria to protect the bacterial genome from invasion by foreign DNA (Bullas et al., 1980). Foreign sequences entering the cell are cleaved by restriction enzyme(s), while the bacterial DNA itself is modified by methylase(s), thus providing protection from its own restriction enzyme (Murray, 2000). R-M systems are composed of genes that encode a specific restriction endonuclease and modification methylase. There are several types of R-M systems, namely type I (*e.g.*, *Eco*KI), type II (*e.g.*, *Eco*RI), and type III (*e.g.*, *Sty* LTI) (Barcus et al., 1995). Types of R-M systems are classified on the basis of their composition and cofactor requirements, the nature of the target sequence, and the site of DNA cleavage with respect to the target sequence (Murray, 2000; Naderer et al., 2002).

Compatibility of R-M systems among strains was proposed as one explanation to account for contrasting recombination rates (Brown et al., 2003). In this model, compatible R-M complexes would permit the successful transfer of larger gene segments among closely related *Salmonella* pathogens; crosses between strains with identical R-M systems would not be subject to restriction (Sharp et al., 1992). A gradation in the size limits of DNA segments exchanged would depend on the polymorphic character of R-M systems in natural strains.

Here, we investigate this model by examining the molecular evolutionary relationships of *hsd* genes encoding R-M complexes among closely related pathogenic *Salmonella* strains (*i.e.,* the 'Typhimurium' complex). If, indeed, *hsd* alleles are freely exchanged themselves among strains that display a substantial tolerance for HGT and recombination of diverged DNA sequences, then an explanation accounting for observed tolerance to extensive HGT begins to emerge for *S. enterica* group I serovars.

4.1 Evidence for HGT of R-M alleles among *Salmonella enterica* group I strains

DNA sequences from three *hsd* type I R-M genes were subjected to cladistic analysis. The resultant invasion gene phylogenies were then compared to phylogenetic groupings from the *mdh* gene and from the *Salmonella* MLEE data. Cladistic comparisons of *hsd* genes to markers of stable *Salmonella* chromosome evolution revealed several findings, and the data for *hsdS* is shown (Fig. 8). For *hsdS* section S1, SARA 56 is removed from neighboring strains when compared to *mdh* or *mutS*. For *hsdS* section S2, the collapsing of numerous clades into a single conserved clade was observed. It should be noted that such collapsing was observed in many of the trees reported here and suggests that HGT may be homogenizing *hsd* alleles. Moreover, a distinct allele that has no homology with its sister allele in a neighboring clade can be seen on the tree. Finally, the *hsdS* section S3 tree breaks up clades from both MLEE and *mdh*. In addition, this tree has three distinct allele types that can be seen phylogenetically, as in the case of S2.

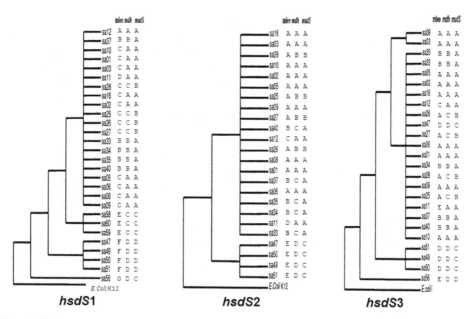

Fig. 8. Phylogenetic trees constructed for three segments comprising the *hsdS* gene, *hsdS1*, *hsdS2*, and *hsdS3*. Each respective gene tree is rooted with an *E. coli* outgroup and compared to MLEE, *mdh* and *mutS* clade patterns. Identical letters signal a common clade origin in the *mdh*, MLEE, or *mutS* datasets. Note that the letter designations from the *mdh*, MLEE, and *mutS* bar columns are independent of each other.

Compatibility among R-M systems has been proposed to account for the extensive levels of HGT documented among subspecies I *Salmonella* pathogens. Since the *mutS* gene appears to have been shuffled among this group of strains, we examined the phylogenetic relationship of *mutS* to type I R-M *hsd* genes. Incongruence was observed between *hsd* genes and *mutS* phylogeny, suggesting that patterns of HGT for *hsd* alleles differ from those for *mutS* alleles. *hsd* segments S1 and S3 each retained at least one incongruent strain between these gene phylogenies. In addition, several *hsd* genes collapsed divergent *mutS* clades into single *hsd* lineages in the trees. For instance, three *mutS* clades composed a single *hsdM* clade, a pattern that held true for other *hsd* genes, including *hsdS* segments 2 and 3. These data indicate distinct roles for HGT between most R-M genes and *mutS*. Nonetheless, observed homogenization of type I R-M loci among subspecies I *Salmonella* strains suggests they are compatible systems, allowing additional genes like *mutS* to be transferred within this population in its entirety.

4.2 Evidence for intra-operon HGT of R-M alleles

Intra-operon evolutionary incongruence between *hsd* genes was further examined using the ILD (incongruence length difference) test, which evaluates the likelihood of a common evolutionary history between genes. The ILD comparisons yielded more notable incongruence between genes than did the tanglegram analysis, suggesting that small patches of sequence within individual genes may be responsible for much of the observed incongruence. Intragenic patterns of HGT have been noted previously for more diverse subspecies (Brown et al., 2002). In the ILD comparisons (Fig. 9), eight of the ten *hsd* data set comparisons yielded significant

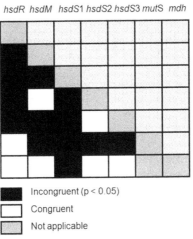

Fig. 9. Pairwise incongruence length difference (ILD) test results for several genes of the Type I Restriction – Modification system in *Salmonella*. ILD tests were performed with 1000 partitions using the Partition Heterogeneity command in PAUP*v.10 (Swofford et al., 2002). Pairwise ILD comparisons were made among the three *hsd* genes R, M, and S including the three sub-regions that were amplified from *hsdS* (i.e., S1, S2, and S3). As in Fig. 3, a p-value of 0.05 or less allows for a rejection of the null hypothesis of congruence (vertical evolution) and accepts the alternative hypothesis of incongruence which is interpreted among bacterial phylogeny as evidence for HGT (LeCointre et al., 1998).

ILD scores (p < 0.05) such that a hypothesis of congruence could be rejected for these intragene comparisons. The only exceptions were the *hsdS2-hsdM* comparison (p = 1.00) and the *hsdS3-hsdS2* comparison (p = 1.00). It is noted that, with the exception of *hsdR*, all of the *hsd* data matrices were also incongruent with *mutS*. When examined in total, the data suggest that the Type-1 R-M operon is a mosaic comprising *hsd* gene sequences that have converged evolutionarily within this operon, but with each possessing a unique phylogenetic path.

It was also noteworthy that *hsdS* segments S2 and S3, however, retained groups of alleles that shared little or no identified homologies. That is, *hsdS2* yielded two unique sequence cassettes, one of which was found in strains of serovar *S*. Paratyphi B in the SARA complex. *hsdS3* yielded three distinct cassette types within the alignment, all of which shared no homology with their counterparts. A cassette retained by SARA strain 56 showed homology to an *hsdS* variant in *E. coli*, suggesting that this sequence has resulted from HGT between these lineages. The other unique cassette, retained by SARA strains 49, 50, and 51, showed no homology to any other *hsd* sequence, indicating that it may be been transferred into *S*. Paratyphi B from a yet unidentified source. The examples of unique cassette formation within this gene reinforce the role that HGT has played in the intra-operon and intragenic evolution of the Type-1 R-M gene system. These data also reveal the exchangeable nature of *hsd* gene sequences in these loci as a result of HGT.

4.3 Key observations

i.   These findings demonstrate several instances for the three *hsd* loci encoding the type I R-M operon in *Salmonella* to be decoupled phylogenetically from the chromosomes of group I *Salmonella* strains (*i.e., mdh* and MLEE), suggesting that certain alleles from these genes have been shuffled by HGT between closely related *S. enterica* strains.

ii.   The *hsd* operon of S. enterica subspecies 1 appears to have undergone intra-operon structuring due to HGT, producing an evolutionary mosaic in the *hsd* region.

iii. The lack of homology within *hsdS* indicates that these specific segments may have been acquired from distantly related bacterial species. An aberrant GC content for *hsdS* of 41%, a value far removed from an average value for enteric bacterial genomes of 56%, reinforces this conclusion.

iv.   The data demonstrate that HGT has been a common occurrence in *hsd* gene evolution and point to a genetic compatibility among closely-related salmonellae for exchange of *hsd* alleles that appears to resemble a panmictic genetic structure among these closely related strains. This may explain, in part, why *Salmonella* known to share homologous genomes and common niches more freely exchange DNA.

5. Discussion and conclusions

In summary, substantial phylogenetic evidence has been presented for the horizontal transfer of *mutS* alleles within a pathogenically homogeneous group of subspecies I *Salmonella enterica* pathogens. Of note, is the observation that *mutS* clades appear to be undergoing homogenization within the 'Typhimurium' strain complex as a result of the repeated HGT of only a few preferred alleles. Moreover, examination of R-M loci revealed that subspecies I *Salmonella* readily exchange *hsd* genes. These findings support the notion that R-M compatibility may be, in part, responsible for the substantial tolerance of HGT and recombined DNA between subspecies I strains.

An overwhelming body of evidence has been compiled that documents the reticulate evolutionary nature of the *Salmonella mutS* gene and its surrounding sequences. In an analysis of nearly 200 different strains documented here and in several previous reports over the past decade (LeClerc et al., 1998; Brown et al., 2002; 2003; Kotewicz et al., 2002; 2003), our laboratory has demonstrated the extent, the chromosomal effects, and the evolutionary history of HGT events that have scrambled this part of the genome in *S. enterica*. Exhaustive phylogenetic comparisons have been brought to bear on *mutS* sequences using various chromosomal markers including MLEE, rDNA, several individual housekeeping genes, and an *a priori* prior agreement MLST data based on concatenation of a three-gene supermatrix (Brown et al., 2002). Puzzling then was a later report that argued a more nonremarkable evolutionary pattern for *mutS* stating, "*mutS* is not more recombinogenic than the other genes" (Octavia and Lan, 2006). The authors based this conclusion solely on a modest 15 strain set of subspecies I salmonellae. Albeit, it remains to be seen to what extent additional homogeneous *Salmonella* populations retain the phylogenetic vestiges of horizontally transferred *mutS* and *hsd* alleles. Whatever the final outcome, it is apparent that horizontal transfer has played a prominent role in the current evolutionary structure of *mutS* and many other genes with virulence, resistance, stress-response, and general housekeeping function, all underscoring recombination as a key mechanism in the generation of genetic diversity among these closely related salmonellae.

Roughly two decades ago, *Salmonella enterica* was regarded as one of only a few eubacterial species that maintained a "truly clonal" evolutionary structure (Selander *et al.*, 1990; 1996; Reeves et al., 1989). Today, armed with whole-genomic analysis, it is now clear that horizontal transfer has shaped and honed unique evolutionary histories for numerous genes, operons, and islands within the *Salmonella* chromosome. With the complete genome sequences of dozens of *Salmonella* now available and countless more underway, such analyses of congruence should aid in determining the extent to which recombination has disrupted clonality throughout the entire *Salmonella* chromosome. Certainly, a greater recognition of precisely how HGT has forged the genomes of *Salmonella* pathogens should enhance the accuracy of risk assessment strategies for these bacteria as well as provide avenues for better detection and characterization of this devastating foodborne pathogen.

6. Acknowledgments

The authors would like to thank Mr. David Weingaertner for repeated and excellent graphical assistance. We would also like to acknowledge Drs. M. Kotewicz, B. Li, A. Mukherjee, A. Shifflet, J. Zheng, S. Jackson, and J. Meng for numerous helpful discussions over many years.

7. References

Allard, M.W., Farris, J.S., and Carpenter, J. M. (1999). Congruence among mammalian mitochondrial genes. Cladistics 15, 75-84.

Barcus, V.A., Titheradge, A.J.B., and Murray, N.E., (1995). The diversity of alleles at the *hsd* locus in natural populations of *Escherichia coli*. Genetics 140, 1187-1197.

Bäumler, A.J. (1997). The record of horizontal gene transfer in *Salmonella*. Trends Microbiol. 5, 318-322.

Bell, R.L., Gonzalez-Escalona, N., Stones, R., and Brown, E.W. (2011). Phylogenetic evaluation of the 'Typhimurium' complex of Salmonella strains using a seven-gene multi-locus sequence analysis. Infect. Genet. Evol. 11, 83-91.

Beltran, P., Plock, S.A., Smith, N.H., Whittam, T.S., Old, D.C., and Selander, R.K. (1991). Reference collection of strains of the *Salmonella typhimurium* complex from natural populations. J. Gen. Microbiol. 137, 601-606.

Boyd, E. F., and Hartl, D. L. (1999). Analysis of the type 1 pilin gene cluster *fim* in *Salmonella*: Its distinct evolutionary histories in the 5' and 3' regions. J. Bacteriol. 181, 1301-1308.

Boyd, E. F., Li, J., Ochman, H., and Selander, R. K. (1997). Comparative genetics of the *inv-spa* invasion complex of *Salmonella enterica*. J. Bacteriol. 179, 1985-1991.

Boyd, E. F., Nelson, K., Wang, F-S., Whittam, T. S., and Selander, R. K. (1994). Molecular genetic basis of allelic polymorphism in malate dehydrogenase (*mdh*) in natural populations of *Escherichia coli* and *Salmonella enterica*. Proc. Natl. Acad. Sci. (USA) 91, 1280-1284.

Boyd, E. F., Wang, F- S., Beltran, P., Plock, S. A., Nelson, K., and Selander, R. K. (1993). *Salmonella* reference collection B (SARB): strains of 37 serovars of subspecies I. J. Gen. Microbiol. 139, 1125-1132.

Boyd, E. F., Wang, F.-S., Whittam, T. S., and Selander, R.K. (1996). Molecular genetic relationships of the salmonellae. Appl. Env. Microbiol. 62:804-808.

Brown, E.W., Kotewicz, M.L., and Cebula, T.A. (2002). Detection of recombination among *Salmonella enterica* strains using the incongruence length difference test. Mol. Phylogenet. Evol. 24, 102-120.

Brown, E.W., LeClerc, J.E., Kotewicz, M.L., and Cebula, T.A. (2001a). Three R's of Bacterial Evolution: How Replication, Repair, and Recombination Frame the Origin of Species. Environ. Molec. Mutagen. 38, 248-260.

Brown, E.W., LeClerc, J.E., Li, B., Payne, W.L., and Cebula, T.A. (2001b). Phylogenetic evidence for horizontal transfer of *mutS* alleles among naturally occurring *Escherichia coli* strains. J. Bacteriol. 183, 1631-1644.

Brown, E.W., Mammel, M.K., LeClerc, J.E., and Cebula, T.A. (2003). Limited boundaries for extensive horizontal gene transfer among *Salmonella* pathogens. Proc. Natl. Acad. Sci. (USA) 100, 15676-15681.

Bullas, L.R., Colson, C., and Neufeld, B. (1980). Deoxyribonucleic acid restriction and modification systems in *Salmonella*: chromosomally located systems of different serotypes. J. Bacteriol. 141, 275-292.

CDC (2007). Preliminary FoodNet data on the incidence of infection with pathogens transmitted commonly through food--10 states, 2006. MMWR Morb. Mortal. Weekly Rep. 56, 336-339.

CDC (2008). Cases infected with the outbreak strain of *Salmonella* Saintpaul, United States, by state, as of August 25, 2008 9pm EDT. Centers for Disease Control and Prevention, *www.cdc.gov*.

Cebula, T. A., and LeClerc, J. E. (1997). Hypermutability and homeologous recombination: ingredients for rapid evolution. Bull. Inst. Past. 95, 97-106.

Cebula, T. A., and LeClerc, J. E. (2000). DNA repair and mutators: effects on antigenic variation and virulence of bacterial pathogens, In *"Virulence Mechanisms of Bacterial Pathogens, 3rd. Ed."* (K. A. Brogden et al., Eds.), pp.143-159, ASM Press, Washington DC.

Cebula, T. A., Levy, D. D., and LeClerc, J. E. (2001). Mutator bacteria and resistance development. In *"Antibiotic Resistance and Antibiotic Development "* (D. Hughes and D. Andersson, Eds.), pp. 107-116, Harwood Academic Publ., Amsterdam.

Christensen, H., and Olsen, J. E. (1998). Phylogenetic relationships of Salmonella based on DNA sequence comparison of *atpD* encoding the beta subunit of ATP synthase. FEMS Microbiol. Lett. 161, 89-96.

Denamur, E., Lecointre, G., Darlu, P., Tenaillon, O., Acquaviva, C., Sayada, C., Sunjevaric, I., Rothstein, R., Elion, J., Taddei, F., Radman, M., and Matic, I. (2000). Evolutionary implications of the frequent horizontal transfer of mismatch repair genes. Cell 103, 711-721.

Dykhuizen, D.E., and Green, L. (1991). Recombination in *Escherichia coli* and the definition of biological species. J. Bacteriol. 173, 7257-7268.

Farris, J.S., Kallersjo, M., Kluge, A.G., and Bult, C. (1995). Testing significance of incongruence. Cladistics 10, 315-319.

Felsenstein, J. (1985). Confidence limits on phylogenies: an approach using the bootstrap. Evolution 39, 783-791.

Forey, P.L., Humphries, C.J., Kitching, I., Scotland, R.W., Seibert, D.J., and Williams, D. (1992). *Cladistics: A practical course in systematics.* Clarendon Press, Oxford.

Fricke, W.F., Mammel, M.K., McDermott, P.F., Tartera, C., White, D.G., Leclerc, J.E., Ravel, J., Cebula, T.A. (2011) Comparative genomics of 28 *Salmonella enterica* isolates: evidence for CRISPR-mediated adaptive sublineage evolution. J Bacteriol. 193, 3556-68.

Funchain, P., Yeung, A., Stewart, J. L., Lin, R., Slupska, M. M., and Miller, J. H. (2000). The consequences of growth of a mutator strain of Escherichia coli as measured by loss of function among multiple gene targets and loss of fitness. Genetics 154, 959-970.

Galan, J. E., and Curtiss, R. (1989). Cloning and molecular characterization of genes whose products allow *Salmonella typhimurium* to penetrate tissue culture cells. Proc. Natl. Acad. Sci. (USA) 86, 6383-6387.

Galan, J. E., and Collmer, A. (1999). Type III secretion machines: bacterial devices for protein delivery into host cells. Science 284, 1322-1328.

Groisman, E. A., and Ochman, H. (2000). The path to *Salmonella*. ASM News 66, 21-27.

Groisman, E. A., Saier, M. H. Jr, and Ochman, H. (1992). Horizontal transfer of a phosphatase gene as evidence for mosaic structure of the *Salmonella* genome. EMBO J. 11, 1309-1316.

Haber, L. T., Pang, P. P., Sobell, D. I., Mankovich, J. A., and Walker, G. C. (1988). Nucleotide sequence mismatch of the *Salmonella typhimurium mutS* gene required for mismatch repair: homology of *mutS* and *hexA* of *Streptococcus pneumoniae*. J. Bacteriol. 170, 197-202.

Haber, L. T., and Walker, G. C. (1991). Altering the conserved nucleotide-binding motif in the *Salmonella typhimurium MutS* mismatch repair protein affects both its ATPase and mismatch binding activities. EMBO J. 10, 2707-2715.

Hacker, J., and Kaper, J. B. (2000). Pathogenicity islands and the evolution of microbes. Ann. Rev. Microbiol. 54, 641-679.

Hall, R.M. (2010). *Salmonella* genomic islands and antibiotic resistance in *Salmonella enterica*. Future Microbiol. 5, 1525-1538.

Hohmann, E. L. (2001). Nontyphoidal salmonellosis. Clin. Infect. Dis. 32, 263-269.

Kelly, B.G., Vespermann, A., and Bolton, D.J. (2009). The role of horizontal gene transfer in the evolution of selected foodborne bacterial pathogens. Food Chem. Toxicol. 47, 951-968.

Kingsley, R. A., and Baumler, A. J. (2000). Host adaptation and the emergence of infectious disease: the *Salmonella* pathogen. Mol. Microbiol. 36, 1006-1014.

Kotewicz, M.L., Brown, E.W., LeClerc, J.E., and Cebula, T.A. (2003). Genomic variability among enteric pathogens: the case of the *mutS–rpoS* intergenic region. Trends Microbiol. 11, 2-6.

Kotewicz, M.L., Li, B., Levy, D.D., LeClerc, J.E., Shifflet, A.W., and Cebula, T.A. (2002). Evolution of multi-gene segments in the *mutS-rpoS* intergenic region of *Salmonella enterica* serovar Typhimurium LT2. Microbiology 148, 2531-2540.

Lan, R., Reeves, P.R., and Octavia, S. (2009). Population structure, origins and evolution of major *Salmonella enterica* clones. Infect. Genet. Evol. 5, 996-1005.

Lawrence, J. G. and Roth, J. R. (1996). Selfish operons: horizontal transfer may drive the evolution of gene clusters. Genetics 143, 1843-1860.

LeClerc, J.E., Li, B., Payne, W.L., and Cebula, T.A. (1996). High mutation frequencies among *Escherichia coli* and *Salmonella* pathogens. Science 274, 1208-1211.

LeClerc, J. E., Li, B., Payne, W. L., and Cebula, T. A. (1999). Promiscuous origin of a chimeric sequence in the *Escherichia coli* O157:H7 genome. J. Bacteriol. 181, 7614-7617.

LeClerc, J. E., Payne, W. L., Kupchella, E., and Cebula, T. A. (1998). Detection of mutator subpopulations in *Salmonella typhimurium* LT2 by reversion of his alleles. Mutation. Res. 400, 89-97.

Lecointre, G. L. Rachdi, P. Darlu, and Denamur, E. (1998). *Escherichia coli* molecular phylogeny using the incongruence length difference test. Mol. Biol. Evol. 15, 1685-1695.

Li, J., Nelson, K., McWhorter, A. C., Whittam, T. S., and Selander, R. K. (1994). Recombinational basis of serovar diversity in *Salmonella enterica*. Proc. Natl. Acad. Sci. (USA) 91, 2552-2556.

Li, J., Ochman, H., Groisman, E. A., Boyd, E. F., Solomon, F., Nelson, K., and Selander, R. K. (1995). Relationship between evolutionary rate and cellular location among Inv/Spa invasion proteins of *Salmonella enterica*. Proc. Natl. Acad. Sci. (USA) 92, 7252-7256.

Liu, S. L. and Sanderson, K. E. (1996). Highly plastic chromosomal organization in *Salmonella typhi*. Proc. Natl. Acad. Sci. (USA). 93, 10303-10308.

Matic, I., Radman, M., Taddei, F., Picard, B., Doit, C., Bingen, E., Denamur, E., and Elion, J. (1997). Highly variable mutation rates in commensal and pathogenic *Escherichia coli*. Science 277, 1833-1834.

McQuiston, J.R., Herrera-Leon, S., Wertheim, B.C., Doyle, J., Fields, P.I., Tauxe, R.V., and Logsdon, J.M. Jr. (2008). Molecular phylogeny of the salmonellae: relationships among *Salmonella* species and subspecies determined from four housekeeping genes and evidence of lateral gene transfer events. J. Bacteriol. 190, 7060-7067.

Murray, N.E. (2000). Type I restriction systems: Sophisticated molecular machines (a legacy to Bertani and Weigle). Microbiol. Molec. Biol. Rev. 64, 412-434.

Naderer, M., Brust, J.R., Knowle, D., and Blumenthal, R.M. (2002). Mobility of a Restriction-Modification System Revealed by Its Genetic Contexts in Three Hosts. J. Bacteriol. 184, 2411-2419.

Nelson, K., and Selander, R. K. (1992). Evolutionary genetics of the proline permease gene (*putP*) and the control region of the proline utilization operon in populations of *Salmonella* and *Escherichia coli*. J. Bacteriol. 174, 6886-6895.

Nelson, K. and Selander, R. K. (1994). Intergeneric transfer and recombination of the 6-phosphogluconate dehydrogenase gene (*gnd*) in enteric bacteria. Proc. Natl. Acad. Sci. (USA) 91, 10227-10231.

Nelson, K., Wang, F -S., Boyd, E. F., and Selander, R. K. (1997). Size and sequence polymorphism in the isocitrate dehydrogenase kinase/phosphatase gene (*aceK*) and flanking regions in *Salmonella enterica* and *Escherichia coli*. Genetics 147, 1509-1520.

Ochman, H., Lawrence, J. G., and Groisman, E. A. (2000). Lateral gene transfer and the nature of bacterial innovation. Nature 405, 299-304.

Octavia, S., and Lan, R. (2006). Frequent recombination and low level of clonality within *Salmonella enterica* subspecies I. Microbiol. 152, 1099-1108.

Patel, P.H., and Loeb, L.A. (2000). DNA polymerase active site is highly mutable: evolutionary consequences. Proc. Natl. Acad. Sci. USA 97, 5095–5100.

Porwollik, S., and McClelland, M. (2003). Lateral gene transfer in *Salmonella*. Microbes and Infect. 5, 977-989.

Radman, M., Matic, I., and Taddei, F. (1999). Evolution of evolvability. Ann. NY Acad.Sci. 870, 146-155.

Reeves, M. W., Evins, G. M., Heiba, A. A., Plikaytis, B. D., and Farmer III, J. J. (1989). Clonal nature of *Salmonella typhi* and its genetic relatedness to other salmonellae as shown by multilocus enzyme electrophoresis, and proposal of *Salmonella bongori* comb. Nov. J. Clin. Microbiol. 27, 313-320.

Selander, R. K., Li, J., and Nelson. K. (1996). Evolutionary genetics of *Salmonella enterica*. In "*Escherichia coli and Salmonella Cellular and Molecular Biology*," (F. C. Neidhardt, Ed.), pp.2691-2707, ASM Press, Washington, DC.

Selander, R.K., Beltran, P., Smith, N.H., Helmuth, R., Rubin, F.A., Kopecko, D.J., Ferris, K., Tall, B.D., Cravioto, A., and Musser, J.M. (1990). Evolutionary genetic relationships of clones of *Salmonella* serovars that cause human typhoid and other enteric fevers. Infect. Immun. 58, 2262-2275.

Sharp, P.M., Kelleher, J.E., Daniel, A.S., Cowan, G.M., and Murray, N.E. (1992). Roles of selection and recombination in the evolution of type-I restriction-modification systems in enterobacteria. Proc. Natl. Acad. Sci. (USA) 89, 9836-9840.

Sukhnanand, S., Alcaine, S., Warnick, L.D., Su, W.L., Hof, J., Craver, M.P., McDonough, P., Boor, K.J., and Wiedmann, M. (2005). DNA sequence-based subtyping and evolutionary analysis of selected *Salmonella enterica* serotypes. J. Clin. Microbiol. 43, 3688-3698.

Swofford, D.L. (2002). Phylogenetic analysis using parsimony (PAUP* 4.0b) program and documentation. The Smithsonian Institution, Washington, DC.

Thampapillai, G., Lan, R., and P. R. Reeves. (1994). Molecular evolution of the *gnd* locus of *Salmonella enterica*. Mol. Biol. Evol. 11, 813-828.

Thompson, J. D., Gibson, T. J., Plewniak, F., Jeanmougin, F., and Higgins, D. G. (1997). The CLUSTAL X windows interface: flexible strategies for multiple sequence alignment aided by quality analysis tools. Nucl. Acids Res. 25, 4876-4882.

Tröbner, W., and Piechocki, R. (1984). Selection against hypermutability in *Escherichia coli* during long term selection. Molec. Gen. Genet. 198, 177-178.

14

The Importance of Mobile Genetic Elements in the Evolution of *Salmonella*: Pathogenesis, Antibiotic Resistance and Host Adaptation

Claudia Silva, Magdalena Wiesner and Edmundo Calva
Departamento de Microbiología Molecular, Instituto de Biotecnología
Universidad Nacional Autónoma de México, Cuernavaca, Morelos
Mexico

1. Introduction

Since the divergence from *Escherichia coli*, more than 100 million years ago, *Salmonella* has acquired by lateral gene transfer a repertoire of genes that confers a set of physiological features that define its particular ecological niche (Ochman and Groisman 1994; Ochman et al., 2000). Some of these chromosomally-encoded genes can be considered as part of the "core genome" of *Salmonella* (i. e. genes present in all the strains), whereas some other chromosomally-encoded genes are part of the "accessory genome" (i. e. genes present in some of the strains) along with mobile genetic elements such as plasmids, bacteriophages, transposons and integrons. In this chapter we review the role that mobile genetic elements have played in *Salmonella* evolution, particularly in pathogenicity attributes, antibiotic resistance and host adaptation.

2. Pathogenicity islands

Pathogenicity islands are large genomic regions that are present in pathogenic variants but less frequently present in closely related non-pathogenic bacteria. They often carry virulence-associated genes, have a G+C content that differs from that of the rest of the chromosome, are frequently associated with tRNA genes and are flanked by repeated sequences (Dobrindt et al., 2004). *Salmonella* pathogenicity islands (SPIs) are large gene cassettes within the *Salmonella* chromosome that encode determinants responsible for establishing specific interactions with the host, and are required for bacterial virulence in a given animal. Like other pathogenicity islands, SPIs generally have a G+C content lower (between 37 and 47%) than the rest of the bacterial chromosome (about 52%), and are often inserted into tRNA genes. Therefore, SPIs have likely been acquired by horizontal transfer from bacteriophage or plasmids of unknown origin, and they are highly conserved between the different *Salmonella* serovars. It is sometimes unclear how certain DNA regions are designated as pathogenicity islands. Small DNA regions (often single genes), which are acquired horizontally, are numerous and their distinction from pathogenicity islands is in some cases arbitrary (Marcus et al., 2000). More than twenty SPIs have been described (Sabbagh et al., 2010). Most SPIs have become part of the core set of genes of *S. enterica* and encode species-specific traits. A smaller subset of SPIs is limited to certain subspecies or serovars. These SPIs harbor genes associated with DNA mobility and are likely to represent more recent acquisitions (Hensel 2004). SPI1 and SPI2 both encode type III secretion systems (TTSS), which mediate the respective virulence phenotype by translocating bacterially-encoded proteins into the host cell cytoplasm (Hansen-Wester and Hensel 2001). TTSS are used by many bacterial pathogens to deliver virulence factors to the host cell and interfere with or subvert normal host cell signaling pathways. They consist of many components, including more than twenty proteins, some of which are homologous to those involved in flagellar assembly. Effector proteins generally require specific chaperones which prevent incorrect folding, degradation and premature association, and may even aid delivery of the effector into host cells. These systems are highly regulated, and proteins are only secreted when the bacteria sense specific environmental cues (Marcus et al., 2000; Cornelis 2006). A remarkable observation is the functional interaction of SPI1 and SPI2 with further loci encoding effector proteins, including SPI5 and some mobile DNA elements (Hensel 2004).

3. *Salmonella* genomic island

A chromosomal island called *Salmonella* genomic island 1 (SGI1) was initially described in the epidemic multiple-drug-resistant (MDR) *Salmonella enterica* serovar Typhimurium phage-type DT104 (Boyd et al., 2001). In this respect, phage-typing is carried out by infecting a *Salmonella* isolate with a number of phages listed in the phage-typing scheme for a specific serovar (Kropinski et al., 2007). The present phage-typing scheme for Typhimurium consists of 34 phages and identifies 207 phage types, referred to as definitive types (DT) (Anderson et al., 1977). Since the original identification of SGI1 in Typhimurium DT104, variants of SGI1 have been described in a wide variety of *Salmonella* serovars (Levings et al., 2005). The 43 kb SGI1 contains 44 open reading frames (ORFs), many of them without homology to known gene sequences. In all the serovars, SGI1 showed the same chromosomal location. In the first part of the island, a number of ORFs showing homology to plasmid-related genes are present. The 13 kb-antibiotic resistance gene cluster is located near the 3′ end of SGI1 (Levings et al., 2005). The MDR cluster confers resistance to streptomycin and spectinomycin, sulfonamides, chloramphenicol and florfenicol, tetracyclines, and beta-lactam antibiotics. The G+C content for the MDR cluster is 59%, while for the rest of the SGI1 it is 49%, as compared with 52% for the Typhimurium chromosome, suggesting a mosaic structure (Boyd et al., 2001). SGI1 is an integrative mobilizable element which contains a complex class 1 integron (see below) named In104, located within the antibiotic resistance cluster located at the 3′ end of the island (Boyd et al., 2001; Doublet et al., 2005; Mulvey et al., 2006). In 2005, Doublet et al. reported that SGI1 could be conjugally transferred from an *S. enterica* donor to *E. coli* recipient strains where it integrated into the recipient chromosome in a site-specific manner (Doublet et al., 2005). First, an extra-chromosomal circle of SGI1 was formed, and this circular intermediate was transferred in the presence of an IncC helper plasmid, which provided the mating apparatus. This study demonstrated that the mobilization of SGI1 probably contributes to the spread of antibiotic resistance genes between *S. enterica* serovars and possibly to other bacterial pathogens (Doublet et al., 2005). As predicted by Doublet et al. (2005), a variant of SGI1 has been reported in *Proteus mirabilis* (Boyd et al., 2008). Moreover, a recent report showing the ability of IncA/C plasmids to mobilize SGI1 has implications for the worldwide spread of these MDR elements (Douard et al., 2010).

The Importance of Mobile Genetic Elements in the Evolution of Salmonella: Pathogenesis, Antibiotic Resistance and Host Adaptation

255

4. Antimicrobial resistance and virulence plasmids

A substantial amount of the variation in bacteria is due to the presence of plasmids (Levin and Bergstrom 2000). Plasmids are part of the flexible genome, which is defined by the high plasticity and modularity of its genetic elements and high rates of gene acquisition and loss (Heuer et al., 2008). They are typically composed of conserved backbone modules coding for replication, maintenance and transfer functions as well as variable accessory modules. The capture of genetic modules by plasmid backbones can increase phenotypic diversity and thereby increment the chances of responding to uncertain environmental changes or exploit an opportunity for niche expansion (Souza and Eguiarte 1997; Frost et al., 2005; Heuer et al., 2008; Norman et al., 2009). Often, antimicrobial or heavy metal resistance, or virulence factors that allow their bacterial host to adapt to changing environments are encoded by plasmids. Plasmids are classified according to incompatibility (Inc) groups, that are based on the inability of plasmids with the same replication or segregation mechanisms to co-exist in the same cell (Couturier et al., 1988). Plasmids of *Salmonella enterica* vary in size from 2 to more than 200 kb. The best described plasmids are the so-called virulence plasmids present in some serovars. Another group of high molecular weight plasmids are responsible for antibiotic resistance, which are in most of the cases conjugative, contributing to the spread of genes in bacterial populations (Rychlik et al., 2006). The low molecular weight multi-copy plasmids are widespread in *Salmonella*, but are less studied and are referred as cryptic, although some of them have been shown to increase resistance to bacteriophage infection due to the presence of modification systems (Rychlik et al., 2006).

Eight *Salmonella* enterica serovars harbour a large (50-285 kb) plasmid named the *Salmonella* virulence plasmid, containing the *spv* operon, which is a major determinant of virulence in their specific hosts (Gulig et al., 1993; Chiu et al., 2000; Fierer and Guiney 2001). In addition to the *spv* operon, other plasmid genes are involved in virulence. The *rsk* and *rck* genes are required for serum resistance, and *traT*, a surface exclusion protein for plasmid transfer, is also responsible for serum resistance (Chu and Chiu 2006). Within a single serovar some strains can carry the virulence plasmid while others not (Olsen et al., 2004). Despite many common properties shared by these plasmids, each virulence plasmid seems to be specific to its serovar, but the outcome of the infection in different animal hosts may be variable. For example, Typhimurium strains that harbour the virulence plasmid are highly virulent to mouse, but there is lack of evidence of an association between the carriage of virulence plasmid and the bacteremia caused in humans (Chiu et al., 2000). Whether the virulence plasmid is necessary to produce systemic infections in humans has been subject of intense debate. Some authors claim that there is lack of evidence of an association between the carriage of the virulence plasmid and human bacteremia (Chiu et al., 2000). Other authors suggest that *spv* genes promote the dissemination of Typhimurium from the intestine (Fierer 2001). In recent reports contrasting results have been found. We studied more than 100 Typhimurium strains isolated from human and food-animal sources in Mexico, and found that only 30% of the strains harboured the plasmid (Wiesner et al., 2009). The presence of the virulence plasmid was significantly associated with human isolates, but only one of the six isolates recovered from patients with systemic infection had the virulence plasmid. Our data support the notion that the virulence plasmid has a role in host adaptation (Baumler et al., 1998); however, it was not consistent with the view that it is associated with systemic infection in humans (Wiesner et al., 2009). In a recent study, Litrup et al. (2010) analyzed 21 Typhimurium strains isolated from patients with mild and sever infections with the aim of correlate genomic content with the outcome of disease. They used a DNA microarray targeting 281 known virulence factors, and found that the presence or absence of the virulence plasmid did not correspond to disease symptoms (Litrup et al., 2010). On the other hand, Heithoff et al. (2008) found that all the Typhimurium strains isolated from animals or humans with bacteremia possessed the virulence plasmid, while 34% of the strains isolated from human gastroenteritis lacked the plasmid (Heithoff et al., 2008). These contrasting results highlight the complex nature of specific host-pathogen interactions, and call to avoid making generalizations since the diversity of environmental (biotic and abiotic), host, and bacterial genetic conditions may produce different outcomes.

Large antimicrobial resistance plasmids are of public health concern. The global scene is that the therapeutic options for MDR microbes are reduced, periods of hospital care are more extended and costly and, in some cases, the strains have also acquired increased virulence and enhanced transmissibility. Realistically, antibiotic resistance can be considered a virulence factor (Davies and Davies 2010). Resistance mechanisms are pandemic and create an enormous clinical and financial burden on health care systems worldwide (Davies and Davies 2010). The resistance genes found in *Salmonella* are closely related to, or are indistinguishable from, those found in other bacteria, including not only members of the *Enterobacteriaceae* but also distantly related bacteria. It is most likely that *Salmonella* acquired these genes from other bacteria, and probably *Salmonella* strains also play a role in the further dissemination of these resistance genes to other bacteria (Michael et al., 2006). Frequencies of conjugative transmission in nature are probably several orders of magnitude higher that those observed under laboratory conditions, and occur readily in networks of multi-host interactions (Dionisio et al., 2002; Sorensen et al., 2005; Davies and Davies 2010).

The IncA/C plasmids exemplify the problematic of resistance plasmids in *Salmonella*. They have attracted the attention of the research community due to their ability to acquire antimicrobial resistance traits and to mobilize across geographical and taxonomical borders (Fricke et al., 2009). Recent comparative studies have addressed the evolutionary relationships among the IncA/C plasmids from *Salmonella enterica*, *Escherichia coli*, *Yersinia pestis*, *Yersinia ruckeri*, *Vibrio cholera*, *Photobacterium damselae* and *Aeromonas salmonicida* (Welch et al., 2007; Kim et al., 2008; McIntosh et al., 2008; Pan et al., 2008; Fricke et al., 2009; Call et al., 2010). The genomic comparison of seven IncA/C plasmids showed that these plasmids share a common backbone, including the origin of replication and a conjugative plasmid transfer system (Welch et al., 2007; Fricke et al., 2009). Several loci containing antimicrobial resistance determinants are distributed along the plasmids, and are integrated at few sites within the conserved plasmid backbone; they are generally located as resistance gene arrays, composed of resistance genes and mobile genetic elements such as insertion sequences, transposons or integrons (Fricke et al., 2009). For example, in the IncA/C plasmids of *Yersinia pestis* and *Salmonella* Newport, a Tn21 transposon is inserted in a similar location but some nucleotide divergence is evident and its orientation is reversed (Welch et al., 2007; Fricke et al., 2009). These studies suggest an evolutionary model in which each IncA/C plasmid diverged from a common ancestor, through processes of stepwise integration events of horizontally-acquired resistance genes arrays (Welch et al., 2007; Fricke et al., 2009).

Over the last decade, increasing attention has been focused on plasmids that harbour the antimicrobial resistance gene bla_{CMY-2}, which encodes an AmpC-type beta-lactamase that hydrolyzes third-generation cephalosporins (Bauernfeind et al., 1996; Zhao et al., 2001;

The Importance of Mobile Genetic Elements in the Evolution of Salmonella: Pathogenesis, Antibiotic Resistance
and Host Adaptation

257

Carattoli et al., 2002; Wiesner et al., 2009; Wiesner et al., 2011). In *Salmonella enterica*, bla_{CMY-2} is frequently carried by IncA/C or IncI1 plasmids (Bauernfeind et al., 1996; Carattoli et al., 2002; Hopkins et al., 2006; Lindsey et al., 2009). In a recent study, Call et al. (2010) analyzed five *E. coli* and *Salmonella* Newport IncA/C plasmids carrying bla_{CMY-2}, and showed that although they share a common ancestor with the *Yersinia* and *Photobacterium* plasmids, they are genetically distinct (Call et al., 2010). In a population study we found that IncA/C plasmids were associated to the Mexican Typhimurium ST213 genotype. We determined that the bla_{CMY-2} gene was carried in IncA/C plasmids, and genetic variability was observed using a plasmid typing scheme, targeting ten conserved regions in IncA/C plasmids (Wiesner et al., 2009; Wiesner et al., 2011). The Typhimurium bla_{CMY-2}-bearing IncA/C plasmids possessed most of the accessory elements found in other *Salmonella* and *E. coli* plasmids (Call et al., 2010), but also more than half contained a class 1 integron (*dfrA12-orfF-aadA2*). The screening of the total Mexican Typhimurium population showed the presence of another IncA/C plasmid harboured by ST213 strains, yet lacking bla_{CMY-2}. These plasmids also carried antibiotic resistance determinants, but they shared only three of the ten genetic markers used to study the IncA/C plasmids, and were smaller than the bla_{CMY-2}-bearing IncA/C plasmids (100 vs. 150-160 kb). Nevertheless, the nucleotide sequences of the regions shared with the bla_{CMY-2}-bearing IncA/C plasmids were identical, suggesting that the bla_{CMY-2}-bearing plasmids could be the result of the insertion of DNA modules into this smaller precursor plasmid (Wiesner et al., 2009; Wiesner et al., 2011). The general agreement from the analysis of the genetic structure of the IncA/C group,is that plasmid evolution progresses faster through the insertion/deletion of DNA stretches rather than by point mutations (Welch et al., 2007; Kim et al., 2008; Fricke et al., 2009; Call et al., 2010; Wiesner et al., 2011).

Large resistance plasmids circulate among microbial populations in distinct environmental niches, even in the absence of antibiotic selective pressure. In other environments the target of the selective pressure could be, for example, mercury resistance carried by many transposons, such as Tn*21* (Liebert et al., 1999; McIntosh et al., 2008). Regardless of the primary selective agent, the complete battery of resistance determinants will be maintained, imposing a global health risk (Liebert et al., 1999; Frost et al., 2005; Welch et al., 2007; McIntosh et al., 2008; Pan et al., 2008; Davies and Davies 2010). Another worrisome situation is the emergence of *Salmonella* virulence-(antibiotic) resistance plasmids. Several studies had reported large hybrid virulence-resistance plasmids in serovars Typhimurium, Choleraesuis and Enteritidis, isolated from Spain, Italy, Czech Republic, Taiwan and the United Kingdom (Chu et al., 2001; Guerra et al., 2001; Guerra et al., 2002; Guerra et al., 2004; Villa and Carattoli 2005; Chu and Chiu 2006; Herrero et al., 2008a; Herrero et al., 2008b; Hradecka et al., 2008; Herrero et al., 2009; Rodriguez et al., 2011). In some of the studies it was demonstrated that the hybrid plasmids were conjugative, which may lead to their spread to new recipients and allow the co-selection of the antibiotic and virulence genes, representing a hazard to human and animal health (Fluit 2005).

5. Integrons

Integrons are assembly platforms that incorporate genes by site-specific recombination and convert them to functional genes by ensuring their correct expression. They are composed of three key elements: a gene encoding an integrase, a primary recombination site, and a promoter that directs the transcription of the captured genes. The integrase can recombine discrete units of circularized DNA known as "gene cassettes"; they are transcribed only when captured into an integron, since most of them lack a promoter. Integration occurs downstream of the resident promoter, at the primary recombination site, allowing the expression of the genes in the cassette (Figure 1). The integron inserted gene cassettes share specific structural characteristics and contain an imperfect inverted repeat at the 3′ end, called the "59-base element". This site functions as a secondary recognition site for the site-specific integrase, and can further integrate gene cassettes. The ability to capture disparate individual genes and to physically link them in arrays suitable for co-expression is a trait unique to integrons, and theoretically facilities the rapid evolution of new phenotypes (Stokes and Hall 1989; Recchia and Hall 1995; Fluit and Schmitz 1999; Holmes et al., 2003; Fluit and Schmitz 2004; Mazel 2006; Boucher et al., 2007; Joss et al., 2009).

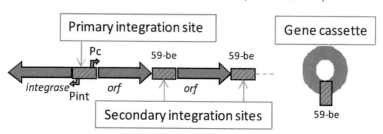

Fig. 1. Diagrammatic representation of the basic features of an integron. Integrons consist of the gene for the integrase, the promoters for the expression of the integrase (Pint) and the gene cassettes (Pc), and the primary recombination site, where the cassettes are integrated. Gene cassettes consist of a single promoter-less gene and a recombination site known as a 59-base element (59-be), which functions as a secondary recognition site for the site-specific integrase, and can further integrate gene cassettes (Stokes and Hall 1989; Hall and Collis 1995; Levesque et al., 1995). An integron carrying an array of two inserted cassettes, and a free circularized gene cassette is shown.

Analyses on the diversity of gene cassettes in environmental samples have shown a great diversity of predicted genes, suggesting that essentially any DNA-encoded function may be contained within a gene cassette (Stokes et al., 2001; Holmes et al., 2003). However, the main part of the gene cassettes found in mobile integrons from cultured bacteria contain antibiotic-resistance determinants (Levesque et al., 1995; Fluit and Schmitz 1999, 2004). Class 1 integrons are found extensively in clinical isolates, and most of the known antibiotic resistance gene cassettes belong to this class (Fluit and Schmitz 2004). Although gene cassettes conferring resistance to nearly every major class of antibiotics have been identified, there are some antibiotic resistance determinants that are preferentially associated to integrons, such as streptomycin, trimethoprim, sulfafurazole, and the early aminoglycosides (White et al., 2001; Fluit and Schmitz 2004). Moreover, it has been reported that MDR is associated significantly with the presence of integrons (Leverstein-van Hall et al., 2003; Fluit 2005; Wiesner et al., 2009). Integrons themselves are not mobile although they can be part of transposons, which are capable of moving from one carrier replicon to another (Fluit and Schmitz 1999; Liebert et al., 1999). A well-known example are the integrons found in Tn21 and related transposons. These transposons generally are located on plasmids, which further enhances the spread of gene cassettes (Fluit and Schmitz 1999; Liebert et al., 1999). The association of integrons with mobile elements and resistance genes has led to their rapid dispersal among various bacteria found in environments exposed to antibiotics (Martinez-Freijo et al., 1999; White et al., 2001; Boucher et al., 2007).

The Importance of Mobile Genetic Elements in the Evolution of Salmonella: Pathogenesis, Antibiotic Resistance
and Host Adaptation

259

Since the discovery of the importance of integrons in the dissemination of antibiotic resistance, many studies have addressed the presence of integrons in *Salmonella*. At the moment of writing there were more than 300 research papers regarding integrons in *Salmonella* (http://www.ncbi.nlm.nih.gov/pubmed). An overview of these studies shows that integrons have been detected in many different *Salmonella* serovars (i. e. Agona, Albany, Anatum, Braenderup, Branderburg, Bredeny, Brikama, Derby, Dublin, Emek, Enteritidis, Eppendorf, Goldcoast, Grumpensis, Hadar, Haifa, Heildelberg, Infantis, Javiana, Kedougou, Kentucky, Kingston, Krefekd, Mbandaka, Muenster, Newport, Panama, Paratyphi B, Rissen, Rough, Saintpaul, Schwarzengrund, Stanley, Senftenberg, Tees, Tshiongwe, Typhimurium, Virchow, Weltevreden, Wien and Worthington), that were isolated from diverse sources (i. e. animal feed, beef, chicken, camel, environment, feline, foodstuff, goat, horse, human, milk, pork and turkey), and from countries all around the world (i. e. Albania, Algeria, Brazil, China, Chile, England, Ethiopia, Germany, Great Britain, Iran, Ireland, Italy, Japan, Lithuania, Mexico, Portugal, Slovak Republic, Spain, Thailand, The Netherlands, United States and Vietnam). Among the most studied cases are the chromosomally-located integrons present in the SGI1. There is a great diversity of integron cassettes detected in *Salmonella*, most of them encoding antibiotic resistance genes, carried in a wide variety of cassette arrays. It is noteworthy that similar cassette arrays are found in different *Salmonella* serovars isolated from different sources and distant countries, and even in other bacterial species. For example, in the study of the integrons present in a Mexican Typhimurium population (Wiesner et al., 2009), we found that the two most abundant integrons (*dfrA12-orfF-aadA2* and *dfrA17-aadA5*) were reported for other *Salmonella* serovars (Anatum, Branderup, Brikama, Enteritidis, Mbandaka, Rissen and Saintpaul) and in other *Enterobacteriaceae* (Lindstedt et al., 2003; Antunes et al., 2006; Su et al., 2006; Molla et al., 2007; Zhao et al., 2007). More surprising was that these integrons were also found in different species of the Gram-positive genera *Staphylococcus* isolated in China (Xu et al., 2008), providing evidence of the successful spread of these integrons around the world and across bacterial phyla (Wiesner et al., 2009).

6. Bacteriophages

Although bacteriophages carrying antibiotic resistance genes have rarely been identified (Davies and Davies 2010), their role in the dissemination of virulence factors has been widely documented (Boyd and Brussow 2002). Bacteriophage-encoded virulence factors can convert their bacterial host, in a process known as phage conversion, from a non-pathogenic strain to a virulent strain or a strain with increased virulence. The phage-encoded proteins involved in lysogenic conversion provide mechanisms to invade host tissues, avoid host immune defenses, and damage host cells (Boyd and Brussow 2002). The extra genes present in prophage genomes which do not have a phage function, but may act as fitness factors for the bacteria, are termed "morons". The moron-encoded genes are not required for the phage life cycle. Their expression is controlled by an autonomous promoter and, thus, can be expressed while the rest of the prophage genes remain silent (Hendrix et al., 2000). Morons enhance phage replication indirectly since moron-encoded functions enhance fitness of the lysogen. This hypothesis provides the theoretical framework for phage-mediated horizontal transfer of fitness factors between bacteria (Hendrix et al., 2000). The ecological success of a lysogenic bacterium contributes to the dissemination of phage genes, providing a case of co-evolution of viruses and bacteria. It has been hypothesized that the driving force behind the emergence of new epidemic clones is the phage-mediated re-assortment of virulence and fitness factors, optimizing the *Salmonella*-host interaction (Mirold et al., 1999; Figueroa-Bossi et al., 2001; Brussow et al., 2004). The contribution of phages to *Salmonella* evolution is one of the best documented cases, and many phage-encoded virulence factors have been documented (Table 1). In particular, the functional biology of many phage-encoded genes has been studied in detail for Typhimurium strains.

Phage	Gene	Protein	Function in virulence
Fels-1	*sodC-III*	Superoxide dismutase	Intracellular survival
Fels-1	*nanH*	Neuraminidase	Intracellular survival
Gifsy-1	*gogB*	Type III efector	Involved in invasion
	gipA	IS-like	Critical for survival in Peyer´s patches
Gifsy-2	*sseI (gtgB)*	Type III efector	Involved in invasion
	sodC-I	Superoxide dismutase	Intracellular survival
	gtgE	Type III efector	Required for full virulence
Gifsy-3	*sspH1*	Type III efector	Involved in invasion
	pagJ	*phoPQ*-activated gene	Bacterial envelope for invasion
Fels-1 and Gifsy-2	*grvA*	Antivirulence gene	Decreases the pathogenicity in the host
P22	*gtrB*	Glucosyl transferase	O-antigen conversion
	gtrA	Flippase	O-antigen conversion
SopEΦ	*sopE*	Type III efector	Involved in invasion
ε34	*rfb*	Glucosyl transferase	Altering antigenicity

Table 1. Bacteriophage-encoded virulence factors of *Salmonella enterica* (Figueroa-Bossi et al., 2001; Boyd and Brussow 2002; Porwollik and McClelland 2003; Ehrbar and Hardt 2005; Kropinski et al., 2007).

Prophages contribute significantly to the diversity among *Salmonella* strains (Boyd and Brussow 2002), and different Typhimurium strains harbor distinct sets of prophages (Figueroa-Bossi et al., 2001; Mirold et al., 2001; Mmolawa et al., 2002). Most of them belong to the P2 family (SopEΦ, Fels-1, and Fels-2) or the lambda family (GIFSY-1, GIFSY-2, GIFSY-3 and P22). Several of the Typhiumurium prophages encode the so-called type three effector proteins, which are injected by the bacterium into animal cells via a type three secretion system (TTSS) (Ehrbar and Hardt 2005). These effector proteins manipulate signal transduction pathways of the cells, which provoke a strong intestinal inflammation and diarrhea. The SopE effector, encoded by SopEΦ, is one of the better studied cases. It is injected into the intestinal cells by the TTSS encoded by SPI-1, and its expression is co-regulated with other genes. The proper timing of SopE expression and delivery into the host cell depends on the regulatory circuits of SPI-1 (Mirold et al., 1999; Brussow et al., 2004; Ehrbar and Hardt 2005). Since the earlier studies, Mirold et al. (1999) demonstrated that SopEΦ is capable of infecting a range of Typhimurium strains (Mirold et al., 1999). In an experimental study, it was demonstrated that the lysogenic conversion of the laboratory Typhimurium strain ATCC14028 with SopEΦ provided increased enteropathogenicity compared with the wild-type strain (Zhang et al., 2002). Thus, it was shown that the horizontal transfer of phage-mediated genes may contribute to the emergence of more pathogenic epidemic clones. Moreover, Mirold et al. (2001) provided evidence for the

The Importance of Mobile Genetic Elements in the Evolution of Salmonella: Pathogenesis, Antibiotic Resistance and Host Adaptation

261

transfer of the SopE cassette between lambda and P2-like phages families (Mirold et al., 2001). They proposed that the transfer of virulence factors between phages increases the flexibility of the re-assortment of effector protein repertories, by circumventing restrictions imposed by immunity functions or the occupancy of the attachment sites by resident prophages (Mirold et al., 2001). By this mean, phages would contribute a great deal to the evolution of bacterial pathogens, and might explain the rapid emergence of new epidemic clones and the ability of *Salmonella* to adapt to a broad range of hosts.

The development of genome-based methods such as microarrays, and the tools to compare complete genome sequences, has opened a new era in the study of *Salmonella* evolution. Several studies have addressed the importance of bacteriophages in the evolution of *Salmonella*, and the role of prophage-encoded virulence factors in pathogenicity. Comparison of complete genomes have pointed out that the prophage content is one of the most important differences between genomes of *Salmonella* serovars, and specially among strains within a single serovar (McClelland et al., 2001; Parkhill et al., 2001; Porwollik et al., 2002; Porwollik et al., 2004; Thomson et al., 2004; Hermans et al., 2005; Cooke et al., 2007; Vernikos et al., 2007; Litrup et al., 2010). The detection of prophage sequences has been recently developed and proposed as a tool for the subtyping of strains (Hermans et al., 2006; Cooke et al., 2007). Recent population studies are supporting the notion that the prophage content is one the most dynamic part of the genome, indicating that phage integration/excision are frequent events shaping *Salmonella* genome evolution (Hermans et al., 2006; Cooke et al., 2007; Drahovska et al., 2007; Matiasovicova et al., 2007; Cooke et al., 2008; Litrup et al., 2010). These observations are in agreement with the phage remnants found in the genomes of *Salmonella* (McClelland et al., 2001; Parkhill et al., 2001; Porwollik et al., 2002; Chan et al., 2003; Porwollik and McClelland 2003; Matiasovicova et al., 2007).

7. Host adaptation

More than a decade ago, Baumler (1998) postulated that the genus *Salmonella* evolved in three phases (Baumler 1997; Baumler et al., 1998). The first phase involved acquisition of SPI1 by an ancestral lineage to all *Salmonella*, since it is present in all phylogenetic lineages of the genus *Salmonella* but absent from *E. coli* and other enterobacteria. In the second phase, the split of *S. enterica* from *S. bongori* involved the acquisition of SPI2, which is not present in *S. bongori* serovars (Ochman and Groisman 1996). Finally, the lineage of *S. enterica* branched into several phylogenetic groups. The formation of *S. enterica* subspecies *enterica* (I) involved a dramatic expansion in host range: while *S. bongori* and *S. enterica* subspecies II, IIIa, IIIb, IV, VI and VII are mainly associated with cold-blooded vertebrates, members of *S. enterica* subspecies I are most frequently isolated from avian and mammalian hosts. The host adaptation of *S. enterica* subspecies I to warm-blooded vertebrates, characterized the third phase in the evolution of virulence in the genus *Salmonella*. The immune system of higher vertebrates is more developed and organized than that of cold-blooded vertebrates. The common ancestor of subspecies I, II, IIIb and VI acquired mechanisms of flagellar antigen shifting (diphasic condition), which is thought to play a role in adaptation to warm-blooded hosts (Li et al., 1995; Baumler et al., 1998; Porwollik and McClelland 2003). The mechanism of phase shifting is amazing and involved the acquisition of the *fljBA* operon, which contains *hin*, encoding for a recombinase that catalyzes the reversible inversion of a segment of the chromosome containing the promoter for the *fljBA* operon. In one orientation, the promoter directs the transcription of the *fljA* (repressor of *fliC*) and *fljB* (phase 2 flagellin) genes, inducing the repression of *fliC* (phase 1 flagellin). In the other orientation, *hin, fljB* and *fljA* are not expressed and *fliC* is expressed (Zieg et al., 1977). A schematic representation of *Salmonella* evolution is presented in Figure 2 (Silva and Wiesner 2009).

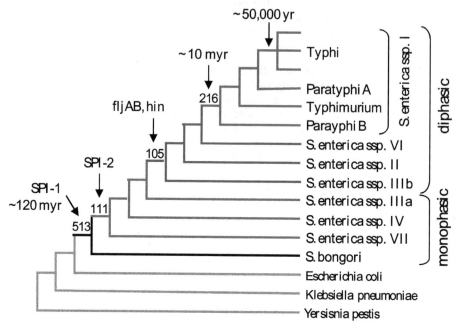

Fig. 2. Schematic representation of *Salmonella* evolution. The cladogram shown is a modified version of that proposed by Porwollik & McClelland (2003), and published by Silva & Wiesner (2009). The number of genes acquired at crucial steps in *Salmonella* evolution is indicated on the nodes along with prominent examples (Porwollik et al., 2002). During the divergence of *E. coli* and *Salmonella*, about 120 million years (myr) ago (Ochman and Wilson 1987), more than 500 genes were acquired, including *Salmonella* pathogenicity island 1 (SPI-1). In the divergence between *S. bongori* and *S. enterica* more than 100 genes were acquired in the *S. enterica* lineage, including those of the SPI-2 (Ochman and Groisman 1996). Along the diversification of the *S. enterica* lineage, about 100 genes were acquired by the common ancestor of subspecies IIIb, II, IV and I, among them were the phase shifting genes (*fljBA* and *hin*), required to confer the diphasic condition to *Salmonella*. During the evolution of subspecies I, which is the most diverse of the subspecies and adapted to warm-blooded vertebrates, more than 200 genes were acquired. Kidgell et al. (2002) estimated that the time of divergence among serovars was around 10 myr, and that the last common ancestor of serovar Typhi existed about 50,000 yr (Kidgell et al., 2002).

However, not everything is said about *Salmonella* evolution. The report of serovar Senftenberg human clinical isolates lacking SPI-1 (Hu et al., 2008) is an example of subsequent loss of genetic determinants during the diversification of *Salmonella*. Another example is the loss of the diphasic trait in some of the serovars of subspecies I (e. g. Typhi) and II. They have reverted to the monophasic condition, usually by loss of expression of phase 2 flagella. In Enteritidis and serovar 4,5,12:i:- the phase 2 flagellin gene (*fljB*) has been

The Importance of Mobile Genetic Elements in the Evolution of Salmonella: Pathogenesis, Antibiotic Resistance and Host Adaptation

263

deleted rather than merely silenced (Selander et al., 1996; Echeita et al., 2001). Likewise, in a study analyzing *E. coli* and *Salmonella* genomes, Retcheless and Lawrence (2007) found that their chromosomes diverged over a 70 million year period, and that the regions flanking SPI1 and SPI2 diverged more recently, suggesting that they did not promote the separation of *E. coli* and *Salmonella* (Retchless and Lawrence 2007).

The distribution of pathogenicity islands, fimbrial operons, and capsular biosynthesis genes among *S. enterica* suggests that during evolution, new combinations of virulence determinants arose through multiple horizontal transfer events, a process which may have driven the development of host adaptation. In addition, deletion events and sequence divergence by point mutations were likely among the events which contributed to changes in the host range of *S. enterica* serovars (Baumler et al., 1998; Porwollik and McClelland 2003). The *S. enterica* subspecies I serovars form a group of pathogens that differ widely in their spectrum of host range within mammals and birds. For classification purposes, they can be categorized into three different groups: broad-host range, host-adapted, and host-restricted serovars. For example, serovars Typhi, Gallinarum, and Abortusovis are host-restricted serovars that are associated with systemic disease in humans, fowl and ovine hosts, respectively. Serovars Dublin and Cholerasuis are host-adapted serovars that are often associated with systemic disease in cattle and pigs, respectively, but can cause disease in other animals. Typhimurium and Enteritidis are broad-host range serovars capable of causing systemic disease in a wide range of animals, but are usually associated with gastroenteritis in a broad range of phylogenetically unrelated host species (Baumler et al., 1998; Kingsley and Baumler 2000; Rabsch et al., 2002). However, even within a single serovar there are differences in host range. For example, Rabsch *et al.* (2002) showed that two Typhimurium variants (DT2 and DT99) were almost exclusively associated with pigeons during decades, over a wide geographic range, indicative of a narrow host range; while other Typhimurium variants, such as DT104, are truly broad-host-range, thus circulating in cattle, swine, poultry and humans. Therefore, it may be more accurate to describe serovar Typhimurium as a collection of variants that vary significantly in their host range and degree of host adaptation. One possible mechanism by which such variants arise is through phage-mediated transfer, of a small number of host-specific virulence factors (Rabsch et al., 2002; Porwollik and McClelland 2003; Brussow et al., 2004; Porwollik et al., 2004; Ehrbar and Hardt 2005; Vernikos et al., 2007). There is currently no genetic explanation for the phenotype of host adaptation; it is unlikely that a single locus will be found to be responsible for this complex biological trait. Instead, a combination of multiple genes is likely to contribute to the overall virulence phenotype (Fierer and Guiney 2001).

8. How much lateral transfer occurs in natural populations?

In this chapter, we have provided extensive evidence on the importance of lateral transfer of genetic information in the evolution of *Salmonella*. Compiling evidence on the role of gene loss and acquisition in the origin of the genus *Salmonella* has been presented in several evolutionary studies (Groisman and Ochman 1997; Lawrence 1999; Porwollik et al., 2002; Lerat et al., 2005; Retchless and Lawrence 2007; Vernikos et al., 2007), exemplified by the acquisition of pathogenicity islands (Ochman and Groisman 1996; Porwollik and McClelland 2003; Hensel 2004). Likewise, the evolutionary processes shaping the genetic structure within serovars and host-adapted ecotypes involve in many cases lateral transfer events, such as prophage insertions (Porwollik and McClelland 2003; Thomson et al., 2004;

Hermans et al., 2006; Cooke et al., 2007; Vernikos et al., 2007). The selective pressure that antimicrobial drugs have imposed on the survival of *Salmonella* probably has increased the acquisition of resistance determinants, often carried by mobile genetic elements that are acquired by lateral transfer. These processes are the result of a long evolutionary history of adaptation to changing environments and hosts. However, we do not want to leave the misconception that the amount of lateral transfer is so rampant that there are no limits to genetic exchange in *Salmonella* populations.

Since more than two decades ago, the clonal nature of Salmonella species was documented by several studies based on multilocus enzyme electrophoresis analysis (Beltran et al., 1988; Reeves et al., 1989; Selander et al., 1990; Boyd et al., 1996; Spratt and Maiden 1999). Evidences for the clonal structure of *Salmonella* include the global distribution of certain genotypes, the congruent relationships between isolates derived from several housekeeping genes, and the robust subspecies structure (Reeves et al., 1989; Boyd et al., 1996; Selander et al., 1996; Falush et al., 2006). In the past decade, multi-locus sequence typing (MLST) studies, analyzing *Salmonella* populations for epidemiological purposes, showed concordant results with the studies based on enzyme electrophoresis, which support the view that *Salmonella* has a clonal population structure. Among these results are the almost strict association between multilocus genotype and serovar, the low genetic diversity within serovars, and the maintenance of old globally-distributed clones (Sukhnanand et al., 2005; Harbottle et al., 2006; Tankouo-Sandjong et al., 2007; Wiesner et al., 2009). The clonal nature of *Salmonella* populations was observed in our study based on a survey of more than 100 Mexican Typhimurium strains. MLST and macrorestriction fingerprints by pulsed-field gel electrophoresis were used to address the core genetic variation, and genes involved in pathogenesis and antibiotic resistance were selected to evaluate the accessory genome. SGI1 was found in a defined subset (16%) of the strains. They were in a compact cluster conformed by strains belonging to the second most abundant genotype (ST19), and in most of the cases they also carried the *Salmonella* virulence plasmid (Figure 3). On the other hand, the strains with the most abundant genotype (ST213) lacked *Salmonella* virulence plasmid or SGI1, but in most of the cases carried a multiple-drug resistant (MDR) IncA/C plasmid. The ST19 isolates carrying the virulence plasmid were significantly associated with the human host, whereas ST213 isolates were more frequently isolated from animal sources, indicating that the distinct accessory genes carried by these genotypes are probably involved in the interaction with the host (Wiesner et al., 2009). No strain carrying both the *Salmonella* virulence plasmid and the MDR IncA/C plasmid, nor hybrid virulence-resistance plasmids, was detected. We concluded that, in the Mexican Typhimurium population, the association between distinct core and accessory genes creates a structure of genetic subgroups within the population, which could be due to the existence of barriers to genetic exchange among subgroups (Wiesner et al., 2009).

It is intriguing why if *Salmonella* evolution is marked by lateral transfer events and genome rearrangements, the genetic structure of populations seems to be extremely clonal. It is possible that in an evolutionary time scale (millions of years) there were several occasions where the mismatch repair system was impaired and large scale recombination events occurred and marked the genomes of diverse *Salmonella* lineages (Taddei et al., 1997; Matic et al., 2000; Didelot et al., 2007). However, in the ecological time scales (thousands of years to decades) the recombination events are rare.

The Importance of Mobile Genetic Elements in the Evolution of Salmonella: Pathogenesis, Antibiotic Resistance and Host Adaptation

265

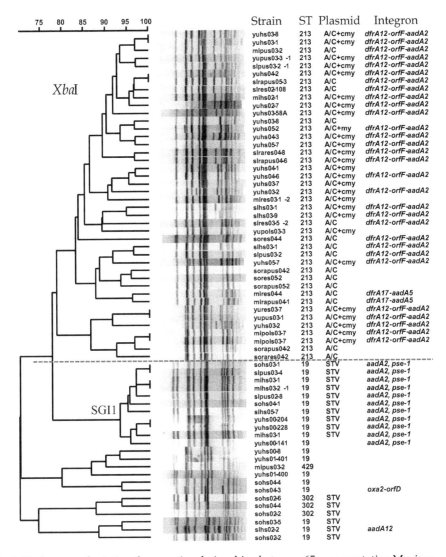

Fig. 3. Dedrogram depicting the genetic relationships between 65 representative Mexican Typhimurium strains, and the associations among core and accessory genes. The data were obtained from Wiesner et al. (2009) and Wiesner et al. (2011). The XbaI restriction fingerprints were clustered by the UPGMA algorithm using Dice coefficients. The first column contains the strain designation. The second column displays the multi-locus sequence type (ST) for each strain. The plasmid column indicates if the strains harbored the Salmonella virulence plasmid (STV), the blaCMY-2-bearing IncA/C plasmid (A/C+cmy) or the smaller blaCMY-2-lacking IncA/C plasmid (A/C). The last column shows the integrons carried by the strains. The horizontal dashed line separates the ST213 strains from the remaining STs. SGI1 indicates the cluster containing the strains with the Salmonella genomic island.

9. Concluding remarks

Mobile genetic elements are key to understanding *Salmonella* evolution and ecology. Actually there is a wealth of information regarding the record of horizontal gene transfer in *Salmonella*; for a revision we recommend the review of Porwollik & McClelland (2003) and the paper of Vernikos *et al.* (2007). In *Salmonella* many of the virulence factors are present as part of pathogenicity islands on the chromosome, yet some virulence factors are encoded on plasmids or bacteriophages. Their composition, presence or absence determines at large differences in pathogenicity and host range between serovars and strains (Fluit 2005). We certainly need to expand our vision, in order to integrate the knowledge about the great variety of genetic virulence determinants in *Salmonella* and the antibiotic-resistance "tool kit". A vision is emerging regarding different molecular routes that determine the plasticity of the accessory genome, which are subject to intricate interactions that we still do not completely understand. Complex interactions among pathogenicity islands, bacteriophages, plasmids and other mobile elements, such as transposons and integrons, are increasingly being evidenced. For example, the transfer of the resistance phenotypes associated with Typhimurium DT104 SGI1, by phage-transduction experiments, suggest the involvement of phages in the mobilization of SGI1 (Lawson et al., 2002b; Fluit 2005). In another example, derivatives of the original Typhimurium DT104 clone have emerged over the more than 20 years after the first report. Different phage-types, such as DT12 or DT120, are indistinguishable from the DT104 clone by genotyping, suggesting that they arose from DT104 through changes that led to different phage susceptibility patterns (Fluit 2005). These shifts in phage-type are probably due to the gain or loss of prophages. More than three decades ago, Threlfall et al. (1978) showed that changes in phage-types can be the result of the loss or acquisition of plasmids (Threlfall et al., 1978). Likewise, in another study related to Typhimurium DT104 and SGI1, some strains showed to be phage-untypeable or DT104B, and it was found that these strains carried a larger plasmid (140 kb) resulting from the recombination between a resistance and virulence plasmid (Guerra et al., 2002). In these latter cases, the changes in the plasmid content could be associated with the loss of determinants required by phages (Fluit 2005), such as surface lipopolysaccharides (Lawson et al., 2002a). Of particular concern is the increasing number of reports of co-integrates of resistance and virulence plasmids, envisioning that new *Salmonella* strains will arise posing a threat to public health (Fluit 2005).

10. References

Anderson, E. S., L. R. Ward, M. J. Saxe, and J. D. de Sa. 1977. Bacteriophage-typing designations of *Salmonella typhimurium*. *J. Hyg. (Lond)* 78:297-300.

Antunes, P., J. Machado, and L. Peixe. 2006. Characterization of antimicrobial resistance and class 1 and 2 integrons in *Salmonella enterica* isolates from different sources in Portugal. *J. Antimicrob. Chemother.* 58:297-304.

Bauernfeind, A., I. Stemplinger, R. Jungwirth, and H. Giamarellou. 1996. Characterization of the plasmidic beta-lactamase CMY-2, which is responsible for cephamycin resistance. *Antimicrob. Agents. Chemother.* 40:221-224.

Baumler, A. J. 1997. The record of horizontal gene transfer in *Salmonella*. *Trends Microbiol.* 5:318-322.

The Importance of Mobile Genetic Elements in the Evolution of Salmonella: Pathogenesis, Antibiotic Resistance and Host Adaptation

267

Baumler, A. J., R. M. Tsolis, T. A. Ficht, and L. G. Adams. 1998. Evolution of host adaptation in *Salmonella enterica*. *Infect. Immun.* 66:4579-4587.

Beltran, P., J. M. Musser, R. Helmuth, J. J. Farmer, 3rd, W. M. Frerichs, I. K. Wachsmuth, K. Ferris, A. C. McWhorter, J. G. Wells, A. Cravioto, and et al. 1988. Toward a population genetic analysis of *Salmonella*: genetic diversity and relationships among strains of serotypes S. choleraesuis, S. derby, S. dublin, S. enteritidis, S. heidelberg, S. infantis, S. newport, and S. typhimurium. *Proc. Natl. Acad. Sci. USA* 85:7753-7757.

Boucher, Y., M. Labbate, J. E. Koenig, and H. W. Stokes. 2007. Integrons: mobilizable platforms that promote genetic diversity in bacteria. *Trends Microbiol.* 15:301-309.

Boyd, D., G. A. Peters, A. Cloeckaert, K. S. Boumedine, E. Chaslus-Dancla, H. Imberechts, and M. R. Mulvey. 2001. Complete nucleotide sequence of a 43-kilobase genomic island associated with the multidrug resistance region of *Salmonella enterica* serovar Typhimurium DT104 and its identification in phage type DT120 and serovar Agona. *J. Bacteriol.* 183:5725-5732.

Boyd, D. A., X. Shi, Q. H. Hu, L. K. Ng, B. Doublet, A. Cloeckaert, and M. R. Mulvey. 2008. *Salmonella* genomic island 1 (SGI1), variant SGI1-I, and new variant SGI1-O in *Proteus mirabilis* clinical and food isolates from China. *Antimicrob. Agents Chemother.* 52:340-344.

Boyd, E. F., and H. Brussow. 2002. Common themes among bacteriophage-encoded virulence factors and diversity among the bacteriophages involved. *Trends Microbiol.* 10:521-529.

Boyd, E. F., F. S. Wang, T. S. Whittam, and R. K. Selander. 1996. Molecular genetic relationships of the salmonellae. *Appl. Environ. Microbiol.* 62:804-808.

Brussow, H., C. Canchaya, and W. D. Hardt. 2004. Phages and the evolution of bacterial pathogens: from genomic rearrangements to lysogenic conversion. *Microbiol. Mol. Biol. Rev.* 68:560-602.

Call, D. R., R. S. Singer, D. Meng, S. L. Broschat, L. H. Orfe, J. M. Anderson, D. R. Herndon, L. S. Kappmeyer, J. B. Daniels, and T. E. Besser. 2010. blaCMY-2-positive IncA/C plasmids from *Escherichia coli* and *Salmonella enterica* are a distinct component of a larger lineage of plasmids. *Antimicrob. Agents Chemother.* 54:590-596.

Carattoli, A., F. Tosini, W. P. Giles, M. E. Rupp, S. H. Hinrichs, F. J. Angulo, T. J. Barrett, and P. D. Fey. 2002. Characterization of plasmids carrying CMY-2 from expanded-spectrum cephalosporin-resistant *Salmonella* strains isolated in the United States between 1996 and 1998. *Antimicrob. Agents Chemother.* 46:1269-1272.

Cooke, F. J., D. J. Brown, M. Fookes, D. Pickard, A. Ivens, J. Wain, M. Roberts, R. A. Kingsley, N. R. Thomson, and G. Dougan. 2008. Characterization of the genomes of a diverse collection of *Salmonella enterica* serovar Typhimurium definitive phage type 104. *J. Bacteriol.* 190:8155-8162.

Cooke, F. J., J. Wain, M. Fookes, A. Ivens, N. Thomson, D. J. Brown, E. J. Threlfall, G. Gunn, G. Foster, and G. Dougan. 2007. Prophage sequences defining hot spots of genome variation in Salmonella enterica serovar Typhimurium can be used to discriminate between field isolates. *J Clin Microbiol* 45:2590-2598.

Cornelis, G. R. 2006. The type III secretion injectisome. *Nat. Rev. Microbiol.* 4:811-825.

Couturier, M., F. Bex, P. Bergquist, and W. K. Maas. 1988. Identification and classification of bacterial plasmids. *Microbiol. Rev.* 52:375-395.

Chan, K., S. Baker, C. C. Kim, C. S. Detweiler, G. Dougan, and S. Falkow. 2003. Genomic comparison of *Salmonella enterica* serovars and *Salmonella bongori* by use of an *S. enterica* serovar typhimurium DNA microarray. *J. Bacteriol.* 185:553-563.

Chiu, C. H., C. Chu, and J. T. Ou. 2000. Lack of evidence of an association between the carriage of virulence plasmid and the bacteremia of *Salmonella typhimurium* in humans. *Microbiol. Immunol.* 44:741-748.

Chu, C., and C. H. Chiu. 2006. Evolution of the virulence plasmids of non-typhoid *Salmonella* and its association with antimicrobial resistance. *Microbes. Infect.* 8:1931-1936.

Chu, C., C. H. Chiu, W. Y. Wu, C. H. Chu, T. P. Liu, and J. T. Ou. 2001. Large drug resistance virulence plasmids of clinical isolates of *Salmonella enterica* serovar Choleraesuis. *Antimicrob. Agents Chemother.* 45:2299-2303.

Davies, J., and D. Davies. 2010. Origins and evolution of antibiotic resistance. *Microbiol. Mol. Biol. Rev.* 74:417-433.

Didelot, X., M. Achtman, J. Parkhill, N. R. Thomson, and D. Falush. 2007. A bimodal pattern of relatedness between the *Salmonella* Paratyphi A and Typhi genomes: convergence or divergence by homologous recombination? *Genome Res.* 17:61-68.

Dionisio, F., I. Matic, M. Radman, O. R. Rodrigues, and F. Taddei. 2002. Plasmids spread very fast in heterogeneous bacterial communities. *Genetics* 162:1525-1532.

Dobrindt, U., B. Hochhut, U. Hentschel, and J. Hacker. 2004. Genomic islands in pathogenic and environmental microorganisms. *Nat. Rev. Microbiol.* 2:414-424.

Douard, G., K. Praud, A. Cloeckaert, and B. Doublet. 2010. The *Salmonella* genomic island 1 is specifically mobilized in trans by the IncA/C multidrug resistance plasmid family. *PLoS One* 5:e15302.

Doublet, B., D. Boyd, M. R. Mulvey, and A. Cloeckaert. 2005. The *Salmonella* genomic island 1 is an integrative mobilizable element. *Mol. Microbiol.* 55:1911-1924.

Drahovska, H., E. Mikasova, T. Szemes, A. Ficek, M. Sasik, V. Majtan, and J. Turna. 2007. Variability in occurrence of multiple prophage genes in *Salmonella* Typhimurium strains isolated in Slovak Republic. *FEMS Microbiol. Lett.* 270:237-244.

Echeita, M. A., S. Herrera, and M. A. Usera. 2001. Atypical, fljB-negative *Salmonella enterica* subsp. *enterica* strain of serovar 4,5,12:i:- appears to be a monophasic variant of serovar Typhimurium. *J. Clin. Microbiol.* 39:2981-2983.

Ehrbar, K., and W. D. Hardt. 2005. Bacteriophage-encoded type III effectors in *Salmonella enterica* subspecies 1 serovar Typhimurium. *Infect. Genet. Evol.* 5:1-9.

Falush, D., M. Torpdahl, X. Didelot, D. F. Conrad, D. J. Wilson, and M. Achtman. 2006. Mismatch induced speciation in *Salmonella*: model and data. *Philos. Trans. R. Soc. Lond. B* 361:2045-2053.

Fierer, J. 2001. Extra-intestinal *Salmonella* infections: the significance of *spv* genes. *Clin. Infect. Dis.* 32:519-520.

Fierer, J., and D. G. Guiney. 2001. Diverse virulence traits underlying different clinical outcomes of *Salmonella* infection. *J. Clin. Invest.* 107:775-780.

The Importance of Mobile Genetic Elements in the Evolution of Salmonella: Pathogenesis, Antibiotic Resistance
and Host Adaptation

269

Figueroa-Bossi, N., S. Uzzau, D. Maloriol, and L. Bossi. 2001. Variable assortment of prophages provides a transferable repertoire of pathogenic determinants in *Salmonella*. *Mol. Microbiol.* 39:260-271.

Fluit, A. C. 2005. Towards more virulent and antibiotic-resistant *Salmonella*? *FEMS Immunol. Med. Microbiol.* 43:1-11.

Fluit, A. C., and F. J. Schmitz. 1999. Class 1 integrons, gene cassettes, mobility, and epidemiology. *Eur. J. Clin. Microbiol. Infect. Dis.* 18:761-770.

Fluit, A. C., and F. J. Schmitz. 2004. Resistance integrons and super-integrons. *Clin. Microbiol. Infect.* 10:272-288.

Fricke, W. F., T. J. Welch, P. F. McDermott, M. K. Mammel, J. E. LeClerc, D. G. White, T. A. Cebula, and J. Ravel. 2009. Comparative genomics of the IncA/C multidrug resistance plasmid family. *J. Bacteriol.* 191:4750-4757.

Frost, L. S., R. Leplae, A. O. Summers, and A. Toussaint. 2005. Mobile genetic elements: the agents of open source evolution. *Nat. Rev. Microbiol.* 3:722-732.

Groisman, E. A., and H. Ochman. 1997. How Salmonella became a pathogen. *Trends Microbiol.* 5:343-349.

Guerra, B., E. Junker, A. Miko, R. Helmuth, and M. C. Mendoza. 2004. Characterization and localization of drug resistance determinants in multidrug-resistant, integron-carrying *Salmonella enterica* serotype Typhimurium strains. *Microb. Drug Resist.* 10:83-91.

Guerra, B., S. Soto, R. Helmuth, and M. C. Mendoza. 2002. Characterization of a self-transferable plasmid from *Salmonella enterica* serotype Typhimurium clinical isolates carrying two integron-borne gene cassettes together with virulence and drug resistance genes. *Antimicrob. Agents Chemother.* 46:2977-2981.

Guerra, B., S. M. Soto, J. M. Arguelles, and M. C. Mendoza. 2001. Multidrug resistance is mediated by large plasmids carrying a class 1 integron in the emergent *Salmonella enterica* serotype [4,5,12:i:-]. *Antimicrob. Agents Chemother.* 45:1305-1308.

Gulig, P. A., H. Danbara, D. G. Guiney, A. J. Lax, F. Norel, and M. Rhen. 1993. Molecular analysis of *spv* virulence genes of the *Salmonella* virulence plasmids. *Mol. Microbiol.* 7:825-830.

Hall, R. M., and C. M. Collis. 1995. Mobile gene cassettes and integrons: capture and spread of genes by site-specific recombination. *Mol. Microbiol.* 15:593-600.

Hansen-Wester, I., and M. Hensel. 2001. *Salmonella* pathogenicity islands encoding type III secretion systems. *Microbes Infect.* 3:549-559.

Harbottle, H., D. G. White, P. F. McDermott, R. D. Walker, and S. Zhao. 2006. Comparison of multilocus sequence typing, pulsed-field gel electrophoresis, and antimicrobial susceptibility typing for characterization of *Salmonella enterica* serotype Newport isolates. *J. Clin. Microbiol.* 44:2449-2457.

Heithoff, D. M., W. R. Shimp, P. W. Lau, G. Badie, E. Y. Enioutina, R. A. Daynes, B. A. Byrne, J. K. House, and M. J. Mahan. 2008. Human *Salmonella* clinical isolates distinct from those of animal origin. *Appl. Environ. Microbiol.* 74:1757-1766.

Hendrix, R. W., J. G. Lawrence, G. F. Hatfull, and S. Casjens. 2000. The origins and ongoing evolution of viruses. *Trends Microbiol.* 8:504-508.

Hensel, M. 2004. Evolution of pathogenicity islands of *Salmonella enterica*. *Int. J. Med. Microbiol.* 294:95-102.

Hermans, A. P., T. Abee, M. H. Zwietering, and H. J. Aarts. 2005. Identification of novel *Salmonella enterica* serovar Typhimurium DT104-specific prophage and nonprophage chromosomal sequences among serovar Typhimurium isolates by genomic subtractive hybridization. *Appl. Environ. Microbiol.* 71:4979-4985.

Hermans, A. P., A. M. Beuling, A. H. van Hoek, H. J. Aarts, T. Abee, and M. H. Zwietering. 2006. Distribution of prophages and SGI-1 antibiotic-resistance genes among different *Salmonella enterica* serovar Typhimurium isolates. *Microbiology* 152:2137-2147.

Herrero, A., M. C. Mendoza, R. Rodicio, and M. R. Rodicio. 2008a. Characterization of pUO-StVR2, a virulence-resistance plasmid evolved from the pSLT virulence plasmid of *Salmonella enterica* serovar Typhimurium. *Antimicrob. Agents Chemother.* 52:4514-4517.

Herrero, A., M. C. Mendoza, E. J. Threlfall, and M. R. Rodicio. 2009. Detection of *Salmonella enterica* serovar Typhimurium with pUO-StVR2-like virulence-resistance hybrid plasmids in the United Kingdom. *Eur. J. Clin. Microbiol. Infect. Dis.* 28:1087-1093.

Herrero, A., M. R. Rodicio, M. A. Echeita, and M. C. Mendoza. 2008b. *Salmonella enterica* serotype Typhimurium carrying hybrid virulence-resistance plasmids (pUO-StVR): a new multidrug-resistant group endemic in Spain. *Int. J. Med. Microbiol.* 298:253-261.

Heuer, H., Z. Abdo, and K. Smalla. 2008. Patchy distribution of flexible genetic elements in bacterial populations mediates robustness to environmental uncertainty. *FEMS Microbiol. Ecol.* 65:361-371.

Holmes, A. J., M. R. Gillings, B. S. Nield, B. C. Mabbutt, K. M. Nevalainen, and H. W. Stokes. 2003. The gene cassette metagenome is a basic resource for bacterial genome evolution. *Environ. Microbiol.* 5:383-394.

Hopkins, K. L., E. Liebana, L. Villa, M. Batchelor, E. J. Threlfall, and A. Carattoli. 2006. Replicon typing of plasmids carrying CTX-M or CMY beta-lactamases circulating among *Salmonella* and *Escherichia coli* isolates. *Antimicrob. Agents Chemother.* 50:3203-3206.

Hradecka, H., D. Karasova, and I. Rychlik. 2008. Characterization of *Salmonella enterica* serovar Typhimurium conjugative plasmids transferring resistance to antibiotics and their interaction with the virulence plasmid. *J. Antimicrob. Chemother.* 62:938-941.

Hu, Q., B. Coburn, W. Deng, Y. Li, X. Shi, Q. Lan, B. Wang, B. K. Coombes, and B. B. Finlay. 2008. *Salmonella enterica* serovar Senftenberg human clinical isolates lacking SPI-1. *J. Clin. Microbiol.* 46:1330-1336.

Joss, M. J., J. E. Koenig, M. Labbate, M. F. Polz, M. R. Gillings, H. W. Stokes, W. F. Doolittle, and Y. Boucher. 2009. ACID: annotation of cassette and integron data. *BMC Bioinformatics* 10:118.

The Importance of Mobile Genetic Elements in the Evolution of Salmonella: Pathogenesis, Antibiotic Resistance and Host Adaptation

271

Kidgell, C., U. Reichard, J. Wain, B. Linz, M. Torpdahl, G. Dougan, and M. Achtman. 2002. *Salmonella* Typhi, the causative agent of typhoid fever, is approximately 50,000 years old. *Infect. Genet. Evol.* 2:39-45.

Kim, M. J., I. Hirono, K. Kurokawa, T. Maki, J. Hawke, H. Kondo, M. D. Santos, and T. Aoki. 2008. Complete DNA sequence and analysis of the transferable multiple-drug resistance plasmids (R Plasmids) from *Photobacterium damselae* subsp. *piscicida* isolates collected in Japan and the United States. *Antimicrob. Agents Chemother.* 52:606-611.

Kingsley, R. A., and A. J. Baumler. 2000. Host adaptation and the emergence of infectious disease: the *Salmonella* paradigm. *Mol. Microbiol.* 36:1006-1014.

Kropinski, A. M., A. Sulakvelidze, P. Konczy, and C. Poppe. 2007. *Salmonella* phages and prophages--genomics and practical aspects. *Methods Mol. Biol.* 394:133-175.

Lawrence, J. G. 1999. Gene transfer, speciation, and the evolution of bacterial genomes. *Curr. Opin. Microbiol.* 2:519-523.

Lawson, A. J., H. Chart, M. U. Dassama, and E. J. Threlfall. 2002a. Heterogeneity in expression of lipopolysaccharide by strains of Salmonella enterica serotype Typhimurium definitive phage type 104 and related phage types. *Lett Appl Microbiol* 34:428-432.

Lawson, A. J., M. U. Dassama, L. R. Ward, and E. J. Threlfall. 2002b. Multiply resistant (MR) *Salmonella enterica* serotype Typhimurium DT 12 and DT 120: a case of MR DT 104 in disguise? *Emerg. Infect. Dis.* 8:434-436.

Lerat, E., V. Daubin, H. Ochman, and N. A. Moran. 2005. Evolutionary origins of genomic repertoires in bacteria. *PLoS Biol.* 3:e130.

Leverstein-van Hall, M. A., M. B. HE, T. D. AR, A. Paauw, A. C. Fluit, and J. Verhoef. 2003. Multidrug resistance among Enterobacteriaceae is strongly associated with the presence of integrons and is independent of species or isolate origin. *J. Infect. Dis.* 187:251-259.

Levesque, C., L. Piche, C. Larose, and P. H. Roy. 1995. PCR mapping of integrons reveals several novel combinations of resistance genes. *Antimicrob. Agents Chemother.* 39:185-191.

Levin, B. R., and C. T. Bergstrom. 2000. Bacteria are different: observations, interpretations, speculations, and opinions about the mechanisms of adaptive evolution in prokaryotes. *Proc. Natl. Acad. Sci. USA* 97:6981-6985.

Levings, R. S., D. Lightfoot, S. R. Partridge, R. M. Hall, and S. P. Djordjevic. 2005. The genomic island SGI1, containing the multiple antibiotic resistance region of *Salmonella enterica* serovar Typhimurium DT104 or variants of it, is widely distributed in other *S. enterica* serovars. *J Bacteriol* 187:4401-4409.

Li, J., H. Ochman, E. A. Groisman, E. F. Boyd, F. Solomon, K. Nelson, and R. K. Selander. 1995. Relationship between evolutionary rate and cellular location among the Inv/Spa invasion proteins of *Salmonella enterica*. *Proc. Natl. Acad. Sci. USA* 92:7252-7256.

Liebert, C. A., R. M. Hall, and A. O. Summers. 1999. Transposon Tn21, flagship of the floating genome. *Microbiol. Mol. Biol. Rev.* 63:507-522.

Lindsey, R. L., P. J. Fedorka-Cray, J. G. Frye, and R. J. Meinersmann. 2009. Inc A/C plasmids are prevalent in multidrug-resistant *Salmonella enterica* isolates. *Appl. Environ. Microbiol.* 75:1908-1915.

Lindstedt, B. A., E. Heir, I. Nygard, and G. Kapperud. 2003. Characterization of class I integrons in clinical strains of *Salmonella enterica* subsp. enterica serovars Typhimurium and Enteritidis from Norwegian hospitals. *J. Med. Microbiol.* 52:141-149.

Litrup, E., M. Torpdahl, B. Malorny, S. Huehn, M. Helms, H. Christensen, and E. M. Nielsen. 2010. DNA microarray analysis of *Salmonella* serotype Typhimurium strains causing different symptoms of disease. *BMC Microbiol.* 10:96.

Marcus, S. L., J. H. Brumell, C. G. Pfeifer, and B. B. Finlay. 2000. *Salmonella* pathogenicity islands: big virulence in small packages. *Microbes Infect.* 2:145-156.

Martinez-Freijo, P., A. C. Fluit, F. J. Schmitz, J. Verhoef, and M. E. Jones. 1999. Many class I integrons comprise distinct stable structures occurring in different species of *Enterobacteriaceae* isolated from widespread geographic regions in Europe. *Antimicrob. Agents Chemother.* 43:686-689.

Matiasovicova, J., P. Adams, P. A. Barrow, H. Hradecka, M. Malcova, R. Karpiskova, E. Budinska, L. Pilousova, and I. Rychlik. 2007. Identification of putative ancestors of the multidrug-resistant *Salmonella enterica* serovar typhimurium DT104 clone harboring the *Salmonella* genomic island 1. *Arch. Microbiol.* 187:415-424.

Matic, I., F. Taddei, and M. Radman. 2000. No genetic barriers between *Salmonella enterica* serovar typhimurium and *Escherichia coli* in SOS-induced mismatch repair-deficient cells. *J. Bacteriol.* 182:5922-5924.

Mazel, D. 2006. Integrons: agents of bacterial evolution. *Nat. Rev. Microbiol.* 4:608-620.

McClelland, M., K. E. Sanderson, J. Spieth, S. W. Clifton, P. Latreille, L. Courtney, S. Porwollik, J. Ali, M. Dante, F. Du, S. Hou, D. Layman, S. Leonard, C. Nguyen, K. Scott, A. Holmes, N. Grewal, E. Mulvaney, E. Ryan, H. Sun, L. Florea, W. Miller, T. Stoneking, M. Nhan, R. Waterston, and R. K. Wilson. 2001. Complete genome sequence of *Salmonella enterica* serovar Typhimurium LT2. *Nature* 413:852-856.

McIntosh, D., M. Cunningham, B. Ji, F. A. Fekete, E. M. Parry, S. E. Clark, Z. B. Zalinger, I. C. Gilg, G. R. Danner, K. A. Johnson, M. Beattie, and R. Ritchie. 2008. Transferable, multiple antibiotic and mercury resistance in Atlantic Canadian isolates of *Aeromonas salmonicida* subsp. *salmonicida* is associated with carriage of an IncA/C plasmid similar to the *Salmonella enterica* plasmid pSN254. *J. Antimicrob. Chemother.* 61:1221-1228.

Michael, G. B., P. Butaye, A. Cloeckaert, and S. Schwarz. 2006. Genes and mutations conferring antimicrobial resistance in *Salmonella*: an update. *Microbes. Infect.* 8:1898-1914.

Mirold, S., W. Rabsch, M. Rohde, S. Stender, H. Tschape, H. Russmann, E. Igwe, and W. D. Hardt. 1999. Isolation of a temperate bacteriophage encoding the type III effector protein SopE from an epidemic *Salmonella typhimurium* strain. *Proc. Natl. Acad. Sci. USA* 96:9845-9850.

Mirold, S., W. Rabsch, H. Tschape, and W. D. Hardt. 2001. Transfer of the *Salmonella* type III effector sopE between unrelated phage families. *J. Mol. Biol.* 312:7-16.

Mmolawa, P. T., R. Willmore, C. J. Thomas, and M. W. Heuzenroeder. 2002. Temperate phages in *Salmonella enterica* serovar Typhimurium: implications for epidemiology. *Int. J. Med. Microbiol.* 291:633-644.

Molla, B., A. Miko, K. Pries, G. Hildebrandt, J. Kleer, A. Schroeter, and R. Helmuth. 2007. Class 1 integrons and resistance gene cassettes among multidrug resistant *Salmonella* serovars isolated from slaughter animals and foods of animal origin in Ethiopia. *Acta Trop.* 103:142-149.

Mulvey, M. R., D. A. Boyd, A. B. Olson, B. Doublet, and A. Cloeckaert. 2006. The genetics of *Salmonella* genomic island 1. *Microbes Infect.* 8:1915-1922.

Norman, A., L. H. Hansen, and S. J. Sorensen. 2009. Conjugative plasmids: vessels of the communal gene pool. *Philos. Trans. R. Soc. Lond. B Biol. Sci.* 364:2275-2289.

Ochman, H., and E. A. Groisman. 1994. The origin and evolution of species differences in *Escherichia coli* and *Salmonella typhimurium. EXS* 69:479-493.

Ochman, H., and E. A. Groisman. 1996. Distribution of pathogenicity islands in *Salmonella* spp. *Infect. Immun.* 64:5410-5412.

Ochman, H., J. G. Lawrence, and E. A. Groisman. 2000. Lateral gene transfer and the nature of bacterial innovation. *Nature* 405:299-304.

Ochman, H., and A. C. Wilson. 1987. Evolution in bacteria: evidence for a universal substitution rate in cellular genomes. *J. Mol. Evol.* 26:74-86.

Olsen, J. E., D. J. Brown, L. E. Thomsen, D. J. Platt, and M. S. Chadfield. 2004. Differences in the carriage and the ability to utilize the serotype associated virulence plasmid in strains of *Salmonella enterica* serotype Typhimurium investigated by use of a self-transferable virulence plasmid, pOG669. *Microb. Pathog.* 36:337-347.

Pan, J. C., R. Ye, H. Q. Wang, H. Q. Xiang, W. Zhang, X. F. Yu, D. M. Meng, and Z. S. He. 2008. *Vibrio cholerae* O139 multiple-drug resistance mediated by *Yersinia pestis* pIP1202-like conjugative plasmids. *Antimicrob. Agents Chemother.* 52:3829-3836.

Parkhill, J., G. Dougan, K. D. James, N. R. Thomson, D. Pickard, J. Wain, C. Churcher, K. L. Mungall, S. D. Bentley, M. T. Holden, M. Sebaihia, S. Baker, D. Basham, K. Brooks, T. Chillingworth, P. Connerton, A. Cronin, P. Davis, R. M. Davies, L. Dowd, N. White, J. Farrar, T. Feltwell, N. Hamlin, A. Haque, T. T. Hien, S. Holroyd, K. Jagels, A. Krogh, T. S. Larsen, S. Leather, S. Moule, P. O'Gaora, C. Parry, M. Quail, K. Rutherford, M. Simmonds, J. Skelton, K. Stevens, S. Whitehead, and B. G. Barrell. 2001. Complete genome sequence of a multiple drug resistant *Salmonella enterica* serovar Typhi CT18. *Nature* 413:848-852.

Porwollik, S., E. F. Boyd, C. Choy, P. Cheng, L. Florea, E. Proctor, and M. McClelland. 2004. Characterization of *Salmonella enterica* subspecies I genovars by use of microarrays. *J. Bacteriol.* 186:5883-5898.

Porwollik, S., and M. McClelland. 2003. Lateral gene transfer in *Salmonella. Microbes Infect.* 5:977-989.

Porwollik, S., R. M. Wong, and M. McClelland. 2002. Evolutionary genomics of Salmonella: gene acquisitions revealed by microarray analysis. *Proc. Natl. Acad. Sci. USA* 99:8956-8961.

Rabsch, W., H. L. Andrews, R. A. Kingsley, R. Prager, H. Tschape, L. G. Adams, and A. J. Baumler. 2002. *Salmonella enterica* serotype Typhimurium and its host-adapted variants. *Infect. Immun.* 70:2249-2255.

Recchia, G. D., and R. M. Hall. 1995. Gene cassettes: a new class of mobile element. *Microbiology* 141 (Pt 12):3015-3027.

Reeves, M. W., G. M. Evins, A. A. Heiba, B. D. Plikaytis, and J. J. Farmer, 3rd. 1989. Clonal nature of Salmonella typhi and its genetic relatedness to other salmonellae as shown by multilocus enzyme electrophoresis, and proposal of Salmonella bongori comb. nov. *J. Clin. Microbiol.* 27:313-320.

Retchless, A. C., and J. G. Lawrence. 2007. Temporal fragmentation of speciation in bacteria. *Science* 317:1093-1096.

Rodriguez, I., B. Guerra, M. C. Mendoza, and M. R. Rodicio. 2011. pUO-SeVR1 is an emergent virulence-resistance complex plasmid of *Salmonella enterica* serovar Enteritidis. *J. Antimicrob. Chemother.* 66:218-220.

Rychlik, I., D. Gregorova, and H. Hradecka. 2006. Distribution and function of plasmids in *Salmonella enterica. Vet. Microbiol.* 112:1-10.

Sabbagh, S. C., C. G. Forest, C. Lepage, J. M. Leclerc, and F. Daigle. 2010. So similar, yet so different: uncovering distinctive features in the genomes of *Salmonella enterica* serovars Typhimurium and Typhi. *FEMS Microbiol. Lett.* 305:1-13.

Selander, R. K., P. Beltran, N. H. Smith, R. Helmuth, F. A. Rubin, D. J. Kopecko, K. Ferris, B. D. Tall, A. Cravioto, and J. M. Musser. 1990. Evolutionary genetic relationships of clones of *Salmonella* serovars that cause human typhoid and other enteric fevers. *Infect. Immun.* 58:2262-2275.

Selander, R. K., J. Li, and K. Nelson. 1996. Evolutionary genetics of *Salmonella enterica*. Pp. 2691-2707 *in* F. C. Neidhardt, R. Curtiss III, J. L. Ingraham, E. C. C. Lin, K. B. Low, B. Magasanik, W. S. Reznikoff, M. Riley, M. Schaechter, and H. E. Umbarger, eds. *Escherichia coli and Salmonella: Celular and Molecular Biology.* American Society of Microbiology, Washington, DC.

Silva, C., and M. Wiesner. 2009. An introduction to systematics, natural history and population genetics of *Salmonella*. Pp. 1-17 *in* C. J. J., and C. E., eds. *Molecular biology and molecular epidemiology of Salmonella infections*. Research Signpost, Singapur.

Sorensen, S. J., M. Bailey, L. H. Hansen, N. Kroer, and S. Wuertz. 2005. Studying plasmid horizontal transfer in situ: a critical review. *Nat. Rev. Microbiol.* 3:700-710.

Souza, V., and L. E. Eguiarte. 1997. Bacteria gone native vs. bacteria gone awry?: plasmidic transfer and bacterial evolution. *Proc. Natl. Acad. Sci. USA* 94:5501-5503.

Spratt, B. G., and M. C. Maiden. 1999. Bacterial population genetics, evolution and epidemiology. *Philos. Trans. R. Soc. Lond. B. Biol. Sci.* 354:701-710.

Stokes, H. W., and R. M. Hall. 1989. A novel family of potentially mobile DNA elements encoding site-specific gene-integration functions: integrons. *Mol. Microbiol.* 3:1669-1683.

Stokes, H. W., A. J. Holmes, B. S. Nield, M. P. Holley, K. M. Nevalainen, B. C. Mabbutt, and M. R. Gillings. 2001. Gene cassette PCR: sequence-independent recovery of entire genes from environmental DNA. *Appl. Environ. Microbiol.* 67:5240-5246.

Su, J., L. Shi, L. Yang, Z. Xiao, X. Li, and S. Yamasaki. 2006. Analysis of integrons in clinical isolates of *Escherichia coli* in China during the last six years. *FEMS Microbiol. Lett.* 254:75-80.

Sukhnanand, S., S. Alcaine, L. D. Warnick, W. L. Su, J. Hof, M. P. Craver, P. McDonough, K. J. Boor, and M. Wiedmann. 2005. DNA sequence-based subtyping and evolutionary analysis of selected *Salmonella enterica* serotypes. *J. Clin. Microbiol.* 43:3688-3698.

Taddei, F., I. Matic, B. Godelle, and M. Radman. 1997. To be a mutator, or how pathogenic and commensal bacteria can evolve rapidly. *Trends Microbiol.* 5:427-428.

Tankouo-Sandjong, B., A. Sessitsch, E. Liebana, C. Kornschober, F. Allerberger, H. Hachler, and L. Bodrossy. 2007. MLST-v, multilocus sequence typing based on virulence genes, for molecular typing of *Salmonella enterica* subsp. *enterica* serovars. *J. Microbiol. Methods.* 69:23-36.

Thomson, N., S. Baker, D. Pickard, M. Fookes, M. Anjum, N. Hamlin, J. Wain, D. House, Z. Bhutta, K. Chan, S. Falkow, J. Parkhill, M. Woodward, A. Ivens, and G. Dougan. 2004. The role of prophage-like elements in the diversity of *Salmonella enterica* serovars. *J. Mol. Biol.* 339:279-300.

Threlfall, E. J., L. R. Ward, and B. Rowe. 1978. Spread of multiresistant strains of *Salmonella typhimurium* phage types 204 and 193 in Britain. *Br. Med. J.* 2:997.

Vernikos, G. S., N. R. Thomson, and J. Parkhill. 2007. Genetic flux over time in the *Salmonella* lineage. *Genome Biol.* 8:R100.

Villa, L., and A. Carattoli. 2005. Integrons and transposons on the *Salmonella enterica* serovar Typhimurium virulence plasmid. *Antimicrob. Agents Chemother.* 49:1194-1197.

Welch, T. J., W. F. Fricke, P. F. McDermott, D. G. White, M. L. Rosso, D. A. Rasko, M. K. Mammel, M. Eppinger, M. J. Rosovitz, D. Wagner, L. Rahalison, J. E. Leclerc, J. M. Hinshaw, L. E. Lindler, T. A. Cebula, E. Carniel, and J. Ravel. 2007. Multiple antimicrobial resistance in plague: an emerging public health risk. *PLoS One* 2:e309.

White, P. A., C. J. McIver, and W. D. Rawlinson. 2001. Integrons and gene cassettes in the *Enterobacteriaceae. Antimicrob. Agents Chemother.* 45:2658-2661.

Wiesner, M., E. Calva, M. Fernandez-Mora, M. A. Cevallos, F. Campos, M. B. Zaidi, and C. Silva. 2011. *Salmonella* Typhimurium ST213 is associated with two types of IncA/C plasmids carrying multiple resistance determinants. *BMC Microbiol.* 11:9.

Wiesner, M., M. B. Zaidi, E. Calva, M. Fernandez-Mora, J. J. Calva, and C. Silva. 2009. Association of virulence plasmid and antibiotic resistance determinants with chromosomal multilocus genotypes in Mexican *Salmonella enterica* serovar Typhimurium strains. *BMC Microbiol.* 9:131.

Xu, Z., L. Shi, M. J. Alam, L. Li, and S. Yamasaki. 2008. Integron-bearing methicillin-resistant coagulase-negative staphylococci in South China, 2001-2004. *FEMS Microbiol. Lett.* 278:223-230.

Zhang, S., R. L. Santos, R. M. Tsolis, S. Mirold, W. D. Hardt, L. G. Adams, and A. J. Baumler. 2002. Phage mediated horizontal transfer of the sopE1 gene increases enteropathogenicity of *Salmonella enterica* serotype Typhimurium for calves. *FEMS Microbiol. Lett.* 217:243-247.

Zhao, S., P. F. McDermott, D. G. White, S. Qaiyumi, S. L. Friedman, J. W. Abbott, A. Glenn, S. L. Ayers, K. W. Post, W. H. Fales, R. B. Wilson, C. Reggiardo, and R. D. Walker. 2007. Characterization of multidrug resistant *Salmonella* recovered from diseased animals. *Vet. Microbiol.* 123:122-132.

Zhao, S., D. G. White, P. F. McDermott, S. Friedman, L. English, S. Ayers, J. Meng, J. J. Maurer, R. Holland, and R. D. Walker. 2001. Identification and expression of cephamycinase *bla*(CMY) genes in *Escherichia coli* and *Salmonella* isolates from food animals and ground meat. *Antimicrob. Agents Chemother.* 45:3647-3650.

Zieg, J., M. Silverman, M. Hilmen, and M. Simon. 1977. Recombinational switch for gene expression. *Science* 196:170-172.

Permissions

The contributors of this book come from diverse backgrounds, making this book a truly international effort. This book will bring forth new frontiers with its revolutionizing research information and detailed analysis of the nascent developments around the world.

We would like to thank Yashwant Kumar, for lending his expertise to make the book truly unique. He has played a crucial role in the development of this book. Without his invaluable contribution this book wouldn't have been possible. He has made vital efforts to compile up to date information on the varied aspects of this subject to make this book a valuable addition to the collection of many professionals and students.

This book was conceptualized with the vision of imparting up-to-date information and advanced data in this field. To ensure the same, a matchless editorial board was set up. Every individual on the board went through rigorous rounds of assessment to prove their worth. After which they invested a large part of their time researching and compiling the most relevant data for our readers. Conferences and sessions were held from time to time between the editorial board and the contributing authors to present the data in the most comprehensible form. The editorial team has worked tirelessly to provide valuable and valid information to help people across the globe.

Every chapter published in this book has been scrutinized by our experts. Their significance has been extensively debated. The topics covered herein carry significant findings which will fuel the growth of the discipline. They may even be implemented as practical applications or may be referred to as a beginning point for another development. Chapters in this book were first published by InTech; hereby published with permission under the Creative Commons Attribution License or equivalent.

The editorial board has been involved in producing this book since its inception. They have spent rigorous hours researching and exploring the diverse topics which have resulted in the successful publishing of this book. They have passed on their knowledge of decades through this book. To expedite this challenging task, the publisher supported the team at every step. A small team of assistant editors was also appointed to further simplify the editing procedure and attain best results for the readers.

Our editorial team has been hand-picked from every corner of the world. Their multi-ethnicity adds dynamic inputs to the discussions which result in innovative outcomes. These outcomes are then further discussed with the researchers and contributors who give their valuable feedback and opinion regarding the same. The feedback is then collaborated with the researches and they are edited in a comprehensive manner to aid the understanding of the subject.

Apart from the editorial board, the designing team has also invested a significant amount of their time in understanding the subject and creating the most relevant covers. They scrutinized every image to scout for the most suitable representation of the subject and create an appropriate cover for the book.

The publishing team has been involved in this book since its early stages. They were actively engaged in every process, be it collecting the data, connecting with the contributors or procuring relevant information. The team has been an ardent support to the editorial, designing and production team. Their endless efforts to recruit the best for this project, has resulted in the accomplishment of this book. They are a veteran in the field of academics and their pool of knowledge is as vast as their experience in printing. Their expertise and guidance has proved useful at every step. Their uncompromising quality standards have made this book an exceptional effort. Their encouragement from time to time has been an inspiration for everyone.

The publisher and the editorial board hope that this book will prove to be a valuable piece of knowledge for researchers, students, practitioners and scholars across the globe.

List of Contributors

Cynthia L. Sheffield and Tawni L. Crippen
United States Department of Agriculture, Agriculture Research Service, Southern Plains Agricultural Research Center, USA

Asit Mazumder
Water and Aquatic Sciences Research Lab Department of Biology, University of Victoria, Victoria, Canada

Chandran Abhirosh
Water and Aquatic Sciences Research Lab Department of Biology, University of Victoria, Victoria, Canada
School of Environmental Sciences, Mahatma Gandhi University, Kottayam, Kerala, India

Sherin Varghese and A.P Thomas
School of Environmental Sciences, Mahatma Gandhi University, Kottayam, Kerala, India

A.A.M Hatha
School of Marine Sciences, Cochin University of Science and Technology, Cochin, Kerala, India

Andreas E. Zautner
Universitätsmedizin Göttingen, Germany

A.A. Aliero
Department of Biological Sciences, Usmanu Danfodiyo University, Sokoto, Nigeria

A.D. Ibrahim
Department of Microbiology, Usmanu Danfodiyo University, Sokoto, Nigeria

Birol Özkalp
Department of Medicinal Laboratory, Vocational School of Health Services of Selçuk University, Konya, Turkey

Lucky H. Moehario
Department of Microbiology Faculty of Medicine University of Indonesia, Jakarta, Indonesia

Enty Tjoa
Department of Microbiology Faculty of Medicine Catholic University of Atmajaya Indonesia, Jakarta, Indonesia

Veronica N. K. D. Kalay
Department of Microbiology Faculty of Medicine Christian University of Indonesia, Jakarta and Division of Microbiology, Siloam Hospital Kebon Jeruk, Jakarta, Indonesia

Angela Abidin
Division of Microbiology, St. Carolus Hospital, Jakarta, Indonesia

Ali Nokhodchi
Medway School of Pharmacy, Universities of Kent and Greenwich, Chatham, UK
Drug Applied Research Center and Faculty of Pharmacy Tabriz University of Medical Sciences, Tabriz, Iran

Taravat Ghafourian
Medway School of Pharmacy, Universities of Kent and Greenwich, Chatham, UK

Ghobad Mohammadi
Faculty of Pharmacy, Kermanshah University of Medical Sciences, Kermanshah

Daniel Arrieta-Baez
Escuela Nacional de Ciencias Biológicas-IPN, Prolongación de Carpio y Plan de Ayala S/N

Rosario Ruiz de Esparza and Manuel Jiménez-Estrada
Instituto de Química, UNAM. Cd. Universitaria, Coyoacán, México

Fernanda Pinheiro de Carvalho Ribeiro, Fernanda Carolina Sousa Fonseca, Isabella Alves Reis, Isabella Santos Araújo, Hélio Mitoshi Kamida and Alexsandro Branco
State University of Feira de Santana, Brazil

Ana Paula Trovatti Uetanabaro
State University of Santa Cruz, Brazil

Saul Ruiz-Cruz and Luis A. Cira-Chávez
Instituto Tecnológico de Sonora, Departamento de Biotecnología y Ciencias Alimentarias, Ciudad Obregón, Sonora, Mexico

Gustavo A. González-Aguilar and J. Fernando Ayala-Zavala
Centro de Investigación en Alimentación y Desarrollo, A. C. (CIAD), Dirección de Tecnología de Alimentos de Origen Vegetal, Hermosillo, Sonora, Mexico

José de Jesús Ornelas-Paz
CIAD Unidad Cuauhtémoc, Chihuahua, México

Renata Albuquerque Costa
Doctoral Student in Fisheries Engineering. C.A.P.E.S. Federal University of Ceará, Brazil

Fátima Cristiane Teles de Carvalho
Doctoral Student in Tropical Marine Sciences. Federal University of Ceará, Brazil

Regine Helena Silva dos Fernandes Vieira
Prof. Dr. Sea Sciences Institute. Federal University of Ceará, Brazil

Mehmet Karadayı, Özlem Barış and Medine Güllüce
Biology Department of Atatürk University, Erzurum, Turkey

Eric W. Brown, Rebecca L. Bell, Marc W. Allard, Narjol Gonzalez-Escalona, Andrei Perlloni, Joseph E. LeClerc and Thomas A. Cebula
Center for Food Safety and Applied Nutrition, Food and Drug Administration, College Park, MD, USA

Claudia Silva, Magdalena Wiesner and Edmundo Calva
Departamento de Microbiología Molecular, Instituto de Biotecnología, Universidad Nacional Autónoma de México, Cuernavaca, Morelos, Mexico

Printed in the USA
CPSIA information can be obtained
at www.ICGtesting.com
JSHW011455221024
72173JS00005B/1079